Microwave Materials

# Microwave Materials

EDITORS
**V.R.K. Murthy**
**S. Sundaram**
**B. Viswanathan**

Springer-Verlag Berlin Heidelberg GmbH

**V.R.K. Murthy**
Department of Physics
IIT Madras, Madras, INDIA

**S. Sundaram**
Microwave Millimeter Wave Wing
Defence Electronics Research Laboratory
Hyderabad, India

**B. Viswanathan**
Department of Chemistry
IIT Madras, Madras, India


Originally published by Springer-Verlag Berlin Heidelberg New York in 1994
Softcover reprint of the hardcover 1st edition 1994



ISBN 978-3-662-08742-8 ISBN 978-3-662-08740-4 (eBook)
DOI 10.1007/978-3-662-08740-4

# Preface

Solid State Materials have been gaining importance in recent times especially in the context of devices which can provide necessary infrastructure and flexibility for various human endeavours. In this context, microwave materials have a unique place especially in various device applications as well as in communication networks. Various technological developments are taking place in fine-tuning these materials for specific applications and in fixed band frequencies. Though the science and technology of these materials has reached an advanced stage, systematic attempts are still lacking in bringing all available information in a single source. The present volume is a modest attempt in this direction, though it cannot be considered to be the one that satisfies completely desired components and information required. The editors have enlisted certain articles of interest in this area, especially those dealing with measurement techniques, chapters dealing with materials like Ferrites, YIGs, Radome and high $T_c$ superconducting materials which are of current interest. The editors are fully aware that the coverages are not comprehensive either in scope or in depth. The purpose of this volume is only to acquaint oneself of certain aspects of a fast developing field. The editors will be grateful for any comments or suggestions in this endeavour.

V.R.K. Murthy
S. Sundaram
B. Viswanathan

# Contents

# 1

# Materials and Processes in Microwave Integrated Circuits Fabrication

T. Rs. Reddy
Hybrid Microelectronics Division
Defence Electronics Research Laboratory
Hyderabad, India

## 1. INTRODUCTION

Microwave Integrated Circuits (MICs), more precisely Hybrid Microwave Integrated Circuits represent an important and significant step towards the integration and miniaturization of the microwave subsystems and systems when compared to those based on the waveguide/coaxial components. But the level of integration is less than that encountered in the Monolithic Microwave Integrated Circuits (MMICs). In spite of the advent of MMICs, however, MICs continued to assert their place in the microwave systems due to some of their unique features which cannot be fully met otherwise. Also MIC technology adopted itself to the integration of MMICs functional chips in realising the microwave systems. MICs span the range from printed circuit board where the integration is limited to realising the conductor pattern, to the hybrid circuits (thin and thick film) where some or all of the passive components are integrated along with the conductor pattern. Since at microwaves frequencies the circuit dimensions tend to be of the order of the wavelength, every element of the microwave circuit influences its performance (Bhat and Koul 1989 and Van Nie, Goedbloed and Kersuzan 1981). This in turn reflects on the choice of the materials that are used to fabricate the MICs. The quality requirements are more demanding for the materials used for the fabrication of MICs than for the low frequency circuits. In the present chapter efforts are made to highlight various aspects related to some of the materials used for the fabrication of MICs.

## 2. TYPES OF MICs

Different types of circuit configuration have been used for the propagation of microwaves and in realising various circuit elements such as inductors, capacitors and resistors. Broadly these can be classified into two categories namely distributed element and lumped element. While in the former case the realisation of inductor and capacitor is related to specific configuration of the transmission line, in the latter case these are realised as lumped elements analogous to those in the low frequency circuits. In order that these behave as lumped elements it should be ensured that their overall sizes do not exceed 1/10 of the wavelength of the microwaves in the medium in which the elements are realised. In real situation however the circuit tends to have mixed features while predominantly tending to be either one of these types.

Most commonly employed circuit configurations include: (1) Suspended line (strip line), (2) Microstrip line, (3) Coplanar line, (4) Slot line and (5) Finline. Typical circuit configurations for some of these are shown in Fig. 1. In almost all cases there is a conductor circuit pattern laid out on a dielectric substrate. In some cases the passive components such as inductors, capacitors and resistors are realised *in situ* and in others are attached to the circuit as discrete components. Invariably the active components (devices) are attached as separate elements and interconnections are made to the circuit lines through one of the several means discussed below. Active devices can

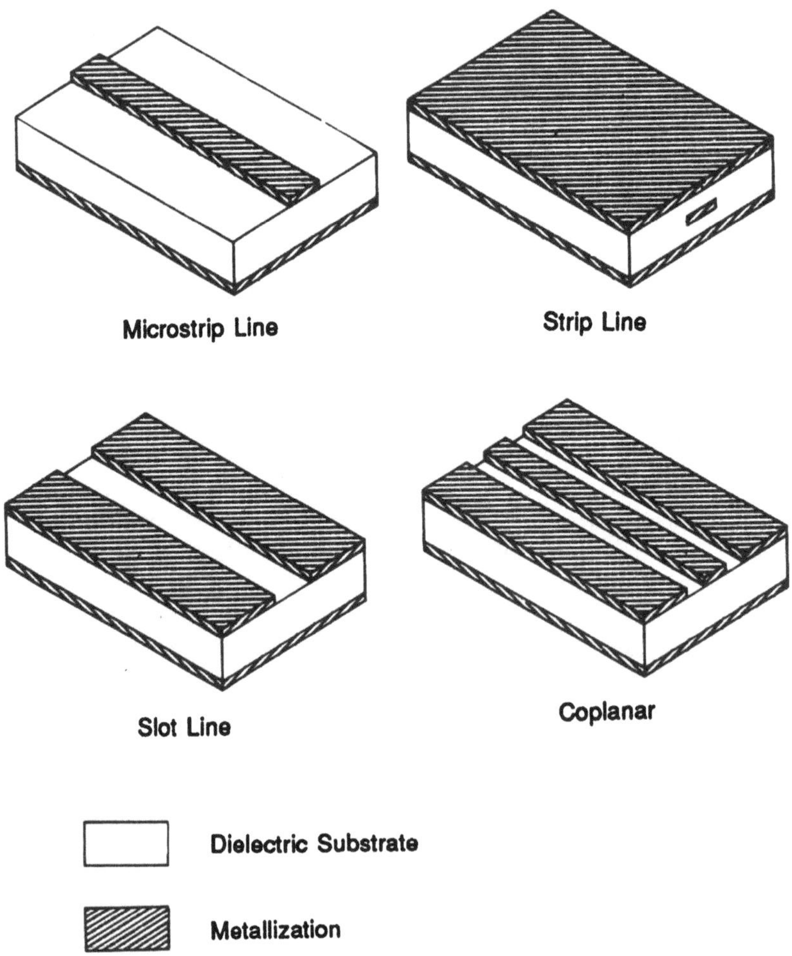


**Fig. 1.** Typical circuit configurations used for microwave integrated circuits.

be discrete components such as diodes, transistors etc., as well as MMICs (Reddy 1988).

During the fabrication of MICs a large variety of materials are used to accomplish various tasks. These include the dielectric substrate, conductor and resistor metallizations, materials for attaching the components to the circuit, interconnecting materials, encapsulants etc. (Grovener 1989, Licari and Enlow 1988, Jowett 1982). Each of these will be discussed in detail in the following sections.

## 3. SUBSTRATE

The substrate supporting the circuit is a dielectric with metallisation on either side. The circuit pattern is printed on one or both sides by photolithographic process. A brief description of the photolithographic fabrication process is in order to appreciate the requirement of the materials for the substrate and the metallization.

### 3.1 Photolithographic Fabrication

Figure 2 shows the flow chart of the photolithographic fabrication of circuit pattern on the substrate. The substrate after proper cleaning is coated with a thin layer of a photoresist. After pre-bake the substrate with the photoresist is exposed to UV light through a photo mask which contains the replica of the desired circuit pattern. After development, the circuit pattern is transferred on to the photoresist exposing the metallization in the non-circuit area. This is then removed by etching either chemically by dissolving the materials in the spray of a suitable chemical reagent or by dry etching technique using plasma with suitable reactive gases or gas mixtures. Chemical milling is a variant of plasma etching in which the reactive ions/molecules/atoms are directed on to the substrate in the form of a beam resulting in steep walls for the circuit lines i.e. high etch factor. Photoresist is finally removed to reveal the delineated circuit pattern on the substrate.

### 3.2 Dielectric Substrate

In contrast to the low frequency circuits the dielectric substrate not only provides a support to the circuit pattern but also actively participates in the functioning of the circuit. The dielectric constant and thickness of the substrate decide the impedance of the transmission lines which are nothing but the conductor lines printed on the dielectric substrates. For a given dielectric substrate the width of the conductor pattern decide the impedance of the transmission line which is typically 50 ohms. In addition, the material has to be of low loss. Tight tolerances are demanded on the uniformity of the dielectric constant and thickness over the substrate area in order to achieve the desired circuit performance. The criticality increases with increase in the frequency.

The basic requirements for the substrate materials are:

1. The dimensional stability over the temperature of the operation.
2. Low dielectric loss.
3. Stable electrical properties such as dielectric constant.

4. High DC resistance.
5. Easy machinability for sizing, making holes etc.
6. Provide smooth surface for metallization.
7. High thermal conductivity for high power applications.
8. Uniformity of these properties over the substrate area.

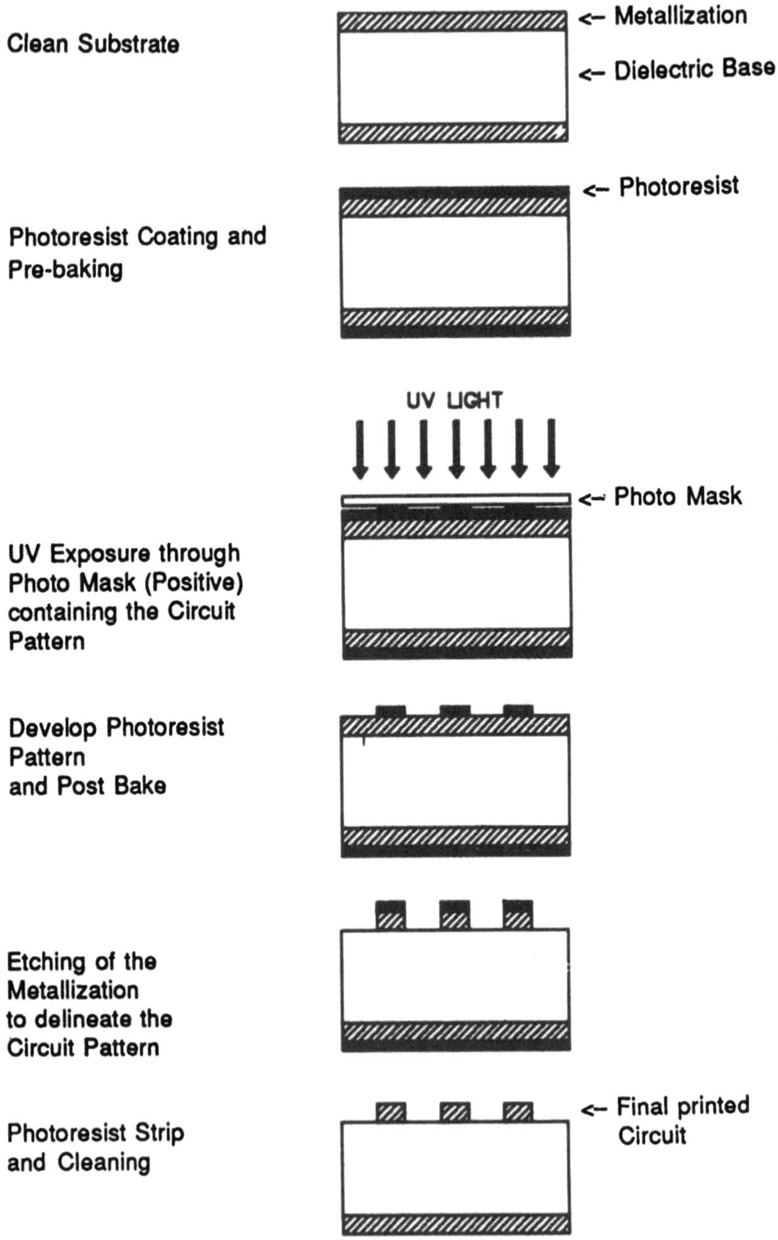


**Fig. 2.** Photolithographic fabrication of microwave integrated circuits.

Substrate materials used in the fabrication of MICs can be broadly classified into soft and hard substrates. Neither type meets all the requirements for the ideal substrate. The selection of one type or the other depends on specific application and the stringency of the requirement. Table 1 lists properties of these two types.

**Table 1. Selected features of hard/soft substrates used for microwave integrated circuits**

| Feature | Hard | Soft |
|---|---|---|
| Dielectric material | Alumina, fused quartz berellia, aluminium nitride | Glass fiber reinforced (woven or non-woven) or ceramic powder filled PTFE |
| Dielectric thickness | 10 to 25 mil | 1 to 250 mil |
| Dielectric constant | 4–10 | 2–10 |
| Minimum dissipation factor | see Table 2 | 0.0009 @ 10 GHz |
| Volume resistivity | see Table 2 | > $10^{15}$ ohm cm |
| Thermal conductivity | see Table 2 | 2.8 BTU in/hr/ft$^2$/deg F |
| Maximum size available | 6" × 6" | 36" × 48" |
| Sizing | Cutting with diamond saw, scribing with diamond tool or laser followed by dicing | Cutting with a shear, chiseling |
| Holes | Ultrasonic or laser drilling | High speed drilling with carbide tools<br>Chiseling or punching |
| Metallization | Large variety of multilayer metallization structures have been evolved. Most common being Cr/Au, Cr/Cu/Au, NiCr/Au, TiW/Au, TaN/TiW/Au | Basically electroless/electroplated or laminated/rolled high purity copper. Gold flash is provided to protect copper from environmental effects and to enable wire bonding |
| Metallization thickness | Overall thickness 3 to 8 micron | 1/8 to 2 oz cu/sq ft (4.5 to 74 micron) |

Soft substrates are typically glass fibre reinforced and/or ceramic filled (loaded) teflon based materials. The fillers or reinforcement materials control the dielectric constant which ranges from 2.2 to 10. The teflon base provides low dielectric loss. Mechanical working with these substrates is relatively easy and lends itself for sizing and for making holes (Taconic Tech Topic). However, when the thickness is small because of their flexible nature these substrates need mechanical support which is normally provided by the aluminium base (in some cases top as well) plate of the package. To provide perfect ground it is necessary to hold the substrate to the base plate with screws at several locations avoiding, of course, the circuit lines.

A variety of ceramic, glass or single crystal materials fall in the category of hard substrates. Contrary to soft substrates hard substrates do not lend themselves to easy mechanical working. Special methods like ultrasonic or laser machining have to be used (Kugler and Culkin 1989). However due to their excellent mechanical and electrical properties these are the preferred substrates in realising high end microwave and millimetric wave circuits. 99.6% alumina is the most common substrate material that is used with glazed or polish surfaces with surface smoothness better than 3 micro inch CLA. Properties of different hard substrates are summarised in Table 2 (Blum and Anzai 1989, Sunahara and Takapataki 1990).

**Table 2. Typical properties of hard substrates used for microwave integrated circuits**

| Property | Alumina | Fused quartz | Berellia | Aluminium nitride |
|---|---|---|---|---|
| Composition | 99.6% $Al_2O_3$ | $SiO_2$ | 99.9% BeO | AlN |
| Dielectric constant | 9.9 | 3.8 | 6.7 | 8.8 |
| Dissipation factor @ 10 GHz | 0.0001 | 0.0001 | 0.0002 | 0.0006 |
| Volume resistivity (ohm cm) | $>10^{14}$ | $>10^{15}$ | $10^{18}$ | $>10^{14}$ |
| Dielectric strength (kV/mm) | 22 | 500 | 14 | 14 |
| Thermal expansion coefficient ($10^{-6}$ $K^{-1}$) | 6.3 | 0.5 | 6.4 | 4.5 |
| Thermal conductivity ($Wm^{-1}$ $K^{-1}$) | 38 | 1.4 | 280 | 170 |
| Fluxural strength (MPa) | 620 | — | 248 | 300 |

### 3.3 Thin Film Metallization

Metallization on the dielectric substrate is meant for realising the circuit pattern or to provide a ground plane. A variety of metallization structures have been employed depending on the requirement of the circuit as well as the nature of the substrate. Important characteristics of the metallization are: (1) good conductivity; (2) good adhesion to the substrate; (3) easy etchability; (4) solderability; (5) wire bondable; (6) good surface finish; and (7) ability to resist environmental degradation such as tarnishing etc.

For soft substrates the metallization is typically copper on both sides with thickness ranging from 1/8 to 2 oz per sq ft i.e., 4.5 to 74 microns. The metallization is applied on to the dielectric either by electroless deposition followed by electroplating or by hot press lamination of a thin copper foil with suitable pre-treatment of the surface of the dielectric followed by the application of adhesive. Copper is of high purity which itself is formed by electroless/ electro deposition on polished lead cylinders to provide smooth surface. Some companies offer special laminations which include integral aluminium backing plate to act as ground plane obviating the need for separate aluminium base plate. This ensures perfect ground plane for the microwave circuits (Fig. 3). To protect copper surface from the tarnishing effects due to the environment about 0.5 to 1 micron thick gold is electroplated onto the copper metallization. This is also necessary to enable wire bonding.

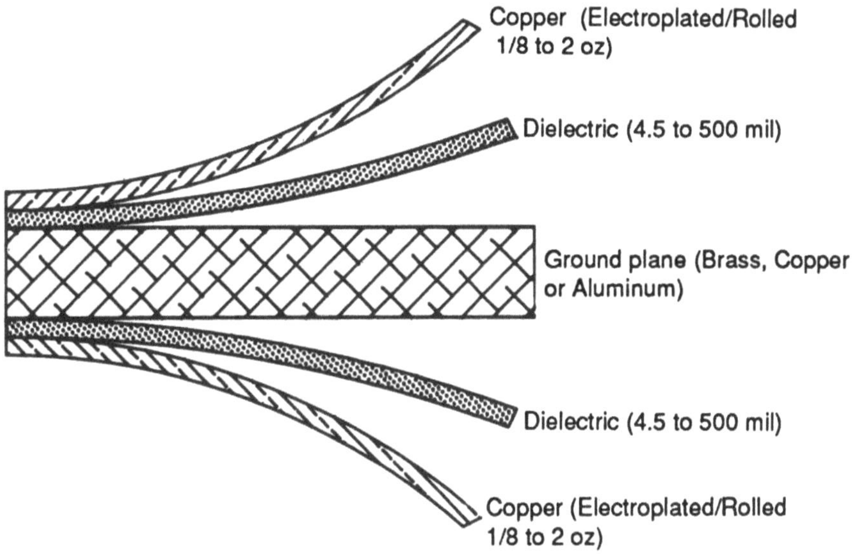


**Fig. 3.** Commercial soft substrate (PTFE based dielectric) with integral ground plane (M/s Rogers Corporation, USA).

Metallization structures employed on hard substrates are more complex for the simple reason that no single metal meets all the requirements (Fig. 4).

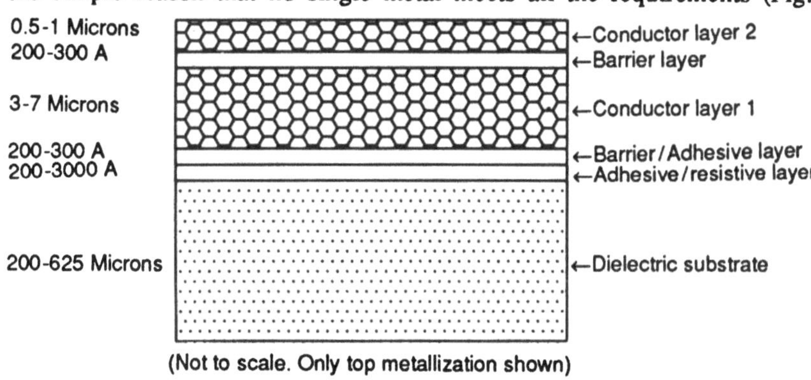


**Fig. 4.** Generalised multilayer metallization scheme for microwave integrated circuits on hard substrates (typical configurations with specifications are given in Table 3).

Adhesion of copper or gold to the substrates is poor. One has to provide an intermediate base layer to improve the adhesion. Additional intermediate barrier layer between the base layer and the top layer sometimes becomes necessary to prevent inter-diffusion between these layers and to prevent the damage to the base layer due to the attack by the moisture in the atmosphere which can easily diffuse through the permeable top conductive layers. Resistance elements in circuits are easily integrated on the hard substrates if a resistive layer is incorporated in the metallization structures. These often play the role of the base layer obviating the need for a separate one. Novel metallization structures employing TiW have been envisaged for enhanced stability and environmental hardness. Variety of metallization structures that are employed on the hard substrates for realising MICs are given in Table 3. The characteristics of different materials employed are summarised in Table 4.

**Table 3. Typical specifications for commonly used metallization structures on hard substrates for microwave integrated circuits**

| Adhesive layer (Angstroms) | Resistive layer (Angstroms) | Barrier/ Adhesive layer (Angstroms) | Conductor layer 1 (Microns) | Barrier layer (Angstroms) | Conductor layer 2 (Microns) |
|---|---|---|---|---|---|
| Cr/250 | — | — | Au/3–8 | — | — |
| Cr/250 | — | — | Cu/3–7 | — | Au/0.5–1 |
| Cr/250 | — | — | Cu/3–7 | Ni/250 | Au/0.5–1 |
| — | NiCr/1500 | — | Au/3–8 | — | — |
| TiW/250 | — | — | Au/3–8 | — | — |
| — | TaN/3000 | TiW/250 | Au/3–8 | — | — |

**Table 4. Properties of thin film materials used for conductor metallization**

| Mate-rial | Functional property | Bulk resis-tivity ($\mu$ ohm cm) | Surface resistance (ohms f in GHz) | Skin depth (Microns f in GHz) | Typical thickness | Interconnections compatibility: Soldering | Interconnections compatibility: Wire bonding |
|---|---|---|---|---|---|---|---|
| Cu | Conductor | 1.7 | $0.85\times10^{-2}\sqrt{f}$ | $2.15/\sqrt{f}$ | 3–7 $\mu$m | Excellent | Possible[@] |
| Au | Conductor | 2.5 | $1.00\times10^{-2}\sqrt{f}$ | $2.54/\sqrt{f}$ | 3–7 $\mu$m | Possible[#] | Excellent |
| Al | Conductor* | 2.7 | — | — | 3–7 $\mu$m | No | Very good |

[@] Gold plating is advised for better results.
[#] Leaching is a problem. Special solders are required.
* Does not require an adhesive layer.

Vacuum evaporation, Sputtering and Electroless/Electro deposition are the common methods employed in the metallization of hard substrates. Choice of the method to be employed is decided by the requirement, throughput and finally the economics of the process (Holland 1965, Berry et al 1968, Maissel and Glang 1970, Chopra and Kaur 1983, Stuart 1983).

Vacuum evaporation is schematically illustrated in Fig. 5. In this, the material to be deposited is heated to its evaporation temperature either by resistive heating or by E–beam in a vacuum better than $10^{-5}$ Torr. The evaporated material is deposited on to the substrate placed above the source material (line of sight). The rate of evaporation is controlled by the heating of the source materials (temperature). E-beam evaporation is superior to resistive heating in several respects.

1. It is clean in the sense that the material itself acts as the container and hence the contamination from the heater element or the container is totally eliminated.
2. E-beam heating affords a better control of the rate of evaporation.

3. Refractory material such as tantalum also can be easily evaporated.
4. Large thicknesses can easily be achieved by the use of large size crucibles holding large quantity of materials to be evaporated.

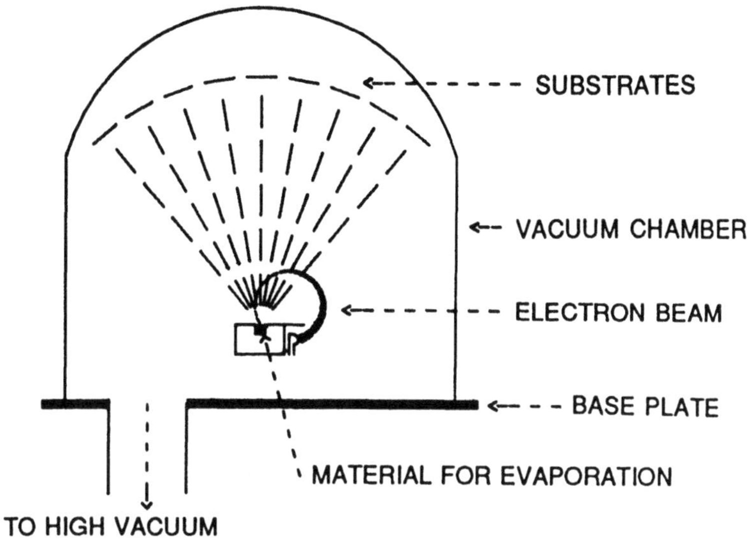


**Fig. 5.** Schematic for E-beam evaporation.

Figure 6 shows a schematic representation of sputtering process. The chamber is evacuated to better than $10^{-6}$ Torr and after thorough degassing, is back-filled with argon to a pressure of $10^{-3}$ Torr. The argon ions produced in the

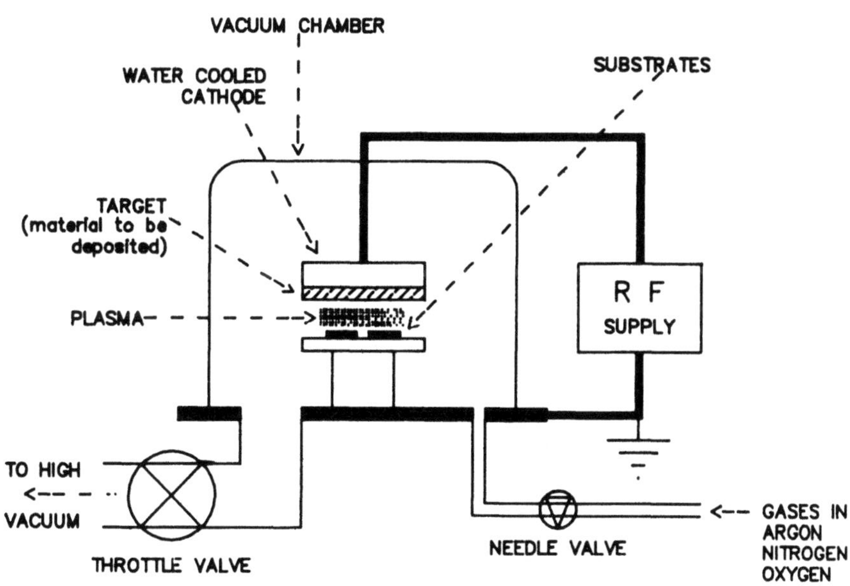


**Fig. 6.** Schematic for sputter deposition.

discharge set up between the two electrodes bombard the source material placed on the cathode and dislodge (sputter) the material which in turn gets deposited on the substrate placed on the anode which is closely placed. There are several variants to this simple arrangement and systems. The discharge can be produced by DC or AC (RF). The discharge can be enhanced by placing the plasma in a magnetic field. In this case, it is referred to as magnetron sputtering. This helps in improving rate of deposition for a given power. Gases other than argon such as nitrogen, oxygen can be employed for reactive sputtering wherein the gaseous ions react with the source (target) material to form a compound/composite film. For example, tantalum target with oxygen results in tantalum oxide film and the desired stoichemetry is achieved by adjusting the oxygen partial pressure and the power.

Extremely well adhering films can be produced by sputtering due to the simple fact that the energy of the sputtered atom is orders of magnitude more than the evaporated ion. In either case, it is essential that the substrates are heated to a temperature anywhere between 150°C and 300°C for improved adhesion and to achieve films with good crystallinity and grain structure. Throughput of the evaporation process is much higher than that in a sputtering process. But this is not a real drawback with the advent of automatic cassette to cassette loading through vacuum load locks. Another specific advantage of sputtering is to be able to deposit composite, compound or alloy films true to the composition existing in the source material, with the added ability to manipulate the composition with reactive sputtering where possible.

Normally the sputtering or vacuum deposition is used to deposit the conductor layer to a thickness of about 1 micron. Build up of this to 5–8 micron is done by electroplating.

Electroless deposition followed by electroplating provide a cheaper alternative for the metallization of the hard substrates. However, the quality of the film is not as good as those prepared by sputtering or vacuum deposition. Good adhesion is one of the main problems which in turn reflects on the difficulty in realising fine line structures in the photolithographic fabrication. Normally thin layer of metallization is deposited by electroless process and the thickness is build up by electroplating. Special formulations of electroless baths are available which obviate the need for electroplating. Most of the commonly used metallization structures can be deposited by the electroless technique. The surface of the substrate is to be suitably prepared to initiate the electroless deposition. Typical process steps for electroless deposition of copper on teflon substrate is given in Table 5 (Prasad and Vijayaraghavan 1987). One of the main drawbacks of this process is the difficulty in controlling the thickness of the film and its texture.

### 3.4 Thick Film Metallization

Thick film technology is a misnomer to the essentially print and fire technology, and historically got its name due to the fact that the thickness of the film realised by print and fire technology is of the order of 25 microns while those obtained by evaporation and sputtering are of the order of 1 micron. Thick film technology is a cheaper alternative to the thin film approach and is extensively used in the

**Table 5. Process steps for electroless deposition**

| Operation | Materials used | Effect / Purpose |
|---|---|---|
| Cleaning | Organic solvents/Detergents | Removal of grease, dirt and other foreign matter |
| Surface preparation | Sodium naphthalene complex | Surface becomes wettable with water |
| Sensitization | Stannous chloride + Concentrated hydrochloric acid | Adsorption of stannous chloride on the surface |
| Activation | Silver nitrate + Ammonium hydroxide | Deposition of silver on the surface through reduction by adsorbed $SnCl_2$ |
| Electroless plating | 1. Copper sulphate + Trisodium citrate<br>2. Sodium hydroxide<br>3. Formaldehyde<br>mix 1, 2 and 3 | Deposition of copper on the surface |

fabrication of hybrid circuits in the lower frequency typically upto 1 GHz. Briefly, the technology involves screen printing a conductive paste/ink on to the substrate through a stainless steel or nylon screen/mesh on to which the circuit pattern has been transferred using a masking material. After suitable drying and pre-baking to drive away the solvents and other organic vehicles the substrates are fired at temperatures of the order of 850–950°C. This leaves the conductor pattern fused on to the substrate. Thick film paste/ink consists of the conductor material in the form of fine particles or flakes suspended in a thixotropic organic vehicle along with glass firt. Organic solvent is added to maintain the desired consistency. Various processes that take place at different stages of processing are indicated in Fig. 7 (Holmes and Loasby 1976, Pitt 1981 and Haskard 1988).

The resulting film consists of the metal particles embedded in glass matrix which binds itself to the substrate during the firing. The conductor particles sinter together to form the conductor tracks but contains voids as well as glass frit intrusions which is the very basis of adhesion. This puts a limit to the suitability of thick film conductor tracks for microwave applications. Secondly, as fired conductor tracks do not have the desired line definition due to the very process of screen printing. Normal pastes can be used to realise circuits operating upto 4 GHz by the use of photolithographic fabrication and etching for delineating the circuit pattern instead of screen printing (Corkhill 1976 and Haskard 1988).

Special new formulations have been developed over the years which enabled to push the frequency to the high end of microwaves region, i.e. upto 18 GHz. These are the fritless pastes and organometallic formulations. The former is conventional thick film paste which contains very little percentage of some oxides instead of the usual glass frit (Corkhill 1976 and Haskard 1988). This forms a very thin binding layer between the metal and substrate leaving behind a better quality conductor track. Organo metallic formulations consist of the metal ions complexed with an organic molecule which when fired is decomposed to leave behind the metallization as a thin film. Contrary to the other thick film formulations the thickness of the film obtained with organo metallic formulations is of the order of 1 micron or less. Building up of the thickness can be achieved by electroplating. As of now organo metallic formulations are available for gold metallization (Haskard 1988).

In either case best results are obtained by screen printing the conductor over the full substrate and realising the circuit pattern photolithographically.

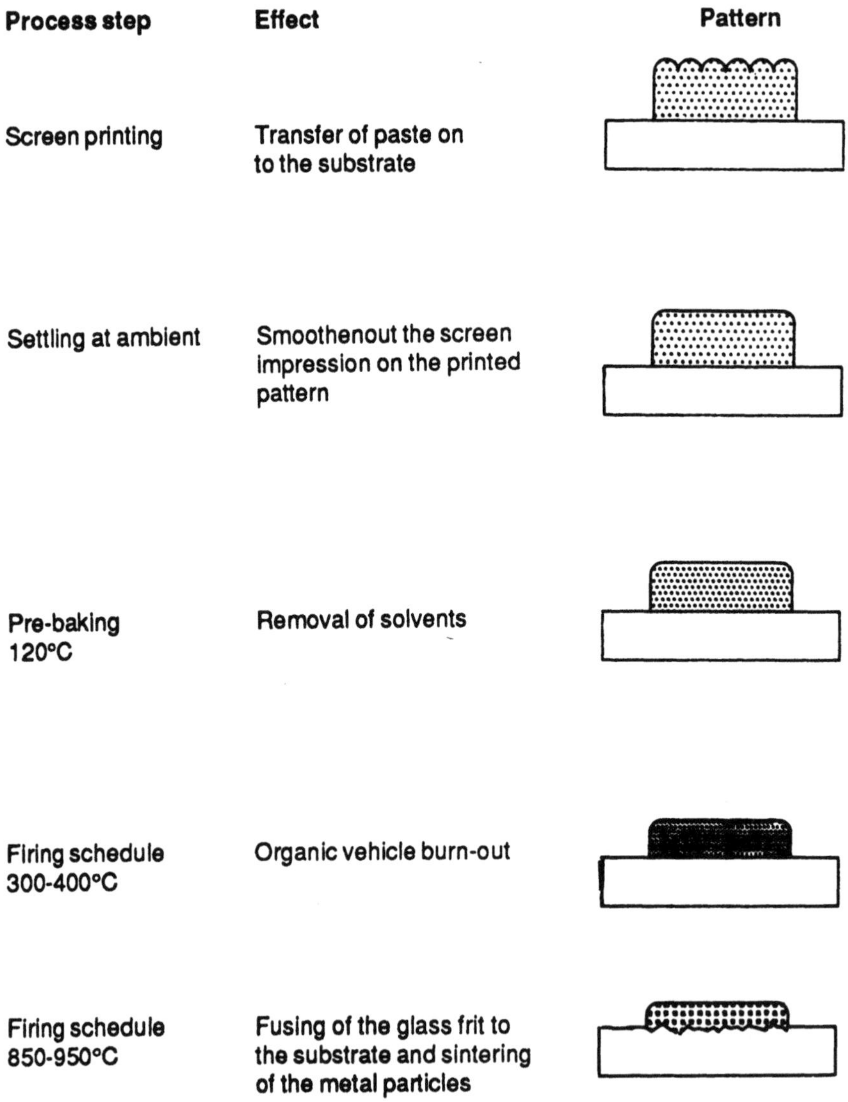


**Fig. 7.** Processes in thick film metal deposition.

## 3.5 Resistive and Dielectric Layers

As already mentioned resistive layers can easily be integrated on to the hard substrates along with the conductor pattern (Siddall 1965 and Hughes 1987).

Historically NiCr (nichrome) is the material used for the resistive layer though to some extent chromium layer itself served the purpose. Tantalum nitride is the newer material offering greater flexibility in controlling the sheet resistance and better stability due to self passivation by the formation of thin oxide layer. Tantalum oxinitride has been experimented with to arrive at low TCR values. Added advantage being that these films as well as the dielectric layer tantalum pentoxide ($Ta_2O_5$) can be deposited by reactive sputtering with nitrogen and oxygen using a single tantalum target. Table 6 summarises the characteristics of different resistive films used in realising MICs.

**Table 6. Properties of thin film resistor materials**

| Material | Sheet resistance (ohm per sq.) | TCR (ppm/°C) | Remarks |
|---|---|---|---|
| Nichrome (Ni:Cr/80:20) | 10–500 | < ±20 to ±150 | Most popular and stable resistance material. But susceptible to contamination |
| Tantalum | upto 100 | ±50 to ±150 | Resistor trimming can be done by partial anodising |
| Tantalum nitride | 30–200 | − 60 to −200 | Resistivity and TCR can be manipulated by incorporating oxygen. Very stable resistors |
| Cr/SiO | upto 600 | < ±20 to ±200 | Stable high resistivity cermets |
| $Al_2O_3$/NiCr/ $Al_2O_3$ | 10–500 | ≈ 0 | Stable resistance and zero TCR sandwiches |

Necessity for a dielectric layer is felt in realising certain circuit elements such as parallel plate capacitors, cross overs particularly while realising a spiral inductor. It is also used as a passivation layer for resistive elements. Silicon dioxide and tantalum pentoxide are the most commonly used dielectric materials while realising the MICs due to their low dielectric loss and ease of deposition. The former can be deposited either by E-beam evaporation or sputtering. The $Ta_2O_5$ layer is more conveniently deposited by reactive sputtering using oxygen gas with tantalum target. Highly stable polymer films such as polyimide and parylene are also used as low dielectric constant alternatives (Grovenor 1989). The characteristics of these materials are given in Table 7.

## 4 ASSEMBLY

Having realised the circuit pattern on the substrate the next step involves populating the printed circuit with discrete devices viz., the active components like transistors, diodes and in some cases the passive components like capacitors, resistors, inductors if these could not be conveniently integrated in the circuit. Conventional leaded package types for these components that are used in the

**Table 7. Properties of thin film dielectric materials**

| Material | Dielectric constant | Dissipation factor | Dielectric strength (V/cm) | Remarks |
|---|---|---|---|---|
| SiO | 2–6 | 0.01–0.04 | $10^6$ | Precise control of composition is difficult |
| $SiO_2$ | 3.4 | 0.005 | $>10^6$ | Most commonly used dielectric for cross-overs, parallel plate capacitor and passivation |
| $Ta_2O_5$ | 21–25 | 0.001–0.01 | $>10^6$ | Not suitable for the cross-overs, but good for realising capacitors. Process integration possible while using TaN resistor elements |

fabrication of the normal Printed Circuit Boards (PCBs) are not suitable for the hybrids to get the full benefits of miniaturization. This is particularly so for MICs which are meant to operate at higher frequencies. All the add on components are to be surface mountable and at the high end of microwaves it is preferable to integrate the passive components in the circuit and to use the active components in the chip form. All the passive components are now available in the surface mountable package or 'chip' form. Typical outline of some of these passive components are shown in Figs. 8 and 9. In addition to the bare chip or 'die', the active components also come in a variety of package types or with leads attached (beam lead type) for use in MICs. Some of these types are shown in Fig. 10.

The process of assembly of these components in the circuit involves: (a) attaching the components on the circuit board to provide mechanical support, and (b) to establish electrical contact between the component terminals and the circuit conductor pattern.

Except in the case where the active components are used in the bare chip form or in some types of passive chip components, the attachment of the component terminals to the conductor circuit pattern for providing electrical contact gives the necessary mechanical support and no additional mechanical fixing is necessary. The bare chip mounted on the metallization provides the mechanical support as well as electrical 'ground' or earthing or in some cases electrical connection to active terminals of the component for example to the collector terminal of transistor. The electrical connection between the connector pads (bonding pads) on the die and the conductor lines on the circuit is done by "wire bonding" a process by which fine dia wires or ribbons are attached to the 'bonding' pads on the die and the conductor lines. Different methods employed for attaching the components to the circuit and for interconnections are summarised in Table 8 (Edwards 1991).

## 4.1 Soldering

This is one of the oldest and most common methods used for attaching the discrete components to the circuit boards and a large varieties of materials and

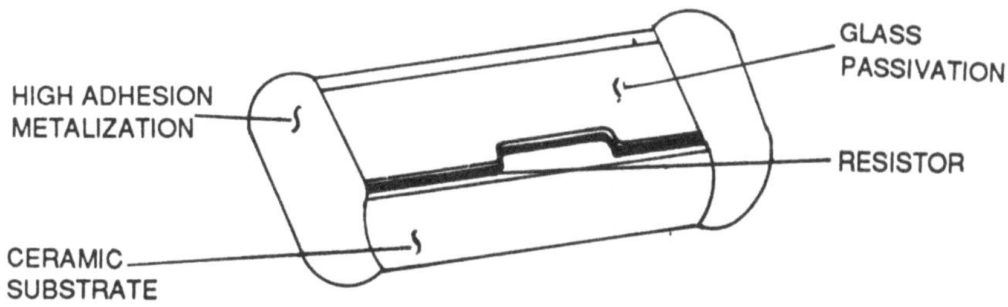


WRAP AROUND TERMINATIONS FOR SOLDER
OR CONDUCTIVE EPOXY

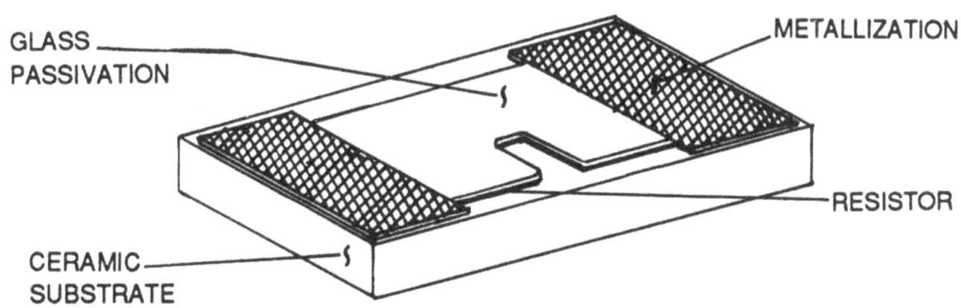


GOLD TOP TERMINATIONS FORWIREBONDING

**Fig. 8.** Outline of surface mount resistors (M/s Mini-Systems, Inc., USA).

techniques have been evolved over the years to suit the specific needs. Soldering is a process by which two metal parts are joined together by the use of a low melting point alloy which when melted forms an intermetallic compound with either of the metal surfaces and bonds the two surfaces when cooled. Tin-lead alloy has been the most commonly used solder for joining copper terminals and lead commonly used in the electronic circuits. Important characteristics of solder materials are:

1. Its melting point should be compatible with the components to be soldered and the environment in which the circuit is to function.
2. It should be able to wet the surfaces to be soldered to enable intermetallic diffusion/bonding to take place to establish good adhesion.
3. It should not leach or scavenge the materials being soldered. In this regard the conventional tin-lead solders are not suitable for use with gold metallization. The problem of leaching is overcome to a great extent with the addition of 2% silver in the solder. Better alternatives are the indium or/and gold based solders but at a higher cost.

A large variety of solder materials have been evolved to meet specific requirements. The characteristics of some typical solders used for the MIC applications are listed in Table 9.

Different techniques used for carrying out soldering in MIC are: (1) Manually, using soldering iron and (2) Reflow with hot gas, hot plate, IR, Vapour Phase, Laser. Such methods as wave soldering and dip soldering, more common in the case of PCB assembly, are not generally used in the fabrication of MICs.

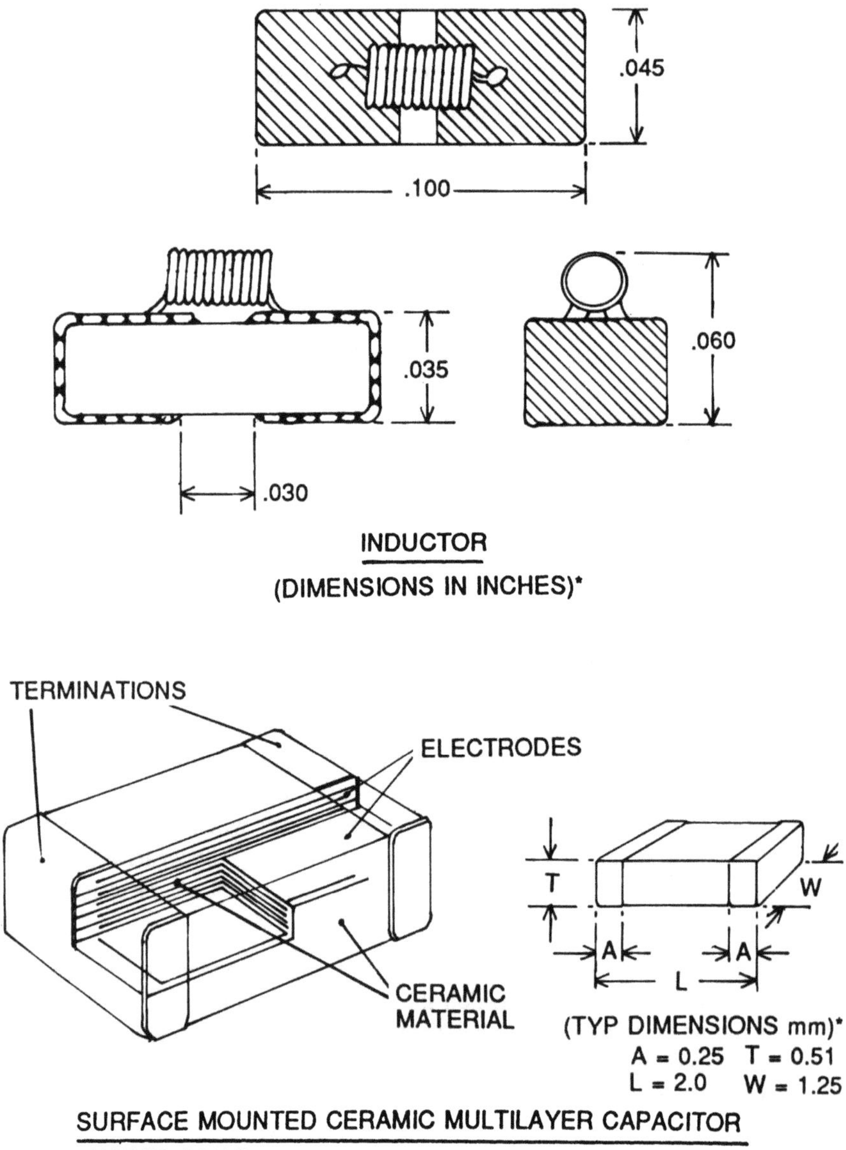


**Fig. 9.** Outline of surface mount inductors and capacitors (M/s Philips Export B.V., Netherlands and M/s Piconics Inc., USA).

For good solder joints it is important that the surfaces to be soldered are clean and free from oxide layers. To help the process of cleaning and to present virgin surface to the solder material, flux materials are used along with the solder material. Both resin based and water soluble (organic or inorganic) fluxes are used. The latter are more active than the resin based fluxes but are easily removed by water cleaning which however is not always possible. Synthetic

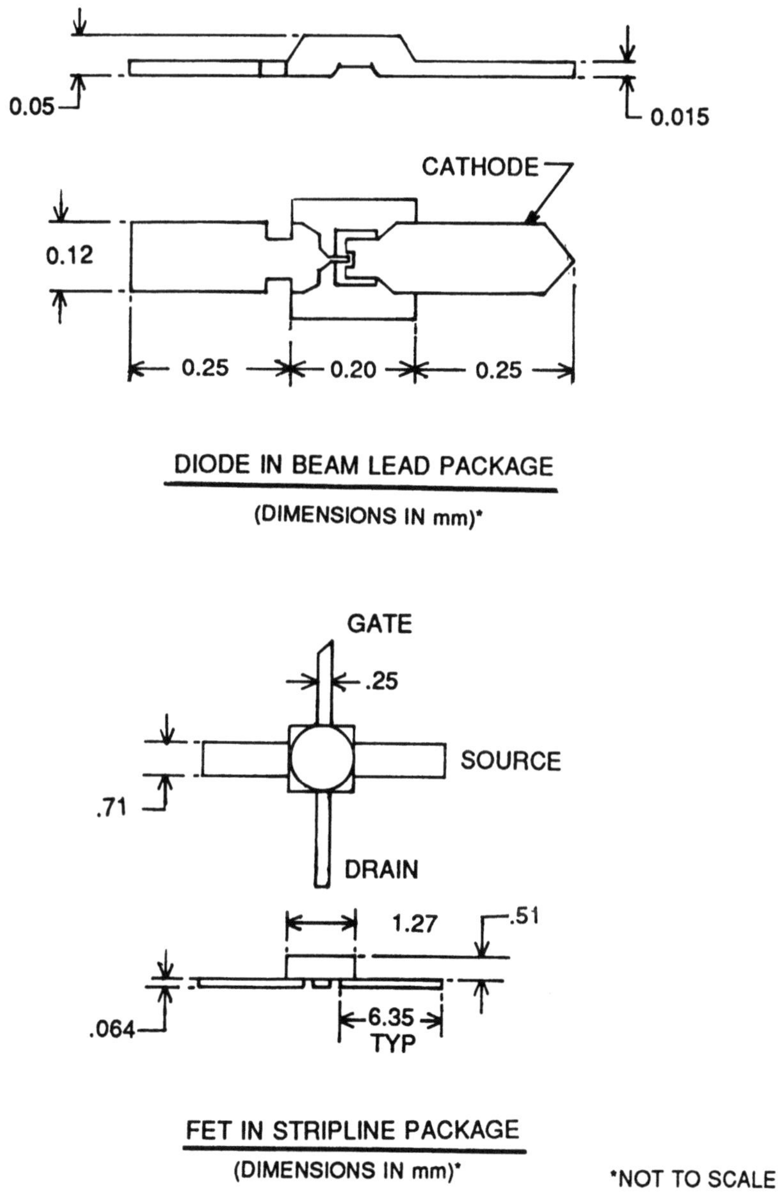


**Fig. 10.** Outline of surface mount active components (M/s California Eastern Laboratories, USA and M/s Avantek Inc., USA).

Table 8. Methods of assembly/interconnection

| Components with leads/terminals | Components in the form of chip/die with bonding pads |
|---|---|
| | (a) Die attachment |
| Solder | Solder |
| | Eutectic |
| Adhesive | Adhesive |
| Welding | (b) Wire/ribbon bonding |

Table 9. Characteristics of some typical solders used for MICs

| Solder material | Melting point (°C) | Remarks |
|---|---|---|
| Au 97%, Si 3% | 370 | Eutectic solder |
| Au 80%, Sn 20% | 280 | Suitable for use with gold metallization |
| In 80%, Pb 15%, Ag 5% | 149 | -do- |
| In 75%, Pb 25% | 163 | -do- |
| In 50%, Pb 50% | 209 | -do- |
| Sn 62%, Pb 36%, Ag 2% | 179 | -do- |
| Sn 63%, Pb 37% | 183 | Standard solder. Not suitable for gold metallization |
| Sn 5%, Pb 95% | 314 | Not suitable for gold metallization |

activated fluxes combine the high activity level typical of water soluble organic fluxes with the benefit of solvent cleanability. The solvent cleaners normally comprise of a blend of polar and nonpolar solvents like propyl alcohol and chlorinated/fluorinated hydrocarbons.

The form in which the solder materials are used depends on the technique used for the soldering. Table 10 summarises the same. The solder preform mainly used for fixing the dies/chips to the substrate is nothing but a disc of solder material with thickness in the range upto 500 microns and is either circular or rectangular (square) in shape with sizes matching those of the dies. It is essential that the die has metallization at the bottom.

Solder paste/cream is similar in characteristics to the thick film paste in the sense that the solder material in the form of fine spherical particles is dispersed in a thixotropic organic polymer along with a solvent to control the viscosity. The paste of course does not contain any glass frit. The flux material is also blended with the paste.

The solder paste can be applied by (1) sharp pin, (2) syringe and (3) screen printing, to the circuit conductor terminations where the components have to be attached.

Application using pin head is essentially a manual process and is only acceptable for laboratory scale activity. There is however absolutely no control over the quantity of solder paste that is being applied and precise control of the location is entirely dependent on the skill of the operator. Application through syringe affords a better control over the quantity of paste. Equipments are available which precisely control the amount of the solder paste to be applied. This can be predetermined by the choice of the nozzle and the other parameters like the pressure in the cylinder. Simpler systems require the nozzle to be moved to the desired location manually and the paste dispensed automatically. More sophisticated systems are available where the nozzle is held on a movable carriage of an X-Y table and its precise location over the substrate is computer controlled. With such systems, the demand on the operator's skill is minimised.

Screen printing affords a precise control of both the quantity of the paste as well as the location. This is a preferred method if a large number of identical circuits are to be fabricated, but is essential if the circuit involves soldering large number of components or components with large number of pins with narrow spacings/gaps. After application of solder cream/paste the components to be soldered are placed in position and the paste is allowed to dry in air to drive off the solvents. The adhesive nature of the paste helps in holding the components in place before it goes for reflow.

Temperature of reflow, time and rate of heating are important in achieving good solder joints. Rapid heating and/or cooling would result in the formation of blisters and voids due to entrapment of volatile organic matter in the paste. For large volume applications conveyerised reflow is convenient. IR reflow is faster without heating too much the other circuit portions. But there is always a danger of excessive heating of some components which have higher absorption in the IR frequencies used for reflow. Vapour phase reflow enables uniform heating of the whole circuit due to the intimate contact of the hot vapour of the liquid. Flourinert fluorinated hydrocarbon with boiling point at 215°C is the most common liquid used for vapour phase reflow. This has the added advantage of cleaning the flux residues of the substrates.

Hot gas reflow involves directing hot gas (air) at the solder points through a specially designed heated tool similar to soldering iron. Special tool tips are provided so that all the leads of the component are heated simultaneously and reflowed. Hot gas reflow has the advantage of heating only the solder points. But basically this is a slow process and useful for repair work wherein one does not heat up the other solder joints in the circuits.

The advantage of laser reflow is also that the laser energy is directed only to the solder joints. The energy and pulse width are so chosen that it is just right for ensuring the reflow without causing the surrounding circuit/components to heat up. Thus any heat sensitive components are protected. Different soldering techniques are compared in Table 10.

### 4.2 Eutectic Die Bonding

This can be considered a special case of soldering in the sense that no external material like solder is involved in joining the two materials. The materials to be bonded together dissolve in each other forming a liquidus at a suitable elevated temperature generally known as eutectic temperature, much below the melting points of individual constituents. The most commonly used system is a silicon die/chip eutectically bonded to gold metallization at about 370°C. In this case the silicon die does not have any metallization on the bottom. The die is brought into intimate contact with the metallization under gentle pressure and rub to aid the eutectic bonding. Eutectic bonding provides very good mechanical strength accompanied with good electrical and thermal conductivity.

**Table 10. Comparision of different soldering techniques**

| Technique | Form of solder material | Automation | Thermal stress to the circuit | Remarks |
|---|---|---|---|---|
| *Melt Soldering* | | | | |
| Soldering iron | Thin wire-plain or flux cored Preform | Strictly manual | Localised heating of the terminals to be soldered | Laboratory scale assembly and rework |
| *Reflow Soldering* | | | | |
| Hot gas | Preform Cream/Paste | Semi-automatic | Only localised heating of the terminals to be soldered | Laboratory scale assembly or batch production. Mainly useful for rework |
| Hot plate | Preform Cream/Paste | Manual to total automation by use of conveyerised system | Whole circuit is raised to the reflow temperature | From batch production to full scale production by integrating with automated dispensing system for application of solder paste and component pick and place equipment |
| IR | Cream/Paste | –do– | Only solder paste heats up for quick reflow. But there is a danger of excessive heating of some components which absorb IR | –do– |
| Vapour phase | Cream/Paste | –do– | Whole circuit is raised to reflow temperature | –do– |
| Laser | Cream/Paste | Best utilised when automated with computer control | Only the solder terminals are heated | Best suited for large production for the investment |

The disadvantages however are: (1) the need to use high temperature for bonding and (2) the strain on the die due to differential thermal expansion between the die material and the substrate—particularly severe for large size dies.

### 4.3 Adhesive Bonding

An effective alternative to soldering is the use of conductive adhesives to attach components to the substrate. This becomes essential if for some reason we cannot subject the components to the temperatures used for soldering. Another area where the adhesives come to rescue is when the two parts cannot be joined together either by use of soldering or by eutectic method. Conductive epoxy adhesives have been first introduced nearly two decades ago. Now it has grown into a mature technology and several other adhesive formulations have come up over the years. Epoxies dominated the field of adhesives used for microelectronic applications. A large variety of formulations have been introduced in the market to meet varied needs of the hybrid manufacturers. Other formulations of adhesives are based on polyimides and the silicones (Jowett 1982 and Soane and Martynenko 1989).

Epoxy resin adhesives are probably the most versatile bonding materials available. These adhesives result in highly cross-linked molecular structures with excellent tensile shear strength but poor peal or cleavage strength. Low cure shrinkage and high resistance to creap under prolonged stress are typical of epoxy adhesives. Their resistance to moisture and solvents is excellent. An epoxy adhesive is primarily composed of an epoxy resin and a curing agent. Secondary ingredients include reactive solvents to adjust viscosity, fillers such as metal particles to make it conductive or ceramics such as alumina powder to increase thermal conductivity. Epoxies are available in solid form or diluted with solvents. The viscosity ranges from sprayable solvent system through thixotropic paste suitable for screen printing. Single component epoxy adhesives incorporate latent curing agents mixed with the epoxy resin, which become active when the adhesive is heated. Variety of other fillers are used to adjust the thermal expansion coefficient, to minimise the cure shrinkage and to adjust the viscosity. The fillers are also used to reduce the exotherm from the cure cycle thus reducing the thermal stress induced in the adhesive bonds.

Silicone adhesives are thermosetting materials. These have excellent heat resistance, however their initial strength is considerably less than that of epoxies. The adhesives are available in the form of viscous solutions either as two component or single component formulations. Curing is normally at elevated temperatures. But RTV (Room Temperature Vulcanizing) silicone adhesives are also available. The desired electrical, mechanical and thermal properties are achieved only after the materials are fully cured. Tensile shear strength is low but their peal strength is relatively high compared to epoxies. Silicone adhesives can withstand temperatures in excess of 300°C for several days.

Polyimide is also a thermosetting material but has the advantage of high tensile shear strength similar to epoxy and high temperature capability like silicone. One disadvantage however arises from the fact that it hardens into a stiff glassy substance and there could be problems with the differential thermal expansion between the component and the substrate during thermal cycling. This is particularly severe if the difference is appreciable and if the component to be mounted is relatively large in size. This material is also available in solution form in consistencies suitable for use with standard dispensing equipments.

When the adhesive has to replace a solder it has to be conductive. This is achieved by loading the adhesive with metal powder, usually silver or gold. Silver-palladium filled adhesives can be used where silver migration is a severe problem at the cost of lower conductivity. The metallic fillers in excess of 70% are used. The adhesives, in particular epoxies can be of two component or single component type; solvent based or solid formulation. Curing time is dependent on temperature. Higher temperature promotes fast curing. Single component adhesives are convenient to use but need special storage conditions to increase the shelf life. Adhesives are available in paste form to enable easy and controlled dispensing through syringe or for screen printing. Important characteristics that decide the choice of a good adhesive apart from providing good adhesion are: (1) minimum shrinkage during curing, (2) minimum out-gassing, (3) high stability over wider operating temperature range and (4) low chloride ion content.

Minimum shrinkage during curing is important otherwise the component will be under great stress. For high reliability, circuits are invariably hermetically sealed in a metal or ceramic package. It is therefore essential that there is minimum out-gassing of harmful substances like water vapour or ionic impurities such as chloride ions which will corrode the aluminium metallization and fine bonding wires. New epoxy formulations which are referred to as 'moisture lock' are being introduced which act as getters to moisture (Estes et al 1985). The requirement of low chloride ion content in the adhesives, as required by new US Military specifications, has been questioned by one of the established epoxy adhesive supplier and is still open to discussion on the acceptable limit (Estes 1986 and Estes and Pernice 1989). Entrapment of the solvents within the adhesive films during curing is one of the greatest problems in attaching the component to the substrate, particularly so when a large size die is to be attached. This not only causes poor performance but also results in reliability problems (Estes 1984). Special epoxy adhesives have been devised which can withstand continuous operating temperature of more than 175°C and intermittent operating temperature of 300–400°C (Estes and Pernice 1989). The main advantage of these over the polyimide formulation is that these epoxies have dual glass transition temperature. The lower Tg value around 80 to 90°C imparts non-brittle or resilient character to the adhesive thereby allowing for thermal cycling from –55 to +155°C without micro cracks developing in the bond line. The higher Tg (160 to 180°C) imparts high temperature strength to the adhesive to enable thermosonic/thermocompression wire bonding to be successfully completed without die floating occurring.

Successful use of adhesives for the Microwave Integrated Circuits operating upto millimeter wave bands has been demonstrated (Hernandez 1974). One point to note is the precise control of the amount of the adhesive to be used

potential and the time. Optimum diameter of the ball is about 1.5 times the wire diameter.

In thermocompression bonding both the bonding tool as well as the substrate are kept at the desired elevated temperature. The bonding tool moves down and the annular rim of the tool squashes the ball against the bonding pad for the preset time leading to bond formation. As already mentioned, the temperature, the bond force (pressure) and the duration of the bonding decide the quality of bond. Excessive force squeezes the material of the wire from underneath the tool leading to weakening of the wire itself. Optimum size of the squashed ball is about 2 to 2.5 times the diameter of the wire. After the bond formation, the tool is withdrawn to a preset height and the substrate is moved to bring the second bond pad into position beneath the capillary tool. The tool is lowered to make the second bond. It may be noted that the second bond cycle does not involve the formation of ball. The wire is squashed by the rim of the capillary to establish the bond. The process of tool lifting after the first bond, translation of the substrate to the second bond position and lowering of the tool for the second bond formation have to be precisely controlled to enable proper 'looping' of the wire bond. This is essential not only to avoid shorting of the wire to the sides of die or chip but also to maintain and control the impedance offered by the bonding wire in the circuit which is important in the microwave integrated circuits. Because of the circular symmetry of the first bond (ball bond) the tool can be moved in any direction to carry out the second bond. The wire is held by a clamp as the tool is moved upto a preset height after the second bond thereby snapping the wire. The wire is cut at the place where it is squashed most.

#### 4.4.1.2 Wedge Bonding

The bonding tool used for this is in the form of a wedge and the wire is bonded to the interconnecting pad by stamping the wire by the foot of the wedge by a preset force for the preset time. For thermocompression bonding here again both the tool and the substrate are kept at the desired temperature. Because of the directional nature of the wedge bond, the tool has to move only backwards to make the second bond. Here again precise control of the tool movement is essential for proper 'looping' of the bonding wire.

Wedge bonding is possible with both gold and aluminium wires whereas ball bonding has not been possible with aluminium because of the difficulty in the formation of the ball. Typical characteristics of the wedge and ball bonding are given in Table 11. Ribbons instead of wires are used particularly for microwave integrated circuits for a better control of the impedance to achieve closer matching to the transmission lines printed on the substrate. Thermocompression bonding is not suitable for aluminium wire bonding and on aluminium metallization due to the contamination of the surface by a thin layer of oxide.

### *4.4.2 Ultrasonic Bonding*

In contrast to thermocompression bonding, the energy required for ensuring bonding between the metallization and the wire is supplied by ultrasonic energy through the bonding tool. The high frequency scrubbing in addition to providing localised heating causes plastic flow of the wire and the contact pad leading to solid state interdiffusion. It also helps in breaking down the surface contaminants if any, in particular the oxide layer on aluminium, helping in intimate contact between the metal surfaces. This is a convenient method where the components cannot be raised to the temperature needed for thermocompression bonding. But it is not useful in bonding highly brittle dies such as GaAs chips.

### *4.4.3 Thermosonic Bonding*

This technique uses both heat as well as ultrasonic energy for the bonding process. This enables one to arrive at the best compromise in the values of substrate/die temperature and ultrasonic energy both of which tend to be lower than in either situations of thermocompression and ultrasonic bonding. In this case the bonding tool is not heated. Thermosonic bonding is versatile, offering wider latitude in the process parameters and is currently the most widely used method for wire bonding. Typical range of parameters in the three cases of wire bonding are given in Table 11.

**Table 11. Comparision of wire/ribbon bonding techniques**

| | Ball bonding | Wedge bonding | |
|---|---|---|---|
| Material | Gold | Gold | Aluminium |
| Form | Wire | Wire/Ribbon | Wire |
| Dimensions | 0.7 to 2 mil (18 to 50 μm) | 0.5 to 3 mil (12.5 to 75 μm)/ 0.5 × 10 mil (12.5 × 250 μm) | 1 to 4 mil (25 to 100 μm) |
| Elongation | 4–6% | 4–6% | 4–6% |
| *Bonding Parameters* | | | |
| Temperature | | | |
| Thermocompression | 300–400°C | 300–400°C | Not possible |
| Ultrasonic | Ambient | Ambient | Ambient |
| Thermosonic | 100–200°C | 100–200°C | 100–20°C |
| Ultrasonic power (maximum) | 2 Watt | 2 Watt | 2 Watt |
| Force (typical) for 1 mil wire | 30 g | 40 g | 30 g |
| Time (sec) | 10–100 | 10–100 | 10–100 |
| *Capabilities* | | | |
| Minimum bond pad size | 50 μm | 20 μm | 20 μm |
| Bond pull strength for 1 mil wire | 8–10 g | 5 g | 5 g |

### *4.4.4 Materials for Bonding Wire/Ribbon*

As already mentioned gold and aluminium in the form of wire/ribbon are the two materials mostly used for interconnection between the bonding pads on the die and the metallised terminations on the substrate. Silver and copper have been considered as alternatives but have not replaced gold and aluminium. The dimensions of the wire and the ribbon commonly employed are indicated in Table 11. One of the important properties of the wire/ribbon that is significant for the wire bonding is ductility. Higher the ductility more plastic flow will occur for any given applied pressure resulting in more intimate interface contact. Oxidation is less important in case of noble metals in determining the bond stability. Normally successful bonding is achieved between metals of similar crystalline structures perhaps due to better mutual solubility in compression bonding. For better bondability therefore most of the metallization structures are given a final layer of gold plating. However, gold wire bonding on aluminium alloy metallization on silicon chip is a common feature. But a potential failure mechanism in such configuration is the formation of brittle intermetallic phases at the interface normally known as the purple plague. This is controlled to some extent though not completely eliminated by alloying small percent of palladium in the gold wire.

Elongation of the wire (the amount the wire stretches before it breaks expressed in %) is very important since it decides the formation of good and stable bond wire loop. Too much elongation results in the sagging of the loop leading to bond failures. Too little elongation would make the formation of loop altogether difficult. Typical values are given in Table 11.

## 5 CONCLUSION

The area of the materials used in the fabrication of Microwave Integrated Circuits is very wide involving materials ranging from polymers, ceramics to special high purity metals and alloys. Each of these serves a specific purpose and possesses special characteristics. In this chapter it was only possible to give an overall picture of the various processes involved in the fabrication of MICs and the materials used therein. No attempt has been made to go into the details of the preparation and characterisation of materials themselves. For more detailed account the reader should consult the books and articles cited under references. Secondly, as regards the characteristics of various materials listed in the tables it should be emphasised that these are meant to be only indicative of their typical values and cannot be used for any design calculations. By the very nature of these materials there is bound to be a variation in these characteristics depending on the detailed composition and the method of preparation. It is therefore advised to use the values specified by the manufacturers of the materials and if necessary and possible carry out inhouse characterisation to confirm the listed specifications.

Another area which needs special attention is the defence and space electronics wherein microwave circuits and systems find wide ranging applications. These applications demand high reliability under adverse environmental conditions. This implies that the system should have the ability to perform over a wide temperature range and under conditions of high humidity, large vibration, high acceleration, great impact etc. Detailed specifications of these and test methods are given in US Military standards 883B and recently introduced standards 1772 or other similar standards. The need for the final subsystem to meet stringent military and space standards implies that the materials used therein have to perform well and meet the stipulated standards. One therefore has to ensure that the materials used for the fabrication of microwave systems for defence or space applications meet the stipulated standards by thorough screening and characterisation as per the methods specified therein.

## REFERENCES


1. Berry RW, Hall PM and Harris MT (1968) 'Thin Film Technology' (Princeton, D. Van Nostrand Co. Inc.).
2. Bhat B and Koul SK (1989) 'Stripline-like Transmission Lines for Microwave Integrated Circuits' (New Delhi: Wiley Eastern Ltd).
3. Blum JB and Anzai K (1989) 'Aluminium Nitride Substrates for Hybrid Microelectronics Application' Hybrid Circuit Technology, 6 No. 8, p. 7–14.
4. Chopra KL and Kaur I (1983) 'Thin Film Device Applications' (New York: Plenum Press).
5. Corkhill JR (1976) 'Thick Films at High Frequencies' in 'Hand Book of Thick Film Technology' Eds. Holmes PJ and Loasby RG (Ayr, Scotland: Electrochemical Publications Ltd.).
6. Edwards PR (1991) 'Manufacturing Technology in the Electronics Industry—An introduction' (London: Chapman & Hall).
7. Estes RH and Pernice RF (1989) "Die Attach Adhesives—Evaluation of VceSAT and θjc Performance in Power Devices" Proceedings of ISHM International Symposium on Microelectronics, Verginia, USA, p. 664–669.
8. Estes RH (1986) 'Adhesives for Military Hybrids' Hybrid Circuit Technology, 3 No. 7, p. 21–24.
9. Estes RH, Kulesza FW and Banfield CE (1985) 'Recent Advances made in Die-Attach Adhesives for Microelectronic Applications' Proceedings of ISHM International Symposium on Microelectronics, Anaheim, USA. p. 391–401.
10. Estes RH (1984) 'The Effect of Porosity on Mechanical, Electrical and Thermal Characteristics of Conductive Die-Attach Adhesives' Solid State Technology, August p. 191–197.
11. Grovenor CRM (1989) 'Microelectronic Materials' (Bristol: Adam Hilger).
12. Haskard MR (1988) 'Thick Film Hybrids—Manufacture and Design' (Sydney: Prentice-Hall of Australia Pty Ltd.).
13. Hernandez L (1974) 'Epoxy Techniques for Hybrid Microwave Integrated Circuits' Proceedings of ISHM International Microelectronics Symposium, Boston, USA, p. 272–5.
14. Holland L (1965) 'Thin Film Microelectronics—The Preparation and Properties of Components and Circuit Arrays' (London: Chapman and Hall Ltd.).
15. Holmes PJ and Loasby RG (1976) 'Hand Book of Thick Film Technology' (Ayr, Scotland: Electrochemical Publications Ltd.).
16. Hughes JE (1987) 'A Review of Thin Film Resistors and their Assembly Problems' Hybrid Circuits, May.
17. Jowett CE (1982) 'Materials and Processes in Electronics' (London: Hutchinson & Co (Publishers) Ltd.).
18. Kugler TR and Culkin TJ (1989) 'Ceramic Processing with Lasers' Hybrid Circuit Technology, 6 No. 8, p. 45–48.
19. Licari JJ and Enlow LR (1988) 'Hybrid Microcircuit Technology Hand Book:

Materials, Processes, Design, Testing and Production' (Park Ridge, New Jersey: Noyes Publications).
20. Maissel LI and Glang R (1970) 'Hand Book of Thin Film Technology' (New York: McGraw-Hill).
21. Pitt KEG (1981) 'An Introduction to Thick Film Component Technology' (Luton, England: Mackintosh Publications Ltd.).
22. Prasad SD and Vijayaraghavan MS (1987) 'Fabricating Hybrid MICs on PTFE based Boards' Hybrid Circuit Technology **4** No. 3, p. 19–21.
23. Reddy TRs (1988) 'MMICs in Hybrids' Proceedings of ISHM (India) '88 Third Annual Conference of International Society for Hybrid Microelectronics (India Chapter), Hyderabad, INDIA p. 31–45.
24. Siddall G (1965) 'The Properties of Passive Circuit Elements' in 'Thin Film Microelectronics—The Preparation and Properties of Components and Circuit Arrays' Ed. Holland L (London: Chapman and Hall Ltd.).
25. Soane DS and Martynenko Z (1989) 'Polymers in Microelectronics: Fundamentals and Application' (Amsterdam: Elsevier Science Publishers B.V.).
26. Stuart RV (1983) 'Vacuum Technology, Thin Film and Sputtering: An introduction' (Orlando, USA: Academic Press Inc.).
27. Sunahara K and Takabatake M (1990) 'Low-Temperature Fired Multilayer Circuit Board for High Frequency Application' Hybrid Circuit Technology, 7 No. 2, p. 35–41.
28. Taconic Tech Topic 'Processing of PTFE/Woven Glass Laminates' Microwave Dielectrics Division (Petersberg, USA: M/s Taconic Plastics Ltd).
29. Van Nie AG, Goedbloed W and Kersuzan G (1981) 'Relaibility and Degradation of Microwave Integrated Circuits' in 'Reliability and Degradation' Eds. Howes MJ & Morgan DV (New York: John Wiley & Sons Ltd) p. 363-439.

# 2

# Materials and Technology for Microwave Integrated Circuits

Bharathi Bhat and Shiban K. Koul
Centre for Applied Research in Electronics
Indian Institute of Technology, Delhi, New Delhi, India

## 1. INTRODUCTION

Advances over the past two decades in the planar techniques and technology and also in the miniature microwave solid state devices compatible with this technology have led to the realization of compact microwave integrated circuits (MICs). These circuits are built on planar transmission lines such as the stripline, microstrip line, suspended stripline, suspended microstrip, inverted microstrip, slot line, coplanar waveguide and coplanar strips. The basic geometries of these lines are shown in Fig. 1. For MICs, these planar transmission lines are formed on substrates made of low loss dielectrics, ferrimagnetics and semiconductors. A typical MIC consists of one or more of the following parts: (a) distributed elements (b) lumped elements in planar/discrete form, (c) semiconductor devices and (d) special elements such as dielectric resonators and ferrite discs. Two different technologies; viz. *hybrid* and *monolithic* are well accepted for the fabrication of MICs.

In the *hybrid* technology, the passive circuit which consists of distributed and/or planar lumped elements is realized as a conductor pattern on the substrate. This is a single level process and involves metallization of the substrate through vacuum evaporation/RF sputtering and defining the desired pattern through accurate photolithographic techniques. Resistors, if required, can be incorporated by depositing resistive films at appropriate places. Semiconductor devices and other discrete passive chip devices (capacitors and resistors) if any, are bonded or soldered to the circuit pattern. Typical operating range of hybrid MICs is 1 to 30 GHz and with the adoption of more sophisticated photolithographic techniques, their operation can be extended into the millimetre wave range up to about 120 GHz. Nearly all coaxial and waveguide components, particularly for operation at low power levels can now be realized in hybrid MIC form. The advent of this technology has resulted in considerable reduction in size and weight, ease of mass production with improved reproducibility and potential low cost of functional circuits. The technology offers the advantage of combining

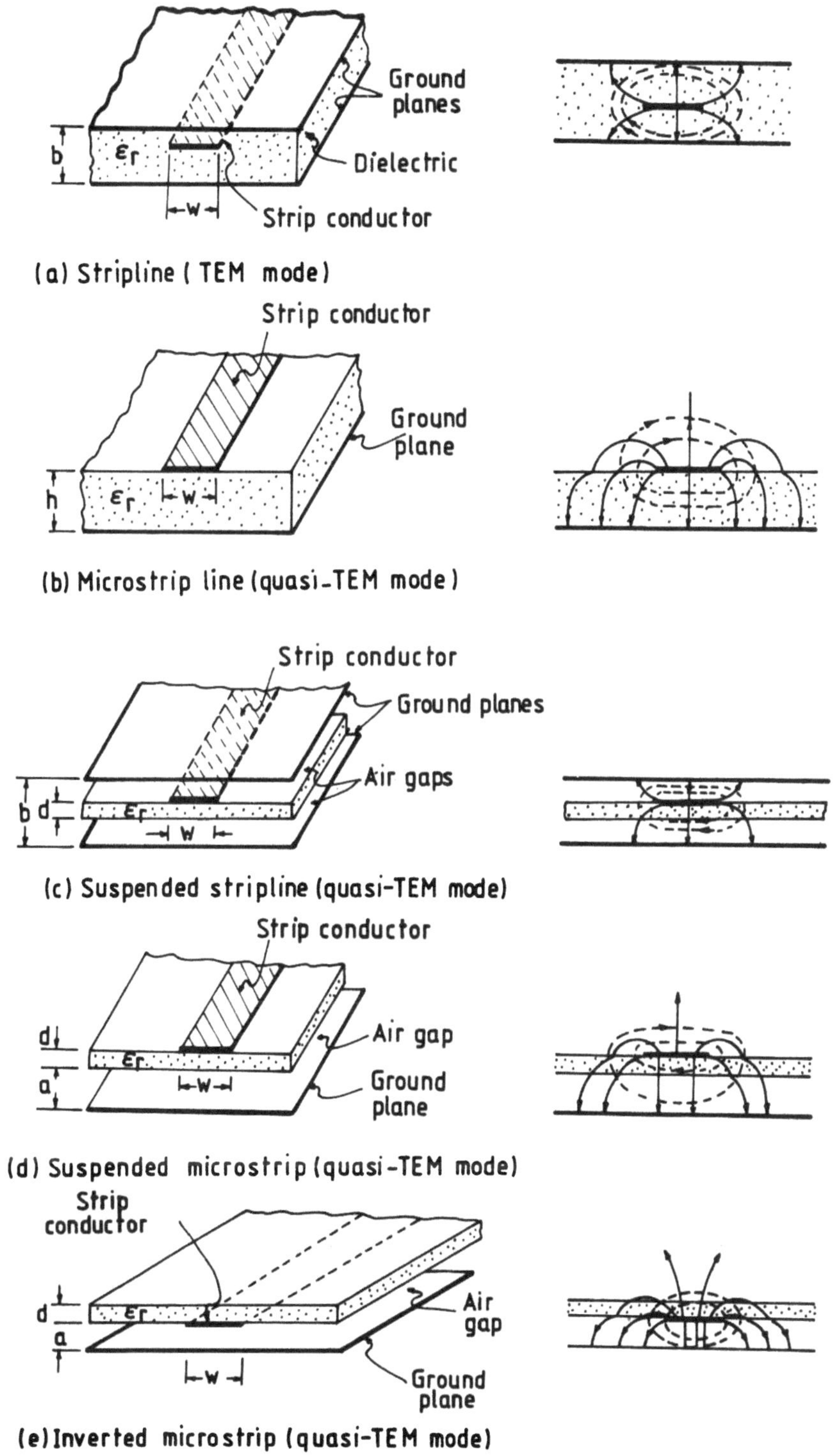
Ground
planes
$\varepsilon_r$
b
Dielectric
W
Strip conductor
(a) Stripline (TEM mode)
Strip conductor
Ground
plane
h
$\varepsilon_r$
W
(b) Microstrip line (quasi-TEM mode)
Strip conductor
Ground planes
Air gaps
b
d
$\varepsilon_r$
W
(c) Suspended stripline (quasi-TEM mode)
Strip conductor
d
Air gap
$\varepsilon_r$
a
W
Ground
plane
(d) Suspended microstrip (quasi-TEM mode)
Strip
conductor
d
$\varepsilon_r$
Air
gap
a
W
Ground
plane
(e) Inverted microstrip (quasi-TEM mode)

**Fig. 1** (Contd.)

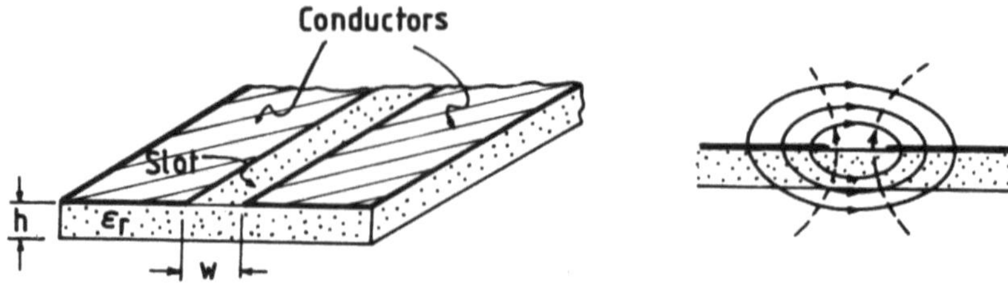


(f) Slotline (non-TEM mode)

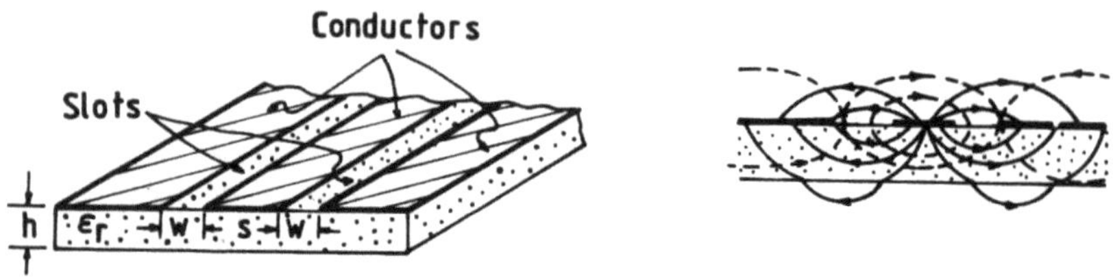


(g)Coplanar waveguide (quasi-TEM mode)

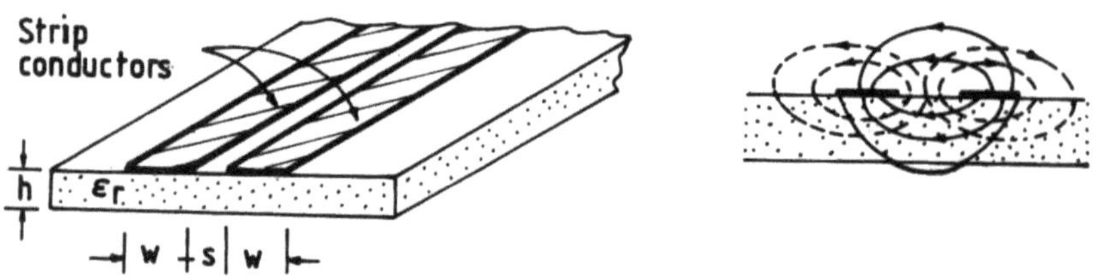


(h) Coplanar strips(quasi-TEM mode)

**Fig. 1.** Planar transmission lines for MICs and typical fields in their cross-sectional planes, ——— E field lines,- - - -H field lines.

multi-function circuits without the interconnecting wires thus permitting realization of compact integrated modules with highly reliable performance. The *monolithic* technology achieves much higher degree of circuit integration than the *hybrid* technology and the circuits are popularly referred to as MMICs (monolithic microwave integrated circuits). In this technology, active devices are grown in-situ on or within a semiconductor substrate, and planar passive circuitry and interconnections are then deposited on the substrate or grown in it. The fabrication involves highly reproducible device technology using diffusion and/or ion implantation, multilevel metallization and a composite process involving photolithography/electron beam lithography. As compared with the *hybrid* technology, the *monolithic* technology offers significant reduction in size and weight; and improved reliability and reproducibility through elimination of wire bonds. Embedding of active devices within the substrate and elimination of wire bonds reduce the undesired parasitics, thereby improving the capability for broadband performance. Furthermore, the ability to incorporate multifunctional capability on a single chip permits the realization of integrated receiver front end and transmit-receive modules.

*Monolithic* technology is capital intensive and involves high cost manufacture. MMICs can become cost-effective only when there is a requirement for large volume production such as for DBS receiver systems and phased array radars. *Hybrid* MICs, on the other hand, have proved to be cost effective and are extensively used in a variety of practical radar, communication, navigation, and sensing systems at microwave and even millimetre wave frequencies. The applications are numerous; microwave test instrumentation, satellite communication, electronic communication, electronic warfare, electronic scanning through phased arrays, and direct broadcast satellite reception.

In the following, we provide a comprehensive review of the basic techniques and technology of MICs. Planar transmission lines, substrate materials, basic elements of planar circuits, and the *hybrid* and *monolithic* technologies are covered.

## 2. PLANAR TRANSMISSION LINES

Planar transmission lines form the basic transmission media for microwave integrated circuits whether the circuit is fabricated in *hybrid* or *monolithic* form. Figure 1 illustrates the basic geometries of the various commonly used planar transmission lines and also the typical field lines for the dominant mode propagation. Besides these, several coupled conductor versions of structures shown in Fig. 1 (a)-(e) and other variants are also used in a variety of circuit applications. The analyses and electrical characteristics of such planar transmission lines are available in several books and reprint volumes [1–7]. In this section, we review the salient characteristics of the commonly used structures shown in Fig. 1. Throughout the section, we use the symbols $Z$ and $\lambda$ to denote the characteristic impedance and guide wavelength, respectively, of the planar transmission line under consideration, $Z^a$ for the characteristic impedance of the same planar transmission line with all dielectrics replaced by air, and $\lambda_0$ to denote the free space wavelength.

### 2.1 Stripline (Fig. 1a) [1, 2, 8, 9]

The basic structure of a stripline consists of a flat strip conductor situated symmetrically between two large ground planes with the intervening space homogeneously filled with a dielectric. The dominant mode of propagation in a stripline is a pure transverse electromagnetic (TEM) mode; that is, the electric and magnetic field components lie entirely transverse to the direction of propagation. The fields concentrate around the strip conductor and decay rapidly with distance away from the strip in the lateral direction. The characteristic impedance $Z$ and the guide wavelength $\lambda$ of the homogeneous stripline are given by

$$Z = Z^a/\sqrt{\varepsilon_r} \tag{1}$$

$$\lambda = \lambda_0/\sqrt{\varepsilon_r} \tag{2}$$

where $\varepsilon_r$ is the relative dielectric constant of the dielectric filling. The following closed from expressions reported in the literature [8] are known to yield virtually exact results for the characteristic impedance:

For $w/b \leq 0.5$

$$Z\sqrt{\varepsilon_r} = 29.979 \ln\left[2\frac{(1+\sqrt{k})}{(1-\sqrt{k})}\right] \tag{3}$$

For $w/b \geq 0.5$

$$Z\sqrt{\varepsilon_r} = \frac{29.979\pi^2}{\ln\left[2\frac{(1+\sqrt{k'})}{(1-\sqrt{k'})}\right]} \tag{4}$$

where

$$k = \text{sech}\ (\pi w/2b) \tag{5a}$$

$$k' = \tanh\ (\pi w/2b) \tag{5b}$$

In the above formulas, $w$ is the strip width and $b$ is the ground plane spacing. The thickness of the strip conductor is assumed to be negligible.

For practical circuits, the dimensions of the stripline are chosen so as to ensure operation in the TEM mode while keeping the losses to a minimum. The attenuation in a stripline decreases with a decrease in $Z$ for a constant ground plane spacing $b$, and also with an increase in $b$ for a constant $Z$ [9]. The maximum value of $b$ is limited by the onset of the TE or TM mode, whichever propagates first. The cut-off wavelengths of the lowest order TE and TM modes, denoted as $\lambda_{c,TE}$ and $\lambda_{c,TM}$, respectively, are given by [10]

$$\lambda_{c,TE} = \sqrt{\varepsilon_r}\ \{2w + (\pi b/2)\} \tag{6}$$

$$\lambda_{c,TM} = 2\sqrt{\varepsilon_r}\ b \tag{7}$$

The stripline is an excellent medium for realizing passive components. Because of its operation in the TEM mode, excellent directivity can be achieved in coupled line directional couplers and high isolation in other coupled line components. The structure, however, is not convenient for incorporating chip devices. Practical stripline circuits are provided with side metallic walls so as to form a complete shielded box as shown in Fig. 2a. The side wall separation $c$ is chosen large enough ($c/w \geq 10$) so that the walls have negligible effect on the propagation characteristics.

## 2.2 Microstrip Line (Fig. 1b) [2-7, 11-21]

Unlike the stripline, the microstrip is an inhomogeneous transmission line. It consists of a dielectric substrate with a strip conductor on one surface and metallization to form a ground plane on the reverse surface. The dominant mode of propagation in this line is quasi-TEM; that is, the mode is dominantly TEM with small field components along the direction of propagation. As shown in Fig. 1b, the fields are confined to the vicinity of the strip conductor with a larger concentration inside the dielectric substrate and less in the air region

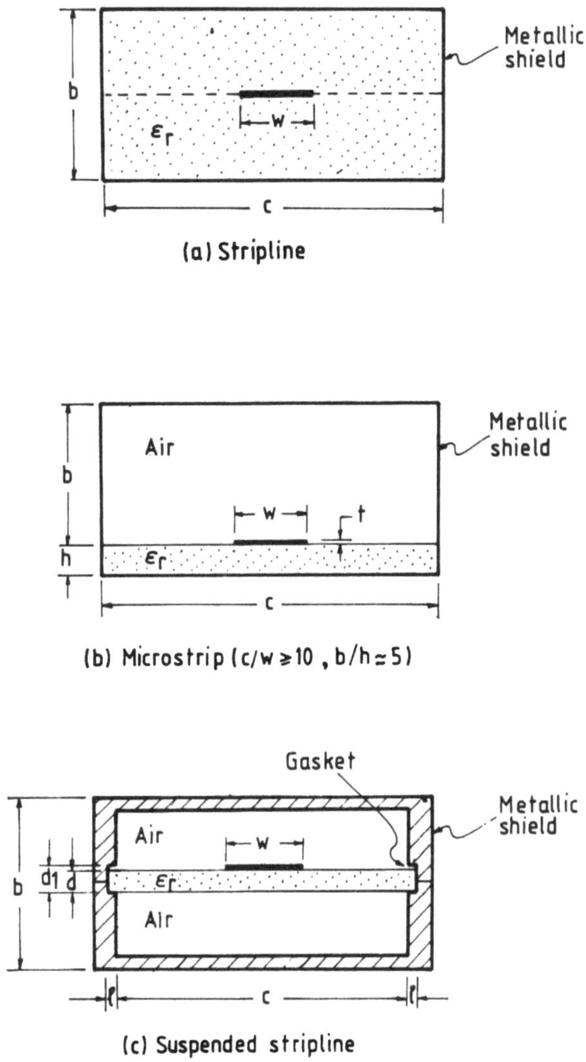


**Fig. 2.** Examples of shielded planar transmission lines.

above. Larger the value of $\varepsilon_r$ of the substrate, the greater is the relative concentration of energy inside the substrate and lesser is the radiation. The characteristic impedance and guide wavelength of a microstrip can be related to $Z^a$ and $\lambda_0$ in the form [11]

$$Z = Z^a/\sqrt{\varepsilon_{eff}} \tag{8}$$

$$\lambda = \lambda_0/\sqrt{\varepsilon_{eff}} \tag{9}$$

where $\varepsilon_{eff}$ is the *effective dielectric constant* (relative) of the microstrip medium. It is defined as the relative dielectric constant of an equivalent homogeneous microstrip which has the same phase velocity as the original (inhomogeneous) microstrip. The value of $\varepsilon_{eff}$ lies in the range

$$0.5(1+\varepsilon_r) \le \varepsilon_{eff} \le \varepsilon_r \tag{10}$$

depending on the value of $w/h$. For small strip widths ($w/h << 1$), the fields are distributed nearly equally in the substrate and air regions and hence $\varepsilon_{eff}$ approaches the lower limit $(1 + \varepsilon_r)/2$. For large strip widths ($w/h >> 1$), the electric field is confined mostly between the strip conductor and the ground plane. The microstrip then resembles a parallel-plate capacitor with $\varepsilon_{eff}$ approaching $\varepsilon_r$.

*Quasi-static formulas for Z and $\varepsilon_{eff}$*

Closed form expressions for evaluating Z and $\varepsilon_{eff}$ of microstrip lines have been reported by several investigators [12-14]. Under the quasi-static approximation and assuming the strip thickness to be negligible, the expression for $Z$ is given by [15]

$$Z = \{60/\sqrt{\varepsilon_{eff}}\} \ln[(h/w)\,F(w/h) + \{1 + (2h/w)^2\}^{1/2}] \tag{11a}$$

where

$$F(w/h) = [6 + (2\pi - 6)\exp\{-(30.666h/w)^{0.7528}\}] \tag{11b}$$

The accuracy of these expressions is reported to be better than 0.01% for $w/h \le 1$ and 0.03% for $w/h << 100$ [14, 15]. The effective dielectric constant is given by [14]

$$\varepsilon_{eff} = \{(\varepsilon_r + 1)/2\} + \{(\varepsilon_r - 1)/2)\}\{1 + (10h/w)\}^{-a.b} \tag{12a}$$

$$a = [1 + (1/49)\ln\left\{\frac{(w/h)^4 + (w/52h)^2}{(w/h)^4 + 0.432}\right\} + (1/18.7)\ln\{1 + (w/18.1h^3)\}] \tag{12b}$$

$$b = 0.564\left(\frac{\varepsilon_r - 0.9}{\varepsilon_r + 3}\right)^{0.053} \tag{12c}$$

The accuracy of this formula is reported to be better than 0.2% for $\varepsilon_r \le 128$ and $0.01 \le w/h \le 100$ [14, 15]. The maximum frequency up to which the above quasi-static formulas can be used is given by [15]

$$f_d = 0.3\left[\frac{Z}{h(\varepsilon_r - 1)^{1/2}}\right]^{1/2} \tag{13}$$

where $h$ is in cm and $f_d$ is in GHz. From (13), it can be seen that the upper frequency limit for quasi-static operation can be extended by choosing substrates of lower thickness and lower dielectric constant.

*Effect of Dispersion*

At higher frequencies, the fields tend to concentrate more within the dielectric substrate. Consequently, $\varepsilon_{eff}$ increases with an increase in frequency, with the value approaching the substrate permittivity $\varepsilon_r$ asymptotically in the limit frequency tends to infinity [16]. The following closed form dispersion formula can be used to compute the frequency dependent $\varepsilon_{eff}$ (denoted as $\varepsilon_{eff}(f)$) in the range $2 \le \varepsilon_r \le 16$, $0.06 \le w/h \le 16$ and $f \le 100$ GHz [17]:

$$\varepsilon_{eff}(f) = \left[\frac{\sqrt{\varepsilon_r} - \sqrt{\varepsilon_{eff}}}{1 + 4F^{-1.5}} + \sqrt{\varepsilon_{eff}}\right]^2 \tag{14a}$$

where

$$F = \frac{4h\sqrt{(\varepsilon_r - 1)}}{\lambda_0}\,[0.5 + \{1 + 2\log(1 + (w/h))\}^2] \tag{14b}$$

and $\varepsilon_{eff}$ is the quasi-static value given by (12). Similarly, the effect of dispersion on Z can be obtained from the closed form expression given by [14]:

$$Z(f) = Z\left(\frac{\varepsilon_{eff}(f) - 1}{\varepsilon_{eff} - 1}\right)\sqrt{\{\varepsilon_{eff}/\varepsilon_{eff}(f)\}} \tag{15}$$

where Z is the quasi-static value given by (11).

*Losses*

The propagation loss (neglecting radiation) in a practical microstrip is primarily due to two types of dissipative losses; conductor loss due to the finite resistivity of the conductor and dielectric loss due to the finite loss tangent of the dielectric substrate. Microstrip losses have been studied by several investigators [18–20]. If we denote $\alpha_c$ as the attenuation constant due to the conductor loss and $\alpha_d$ as the attenuation constant due to the dielectric loss, both expressed in dB/unit length, then the total attenuation constant $\alpha$ is given by

$$\alpha = (\alpha_c + \alpha_d) \text{ dB/unit length} \tag{16}$$

The following closed form expressions due to Pucel et al [18] can be used to compute $\alpha_c$ in dB/unit length:

$$\alpha_c = \begin{cases} 1.38\,\dfrac{R_s'}{hZ}\left[\dfrac{32 - (w'/h)^2}{32 + (w'/h)^2}\right]A & w/h \le 1 \\[2ex] 6.1 \times 10^{-5}\,\dfrac{R_s'\,Z\,\varepsilon_{eff}(f)}{h}\left[\dfrac{w'}{h} + \dfrac{0.667w'/h}{(w'/h) + 1.444}\right]A & w/h \ge 1 \end{cases} \tag{17a}$$

$$A = \begin{cases} 1 + (h/w')\,\{1 + (1.25t/\pi w) + (1.25/\pi)\ln(4\pi w/t)\} & w/h \le 1/2\pi \\ 1 + (h/w')\,\{1 - (1.25t/\pi h) + (1.25/\pi)\ln(2h/t)\} & w/h \ge 1/2\pi \end{cases} \tag{17b}$$

where

$$w'/h = \begin{cases} w/h + (1.25/\pi)(t/h)\,\{1 + \ln(4\pi w/t)\} & w/h \le 1/2\pi \quad (18a) \\ w/h + (1.25/\pi)(t/h)\,\{1 + \ln(2h/t)\} & w/h \ge 1/2\pi \quad (18b) \end{cases}$$

$$R_s' = R_s[1 + (2/\pi)\tan^{-1}\{1.4(\Delta/\delta)^2\}] \tag{18c}$$

$$R_s = \sqrt{(\omega\mu_0/2\sigma)} \tag{18d}$$

$$\delta = (1/R_s\sigma) \tag{18e}$$

In these formulas, $R_s$ is the surface resistivity of the conductor, $\delta$ is the skin depth, $\sigma$ is the conductivity of the metal, $\Delta$ is the rms surface roughness and $t$ is the thickness of the strip conductor.

The expression for $\alpha_d$ in dB/unit length is given by [20]

$$\alpha_d = 27.3\{\varepsilon_r/\sqrt{\varepsilon_{\text{eff}}(f)}\}\left(\frac{\varepsilon_{\text{eff}}(f) - 1}{\varepsilon_r - 1}\right)(\tan\delta/\lambda_0) \tag{19}$$

*Q-factor*

The $Q$-factor of a microstrip corresponding to the conductor and dielectric losses can be obtained from [15]

$$Q = \frac{\pi\sqrt{\varepsilon_{\text{eff}}(f)}}{\lambda_0\alpha} \tag{20}$$

where $\alpha$ is the total attenuation constant specified in dB/unit length.

*Maximum Frequency of Operation*

The maximum frequency of operation of a microstrip is limited primarily by the onset of higher order modes, higher losses and the requirement on stringent fabricational tolerances. The frequency at which significant coupling takes place between the quasi-TEM mode and the TM-surface wave mode is given by [21]

$$f_{c,\text{TM}} = \frac{v_0\tan^{-1}(\varepsilon_r)}{\sqrt{2}\pi h\sqrt{(\varepsilon_r - 1)}} \tag{21}$$

where $v_0$ is the velocity of electromagnetic waves in free space. A major advantage of the microstrip is that its surface is accessible for mounting passive as well as active discrete devices. It is also a versatile medium for realizing a variety of circuit forms and combining several circuit functions. Practical microstrip circuits are housed in a shielded enclosure in order to suppress radiation and provide electromagnetic shielding. Fig. 2b shows the cross-section of a shielded microstrip. The dimensions of the enclosure are selected such that the waveguide modes are below cut-off, and the top and side walls have practically no effect on the propagation characteristics. Referring to the dimensional parameters marked in Fig. 2b, these criteria can be met by choosing $c/w \simeq 10$, $b/h \simeq 5$ and $w, h << \lambda/2$.

## 2.3 Suspended Stripline (Fig. 1c) [2, 5, 6, 22-27]

The suspended stripline is the most useful variant of the stripline. It is essentially an inhomogeneous stripline in which the substrate carrying the strip conductor is placed symmetrically between the two ground planes thereby creating an air gap on either side of the substrate. The dominant mode of propagation is quasi-TEM. The introduction of air gap results in the reduction of the effective dielectric constant of the propagating medium. This configuration, therefore, permits larger circuit dimensions leading to relaxed dimensional tolerances and increased accuracy of circuit fabrication as compared with the microstrip. The presence of the air gap also reduces the conductor loss in the ground plane, because most of the electromagnetic energy gets concentrated in the dielectric substrate. By choosing the substrate sufficiently thin, the effective dielectric constant can be made close to that of air, thereby extending the frequency range of operation in the dominant mode, which reduces to nearly TEM [22]. Because of the above advantages, the suspended stripline is ideally suited for operation in the millimetre wave range. Practical suspended striplines require some provision for supporting the dielectric substrate and also for electromagnetic shielding. Fig. 2c shows the cross-section of a shielded structure with grooves on the side walls to suspend the substrate.

*Closed form expressions for Z and $\varepsilon_{eff}$*

The characteristic impedance and guide wavelength for a suspended stripline (Fig. 2c) can be obtained from the following closed form expressions [27]:

For $0 < w < c/2$

$$\lambda = \lambda_0 / \sqrt{\varepsilon_{eff}} \tag{22a}$$

where

$$\sqrt{\varepsilon_{eff}} = [1 + \{E - F \ln (w/b)\} \ln (1/\sqrt{\varepsilon_r})]^{-1} \tag{22b}$$

and

$$Z = 60\pi \left[ V + R \ln \left\{ \frac{6}{(w/b)} \right\} + \left\{ 1 + \frac{4}{(w/b)^2} \right\}^{1/2} \right] \tag{22c}$$

where

$$\begin{aligned} E &= 0.2077 + 1.2177\,(d/b) - 0.08364(c/b) \\ F &= 0.03451 - 0.1031\,(d/b) + 0.01742(c/b) \\ V &= -\,1.7866 - 0.2035\,(d/b) + 0.4750(c/b) \\ R &= 1.0835 + 0.1007\,(d/b) - 0.09457(c/b) \end{aligned} \tag{22d}$$

For $c/2 < w < c$

$$\lambda = \lambda_0 / \sqrt{\varepsilon_{eff}} \tag{23a}$$

where

$$\sqrt{\varepsilon_{eff}} = [1 + \{E - F \ln (w/b)\} \ln (1/\sqrt{\varepsilon_r})]^{-1} \tag{23b}$$

and

$$Z = 120\pi[V + R\{(w/b) + 1.393 + 0.667 \ln ((w/b) + 1.444)\}^{-1}] \tag{23c}$$

where

$$E = 0.464 + 0.9467\,(d/b) - 0.2063\,(c/b)$$
$$F = 0.1424 + 0.3017\,(d/b) - 0.02411\,(c/b)$$
$$V = -\,0.6301 - 0.07082\,(d/b) + 0.247\,(c/b)$$
$$R = 1.9492 + 0.1553\,(d/b) - 0.5123\,(c/b) \qquad (23d)$$

These expressions are valid for $1 \le c/b \le 2.5$, $1 < \varepsilon_r < 4$, $0.1 < d/b < 0.5$. The accuracy of these expressions is within ± 2% for $0 < w < c/2$ and ± 3% for $c/2 < w < c$ compared to rigorous methods.

In practice, the dimensions of the housing and grooves must be carefully chosen so as to avoid propagation of undesired waveguide modes. The first waveguide mode to appear can be either the $TE_{10}$-type or the distorted $TE_{01}$ mode depending on the dielectric constant and the dimensions of the enclosure. For the $TE_{10}$ type of mode, the cut-off frequency can be obtained from the following relation [23] (refer Fig. 2c):

$$\tan\left(0.5c\,\beta_c\sqrt{\varepsilon_1}\right) . \tan\left(\beta_c\sqrt{\varepsilon_2}\,l\right) = (b/d_1)\sqrt{(\varepsilon_2/\varepsilon_1)} \qquad (24a)$$

where

$$\varepsilon_1 = [1 - (d/b)\{(\varepsilon_r - 1)/\varepsilon_r\}]^{-1} \qquad (24b)$$

$$\varepsilon_2 = [1 - (d/d_1)\{(\varepsilon_r - 1)/\varepsilon_r\}]^{-1} \qquad (24c)$$

and $\beta_c$ is the propagation constant at the cut-off frequency.

For the $TE_{01}$ type of mode, the cut-off frequency can be obtained by solving the following transcendental equation [24]:

$$\sqrt{\varepsilon_r}\,\tan\{(\omega_c\sqrt{\varepsilon_r}\,d)/2v_0\} . \tan(\omega_c c/v_0)^{-1} = 0 \qquad (25)$$

where $\omega_c$ is the cut-off angular frequency and $v_0$ is the free space velocity.

## 2.4 Suspended and Inverted Microstrip Lines (Fig. 1d and 1e) [11, 20, 28, 29]

The suspended microstrip (Fig. 1d) is a variant of the microstrip with an air gap between the substrate and the ground plane. The inverted microstrip (Fig. 1e) differs from the suspended microstrip in that the strip conductor is situated on a lower surface of the dielectric substrate facing the ground plane. Both suspended and inverted microstrip lines can also be viewed as special cases of the suspended stripline with one of the ground planes removed. These structures retain the advantages of the suspended stripline in terms of achieving larger strip dimensions and lower dissipative losses with respect to the microstrip. In comparison with the suspended microstrip, the inverted microstrip has the advantages of reduced radiation loss by virtue of having the strip conductor below the substrate. However, because the air gap involved is too small (typically on the order of 1 mm or less), incorporation of semiconductor devices becomes very difficult in this configuration.

The analysis and propagation characteristics of both these structures are reported in the literature [11, 20, 28, 29]. In the following, we provide closed form expressions for evaluating the characteristic impedance Z and effective dielectric constant $\varepsilon_{eff}$[29]. The guide wavelength is given by $\lambda_0/\sqrt{\varepsilon_{eff}}$.

*Suspended Microstrip* (Fig. 1d)

$$Z = (60/\sqrt{\varepsilon_{eff}}) \ln [f(u)/u + (1 + 4/u^2)^{1/2}] \tag{26a}$$

where

$$f(u) = 6 + (2\pi - 6) \exp [-(30.666/u)^{0.7528}] \tag{26b}$$

$$u = w/(a + b) \tag{26c}$$

$$\sqrt{\varepsilon_{eff}} = [1 + (d/a).(a_1 - b_1 \ln (w/a)) (1/\sqrt{\varepsilon_r} - 1)]^{-1} \tag{27a}$$

where

$$a_1 = [0.8621 - 0.1251 \ln (d/a)]^4 \tag{27b}$$

$$b_1 = [0.4986 - 0.1397 \ln (d/a)]^4 \tag{27c}$$

*Inverted Microstrip* (Fig. 1e)
The expression for the characteristic impedance is the same as (26a) with $f(u)$ given by (26b) and $u = w/a$. The effective dielectric constant is given by

$$\sqrt{\varepsilon_{eff}} = [1 + (d/a).(a_2 - b_2 \ln (w/a)) (1/\sqrt{\varepsilon_r} - 1)]^{-2} \tag{28a}$$

where

$$a_2 = [0.5173 - 0.1515 \ln (d/a)]^2 \tag{28b}$$

$$b_2 = [0.3092 - 0.1047 \ln (d/a)]^2 \tag{28c}$$

The accuracy of (27) and (28) is reported to be within ± 1% for $1 < w/a \leq 8$, $0.2 \leq d/a \leq 1$ and $\varepsilon_r \leq 6$.

As shown in the case of the suspended stripline at Fig. 2c, suspended and inverted microstrip lines can also be housed in shielding enclosures. However, unlike the suspended stripline, in these lines, the top wall is placed sufficiently high so as not to take part in the propagation.

### 2.5 Slotline (Fig. 1f) [3, 30, 31]

The slotline consists of a slot etched from the conducting layer on one surface of a dielectric substrate with the opposite surface being bare. The structure is thus complementary to that of a microstrip. The electric field lines are oriented essentially across the slot whereas the magnetic field lines have both transverse and longitudinal components. The dominant mode of propagation is non-TEM and hence, the characteristic impedance and guide wavelength are dependent on frequency, although at a slow rate. As compared with the microstrip, the slotline is more dispersive. The slotline mode resembles the $TE_{10}$ mode of a rectangular waveguide but it differs from the waveguide in that it has no cut-off frequency.

Based on the analysis of Cohn [30], Garg and Gupta [31] have obtained closed form expressions for the characteristic impedance and guide wavelength of slotline by curve fitting the numerically computed results. Their formulas are given below:

For $0.02 \le w/h \le 0.2$

$$Z = 72.62 - 15.283 \ln(\varepsilon_r) + 50 \frac{\{(w/h) - 0.02\}\{(w/h) - 0.1\}}{w/h}$$
$$+ \ln\{100(w/h)\}\{19.23 - 3.693 \ln(\varepsilon_r)\} - [0.139 \ln(\varepsilon_r) - 0.11$$
$$+ (w/h)\{0.465 \ln(\varepsilon_r) + 1.44\}] \cdot \{11.4 - 2.636 \ln(\varepsilon_r) - (h/\lambda_0) \times 10^2\}^2 \tag{29}$$

$$\lambda/\lambda_0 = 0.923 - 0.195 \ln(\varepsilon_r) + 0.2(w/h) - \{0.126(w/h) + 0.02\} \ln\{(h/\lambda_0) \times 10^2\} \tag{30}$$

For $0.2 \le w/h \le 1.0$

$$Z = 113.19 - 23.257 \ln(\varepsilon_r) + 1.25(w/h)\{114.59 - 22.531 \ln(\varepsilon_r)\}$$
$$+ 20\{(w/h) - 0.2\}\{1 - (w/h)\} - [0.15 + 0.1 \ln(\varepsilon_r) + (w/h)\{-0.79$$
$$+ 0.899 \ln(\varepsilon_r)\}] \cdot \{10.25 - 2.171 \ln(\varepsilon_r) + (w/h)\{2.1 - 0.617 \ln(\varepsilon_r)\}$$
$$- (h/\lambda_0) \times 10^2\}^2 \tag{31}$$

$$\lambda/\lambda_0 = 0.987 - 0.21 \ln(\varepsilon_r) + (w/h)(0.111 - 0.0022\varepsilon_r)$$
$$- \{0.053 + 0.041(w/h) - 0.0014\varepsilon_r\} \ln\{(h/\lambda_0) \times 10^2\} \tag{32}$$

The above expressions (29)-(32) are reported [31] to be accurate within 2% over the range $9.7 \le \varepsilon_r \le 20$ and $0.01 \le h/\lambda_0 \le (h/\lambda_0)_c$, where $(h/\lambda_0)_c$ is the cut-off value for the $TE_{10}$ surface-wave mode on the slot line. The value of $(h/\lambda_0)_c$ can be obtained from

$$(h/\lambda_0)_c = 0.25/\sqrt{(\varepsilon_r - 1)} \tag{33}$$

The slotline is a convenient medium for shunt mounting of discrete devices. It is particularly suitable for ferrite components that require regions of circularly polarized magnetic field. One major disadvantage of the slotline is that its $Q$-factor is low ($\simeq 100$) as compared with other transmission lines.

### 2.6 Coplanar Waveguide and Coplanar Strips (Fig. 1g and 1h) [3, 32-34]

The coplanar waveguide is basically a coupled slotline (Fig. 1g) in which the signal is applied to the centre conductor with respect to the two ground conductors on the two sides. The configuration of coplanar strips (Fig. 1h) is complementary to the coplanar waveguide. Unlike the slotline, which is non-TEM in nature, both the coplanar waveguide and coplanar strips propagate quasi-TEM mode. At higher microwave frequencies, however, the contribution due to the longitudinal magnetic field is sufficiently large resulting in non-TEM mode of propagation.

Both coplanar waveguide and coplanar strips have been analytically studied by several authors [3, 32–34]. The expressions for the characteristic impedance $Z$ and effective dielectric constant $\varepsilon_{eff}$ of both these structures are given below [34]:

*Coplanar Waveguide* (Fig. 1g)

$$Z = \frac{30\,\pi}{\sqrt{\varepsilon_{eff}}} \; \frac{K(k')}{K(k)} \tag{34}$$

$$\varepsilon_{eff} = 1 + \{(\varepsilon_r - 1)/2\} \frac{K(k')\,K(k_1)}{K(k)\,K(k_1')} \tag{35}$$

where

$$k = \{s/(s + 2w)\} \tag{36a}$$

$$k_1 = [\{\sinh(\pi s/4h)\}/\{\sinh(\pi(s + 2w)/4h)\}] \tag{36b}$$

$$k' = \sqrt{(1 - k^2)} \tag{36c}$$

$$k_1' = \sqrt{(1 - k_1^2)} \tag{36d}$$

In (35), $K(k)$ represents a complete elliptic function of the first kind and $K(k')$ with $k' = \sqrt{(1 - k^2)}$ is its complementary function. An approximate expression for $K(k)/K(k')$ which is accurate to 8 ppm is given by

$$\frac{K(k)}{K(k')} = \begin{cases} [(1/\pi)\ln\{2(1 + \sqrt{k'})/(1 - \sqrt{k'})\}]^{-1}, & 0 \le k \le 0.7 \\ [(1/\pi)\ln\{2(1 + \sqrt{k})/(1 - \sqrt{k})\}]^{-1}, & 0.7 \le k \le 1 \end{cases} \tag{36e}$$

*Coplanar Strips* (Fig. 1h)

$$Z = \frac{120\pi}{\sqrt{\varepsilon_{eff}}} \; \frac{K(k)}{K(k')} \tag{37}$$

$$\varepsilon_{eff} = 1 + \{(\varepsilon_r - 1)/2\} \frac{K(k')\,K(k_1)}{K(k)\,K(k_1')} \tag{38}$$

where $k$, $k_1$, $k'$ and $k_1'$ are given by (36a) to (36d), respectively.

One major advantage of both the coplanar waveguide and coplanar strips is that they permit simple integration of hybrid components both for series and parallel connections. In the fabrication of MMICs, these coplanar structures offer a convenient media in addition to the microstrip. Their Q-factors, however, are low ($\simeq$100) as compared with that of a microstrip ($\simeq$250).

## 3. SUBSTRATES AND TRANSMISSION LINE CHARACTERISTICS

The selection of substrates for MICs depends on three major factors-namely, the frequency of operation, the type of planar line used and the fabrication technology; that is hybrid or monolithic. Other factors include, the desired electrical, mechanical and thermal properties and cost. Hybrid MICs use dielectric substrates which are generally isotropic, although anisotropic dielectric and ferrimagnetic substrates are also used for certain special applications.

## 3.1 Substrates for Hybrid MICs

*General Requirements*
The dielectric substrate for hybrid MICs must have the following general properties:

1. *Low dielectric loss*: The dielectric loss in a circuit is directly proportional to the loss tangent (tan $\delta$) of the substrate material at the operating frequency. Typically values of the order of $5 \times 10^{-4}$ or less would be desirable.

2. *Fine surface finish*: The conductor loss due to the metal film at microwave frequencies depends on the degree of surface finish of the substrate (in addition to the resistivity of the metal), with the requirement on fineness of polish becoming increasingly stringent with increasing frequency. Smooth surface is also important for achieving fine line definition of the conducting pattern on the substrate. The surface finish must be at least of the order of one skin depth (typically 1 $\mu$m or better at X-band).

3. *Low temperature coefficient*: In order to ensure stability of circuit performance with temperature, the variation in dielectric constant and loss tangent with temperature must be as low as possible.

4. *Uniformity of thickness and dimensional stability*: Uniformity of thickness in substrates from a batch and also their dimensional stability are important from the point of view of reproducibility of circuit performance. These parameters are of special importance in circuits realized in suspended and inverted microstrips because they reflect on the uniformity of air gap between the substrate and the bottom ground plane which is a critical.parameter in circuit design.

5. *Easy machining*: From the point of view of circuit manufacture, substrates that can be cut and drilled easily are preferred.

6. *High thermal conductivity*: Higher thermal conductivity permits better heat dissipation in the circuit. This is important in circuits operating at high power levels.

7. *Mechanical stability up to high temperature*: This property is important for circuits incorporating discrete components by soldering or thermo-compression bonding.

Table 1 lists the properties of some of the substrate materials used in hybrid MICs. These materials can be classified under three main categories: ceramic, plastic and ferrimagnetic.

*Ceramic Substrates*
Ceramic substrates, in general, are hard and can be produced with good surface polish. These substrates, however, cannot be easily cut or drilled.

*Alumina* possesses a good combination of dielectric, conductive, mechanical

**Table 1. Properties of substrates usable in hybrid MICs**

| Material | $\varepsilon_r$ (approx.) | $\tan\delta \times 10^4$ at 10 GHz | Surface roughness (μm) | Thermal conductivity (W/cm K) | Thermal expansion (ppm/K) | Dielectric strength (KV/cm) | Special features |
|---|---|---|---|---|---|---|---|
| *(a) Ceramic Substrates* | | | | | | | |
| Alumina (99.5%) | 9.6–10.4 | 1 | 0.05 (polished) | 0.37 | 6.3 | $4 \times 10^3$ | Au-coated substrates, widely used in microwave range |
| Fused quartz (99.9%) | 3.78 | 1 | 0.006–0.025 | 0.017 | — | $10 \times 10^3$ | Reproducible substrates, minimum thickness 0.05 mm, smooth surface, useful in mm-range, high cost |
| Sapphire (Single crystal) | $\varepsilon_{r\perp} = 9.6$ $\varepsilon_{r\parallel} = 11.6$ | 1 | 0.005–0.025 | 0.42 | 6.0 | $4 \times 10^3$ | Reproducible substrates, smooth surfaces, anisotropic, high cost |
| Beryllia BeO 98% | 6.3 | 60 | 0.05–1.25 | 2.5 | 6.1 | — | High conductivity, rough surface, compound substrate |
| Rutile ($TiO_2$) | 85 | 40 | 0.25–2.5 | 0.05 | 7.5 | — | Reduced size components, rough surface, $\varepsilon_r$ temperature sensitive |
| *(b) Plastic Substrates* | | | | | | | |
| RT-duroid 5880 | 2.2–2.3 | 3–7 | 0.075-1* 4.25–8.75** | 0.0026 | 16–100 | $1 \times 10^3$ | Cu-coated, large sheet size, minimum thickness 0.1 mm, easily machinable, low cost |
| RT-duroid 6010 | 10.2–10.7 | 10–15 | " | 0.0041 | — | — | Cu-coated, high $\varepsilon_r$, easily machinable, high loss |
| Kapton | 3–3.5 | — | — | — | — | — | Flexible substrates, minimum thickness 0.025 mm, easily machinable, low cost |
| Cu-flon | 2.1 | 4.5 | — | — | — | $1 \times 10^3$ | " |
| *(c) Ferrimagnetic substrates* | | | | | | | |
| Ferrite/garnet | 9–16 | 10 | 0.25 | 0.03 | — | $4 \times 10^3$ | Porous, hard, brittle, anisotropic under magnetic bias, useful for non-reciprocal devices |

*Average peak to valley difference in height (rolled Cu).
**Average peak to valley difference in height (electro-deposited Cu).

and thermal properties. It has very low loss, negligible dispersion between batches but is slightly anisotropic. Alumina substrates for MIC applications are available in standard thicknesses 0.1, 0.254, 0.635, 1.27 and 2.54 mm with dimensions in multiples of one inch (1" × 1" to 4" × 4"). Because of its high dielectric constant, alumina substrates are used extensively at lower microwave frequencies.

*Fused quartz* has a lower dielectric constant than alumina and is available in substrates with thickness as low as 0.05 mm and with a high degree of surface finish ($\simeq$0.006 $\mu$m). With these features and its low loss property, fused quartz offers an excellent medium for use at millimetre wave frequencies. The undesirable features of fused quartz are, its high cost and highly brittle nature leading to easy breakage during machining.

*Sapphire* is the mono-crystalline form of alumina. It is dielectrically anisotropic. The substrate is normally cut with the *c*-axis normal to the ground plane. This gives electrical properties that are independent of the direction of propagation across the substrate. For electric fields along the c-axis, $\varepsilon_{r\parallel} = 11.6$, and for fields perpendicular to the c-axis, $\varepsilon_{r\perp} = 9.6$. Its anisotropy is well defined, with negligible variation from batch to batch, and hence is the most useful of all the anisotropic materials. Sapphire substrates can be polished to achieve optical grade surface finish, thereby offering very uniform and precisely defined conductor patterns. The material is optically transparent. Its thermal conductivity is about 30% higher than that of alumina and hence may be preferred for mounting high power active devices.

In microstrip active devices where high dissipation rates are encountered, materials with a high thermal conductivity are required. *Poly-crystalline BeO* which has a thermal conductivity of the order of 2.5 W/cm k is ideal in such applications. It can also be used as an insert in a combined substrate where metal cannot be used for transfer of heat. The disadvantage of this material is that it is highly toxic demanding special precautionary measures while machining. *Rutile* ($TiO_2$) has a very high dielectric constant but is temperature sensitive.

*Plastic Substrates*
Unlike the ceramic substrates, plastic substrates can be easily machined and drilled. The most commonly used plastic substrate is the glass-fibre reinforced polytetrafluoroethylene (PTFE) trade named as *RT-Duroid 5880*. It has a lower dielectric constant ($\varepsilon_r = 2.2$–$2.3$) than fused quartz and hence can be used at millimetre wave frequencies. The loss tangent of this material is three to four times higher than that of fused quartz, but this would have negligible effect on circuit performance since the overall loss is governed primarily by the conductor loss (which is nearly ten times more than the dielectric loss). The thermal expansion coefficient ($\simeq$10 ppm/°C) of RT-duroid is compatible with the metal compounds which are used for conductor patterns ($\simeq$10–30 ppm/°C for Cu, Au) whereas that of fused quartz is much smaller ($\simeq$0.5 ppm/°C). Duroid is also mechanically stable up to temperatures as high as 350°C. It is a low cost

material and is available in larger sheets with standard thicknesses (1/64″, 1/32″, 1/16″ and 1/4″) and with copper metallization (≈ 10-30 μm) on one or both sides. Although its surface finish is inferior to that of fused quartz, because of the other advantages enumerated above, RT-duroid 5880 is widely used at Ku-band and above.

For higher dielectric constant, ceramic-loaded plastic trade named *RT-duroid 6010* is available. Its dielectric constant is close to that of alumina but the loss tangent is nearly ten times higher. Since machining and drilling is easy, this substrate can substitute alumina during the laboratory prototype development.

*Ferrimagnetic Substrates*

*Ferrites and garnets* have relative dielectric constants in the range 9 to 16 and dielectric loss tangent of the order of $10^{-3}$ at 10 GHz. These materials are hard and brittle. Substrates are normally cut from a bar and ground to the required thickness and surface finish depending on the device application. In the presence of a d.c biasing magnetic field, they exhibit gyromagnetic anisotropy thereby permitting the realization of non-reciprocal microstrip devices.

## 3.2 Substrates for MMICs

Semiconductor substrates of Si and GaAs have dielectric constants close to that of alumina but their dielectric loss tangents are nearly 10 to 100 times higher. Hybrid MICs using these substrates would therefore be very lossy. Their suitability lies in monolithic MICs where the active devices, passive components and interconnections are all formed into the bulk or onto the surface of the same semiconductor material. Table 2 lists the relevant properties of both Si and GaAs. Of the two materials GaAs has emerged as a versatile and superior material for MMICs. The resistivities of semi-insulating GaAs ($10^7$ to $10^9$ Ω-cm) is nearly four orders of magnitude higher than that of semi-insulating Si. A semi-insulating GaAs is therefore more suitable as a low-loss dielectric substrate for realizing passive circuitry and also for providing electrical isolation between active devices in the circuit. The electron mobility in GaAs is more than six times that of Si, thus making it suitable for realizing active semiconductor devices (MESFETs) for operation at higher frequencies even up to 100 GHz. The minority carrier life time for GaAs ($\approx 10^{-8}$ sec) is nearly five orders of magnitude less than that for Si ($\approx 2.5 \times 10^{-3}$ sec), which makes it well suited for realizing fast switching p-i-n diodes. The electrical properties of the substrate can be selected by doping so that it can act either as a semiconducting layer or as an insulating mechanical support. The high-resistivity properties of semi-insulating GaAs substrates are utilized for realizing passive components whereas the high mobility properties of the semiconductor layer that is either epitaxially grown onto the semi-insulating substrate or implanted into it are utilized for realizing active devices.

## 3.3 Selection of Substrates

Considering the properties of various commercially available substrates, it is clear that there is no single substrate which can meet all the requirements. For

**Table 2. Properties of substrates used in MMICs**

| Material | $\varepsilon_r$ (approx.) | $\tan\delta \times 10^4$ at 10 GHz | Surface roughness ($\mu$m) | Thermal conductivity $K$ (W/cm/°C) | Resistivity ($\Omega$-cm) | Electron mobility ($cm^2$/V-sec) | Thermal expansion (ppm/K) | Dielectric strength (KV/cm) |
|---|---|---|---|---|---|---|---|---|
| Si | 11.7–12 | 40–150 | < 0.025 | 0.9 | — | 700* | — | — |
| Semi-insulating Si | 11.7–12 | 10–100 | < 0.025 | 0.9 | $10^3$–$10^5$ | — | 4.2 | 300 |
| GaAs | 12–13 | 10–60 | < 0.025 | 0.3 | — | 4300* | — | — |
| Semi-insulating GaAs | 12–13 | 16 | ≃ 0.025 | 0.3 | $10^7$–$10^9$ | — | 5.7 | 350 |

*At $10^{17}/cm^2$ doping.

a given application, one must choose a substrate which offers the best compromise between the desirable properties.

For hybrid MICs at microwave frequencies, microstrip on alumina substrate is the most extensively used planar transmission line. Alumina, with its high dielectric constant keeps the circuit size small and its low loss tangent keeps the losses low. The choice of substrate thickness depends on the desired impedance level and the operating frequency. For operation up to about 20 GHz, substrates of standard thickness 0.635 mm are employed. Typical characteristics of a microstrip on alumina substrate are discussed in Sec. 3.4. It may be noted from (21) that as the frequency of operation increases, microstrip circuits must adopt progressively thinner substrates preferably with a lower dielectric constant. Another factor which assumes importance as the frequency increases, is the surface roughness of the substrate. As can be seen from (17) the conductor loss increases considerably as the skin depth in the metal reduces to the order of rms surface roughness and lower. Using alumina substrates of smaller thicknesses 0.2 to 0.3 mm a variety of integrated circuit components and subsystems upto about 50 GHz have been reported [35–36].

Fused quartz, with its lower dielectric constant ($\varepsilon_r = 3.8$) and extremely smooth surface finish ($\simeq$0.006 mm which is nearly 50 times smoother than alumina), and availability in sufficiently small thicknesses offers a superior alternative to alumina at frequencies above 20 GHz. By using 0.12 mm thick Z-cut quartz substrates, several microstrip components such as the up-converter and balanced mixer, have been fabricated and integrated into hermetically sealed receivers for operation in the W-band (75-100 GHz) [37]. Fused quartz also has the advantage that it is compatible with most ferrite materials. This property allows insertion of ferrite discs into a quartz substrate thereby enabling complete integration of circulators with other components on a single substrate. Fused quartz substrates, however are very expensive. They are difficult to machine and drill.

An inexpensive alternative to fused quartz at millimetre wave frequencies is the copperclad RT-duroid 5880 ($\varepsilon_r = 2.22$) or Cu-flon ($\varepsilon_r = 2.1$). Unlike fused quartz, these substrates are flexible, and easy to machine and drill. Using these substrates in shielded microstrip configuration, compact receivers have been reported for operation in the V-band (50-75 GHz) [38, 39]. By using 0.127 mm thick RT-duroid substrate in suspended stripline configuration, integrated circuit mixers have been realized at W-band (75-110 GHz) [40] and even at D-band (110-170 GHz) [41].

Among the dielectrically anisotropic materials, sapphire is the most useful because of its well defined and repeatable electrical characteristics, availability in small thickness and optical grade finish. The anisotropic property can be advantageously utilized for equalizing the even-and odd-mode phase velocities in microstrip and suspended stripline coupled lines to achieve high directivity in directional couplers [42]. Microstrip-like transmission lines and coplanar lines with anisotropic substrates also find applications in electro-optic modulators and surface-acoustic wave (SAW) applications [43].

Ferrites exhibit magnetic anisotropy under d.c magnetic bias. Planar transmission lines-namely stripline, microstrip line, slot-line and coplanar lines on magnetised ferrite substrates find applications in electronically variable phase shifters [44]. Reciprocal as well as non-reciprocal phase shifting is possible by suitably orienting the magnetic bias field either parallel or perpendicular to the substrate plane depending on the type of circuit employed. Ferrite based striplines and microstrip lines are also used in the realization of isolators and circulators [45].

The application of semiconductor materials is in MMICs and among the various materials, GaAs offers the best properties. Chrome compensated semi-insulating GaAs offers stable resistivities over the temperature range 850–900°C which is required for the fabrication of microwave devices. The high resistivities of the GaAs substrate allows the realization of active devices by suitably doping either through epitaxial deposition or by ion implantation process. The high electron mobility of GaAs permits short-gate length FET devices to operate at frequencies as high as 100 GHz. The GaAs MESFETs have found wide application in amplifiers, mixers, switches and phase shifters. The versatility of GaAs MESFETs in a variety of such functional circuits has added to the popularity of GaAs substrates in MMICs.

### 3.4 Typical Transmission Line Characteristics

Table 3 provides a qualitative comparison of various planar transmission lines considered in Sec. 2. The range of impedances and $Q$-factors listed are typical values achievable with the commercially available substrates at microwave frequencies. The upper and lower limits of impedances are governed by two major factors. First, the accuracy of the photolithographic technique sets tolerances on the minimum strip and slot widths that can be achieved. The second factor is the possible excitation of the higher order modes when the strip width exceeds about one-quarter wavelength.

Of the various planar transmission lines listed, the most commonly used are, the microstrip at microwave frequencies, and the suspended substrate transmission line configurations—namely the suspended stripline and the suspended microstrip, at millimetre wave frequencies. In the following, we discuss some typical characteristics of the structures:

Figures 3 to 5 illustrate typical variation in the effective dielectric constant, characteristic impedance Z, and the attenuation constant $\alpha$ (equal to the sum of the conductor and dielectric attenuation constants), respectively, of a microstrip. The substrates chosen are, alumina ($\varepsilon_r = 9.6$), fused quartz ($\varepsilon_r = 3.8$) and RT-duroid ($\varepsilon_r = 2.22$). From Fig. 3, we note that $\varepsilon_{eff}$ of the microstrip increases with an increase in $w/h$ and also $\varepsilon_r$. This is an expected result because, in both the cases, the electric field concentration in the substrate region increases thereby increasing $\varepsilon_{eff}$. Consequently both the impedance $Z$ and the guide wavelength $\lambda$ which are inversely proportional to $\sqrt{\varepsilon_{eff}}$ decrease with an increase in $w/h$ and $\varepsilon_r$ (see Figs. 3 and 4). For a given substrate thickness $h$ and impedance level $Z$, the strip width $w$ increases with a decrease in $\varepsilon_r$. The normally encountered impedance range in practical circuits is from 20 to 120 Ω and for these impedance levels a microstrip on alumina substrate of standard thickness 0.635 mm yields convenient strip widths. We also note from (21) that with $h = 0.635$ mm and

**Table 3. Comparison of various planar transmission lines**

| Configuration | Typical impedance range (Ω) | Unloaded $Q$-factor ($Q_u$) | Dispersion | Power handling capability | Commonly used substrates/special features |
|---|---|---|---|---|---|
| Stripline | 20–120 | high (~ 500) | negligible | high | Substrates—RT duroid/for passive components, inconvenient for mounting discrete devices |
| Microstrip line | 20–120 | medium (≃ 250) | small | high | Substrates—alumina, fused quartz, RT-duroid for hybrid MICs; GaAs for MMICs, and ferrites for non-reciprocal devices/most widely used transmission line at microwave frequencies |
| Suspended stripline | 40–150 | high | small | high | Substrates—fused quartz, RT-duroid/well suited for millimetre wave components, inconvenient for shunt mounting of active devices |
| Suspended microstrip | 40–150 | high | small | high | Features similar to suspended stripline |
| Inverted microstrip | 25–130 | high | small | medium | Substrates—alumina, fused quartz, RT-duroid/suitable for higher microwave frequencies, inconvenient for mounting active devices |
| Slot line | 50–200 | low (~ 100) | large | low | Normally used in conjunction with microstrip, suitable for shunt mounting of chip devices. Produces elliptically polarized RF magnetic field, useful for non-reciprocal devices on ferrite substrates |
| Coplanar waveguide | 40–150 | low | medium | medium | Easy connection of series and shunt elements, useful for MMICs on GaAs and non-reciprocal components on ferrite |
| Coplanar Strips | 40–250 | low | medium | medium | Easy connection of series and shunt elements, useful for MMICs on GaAs |

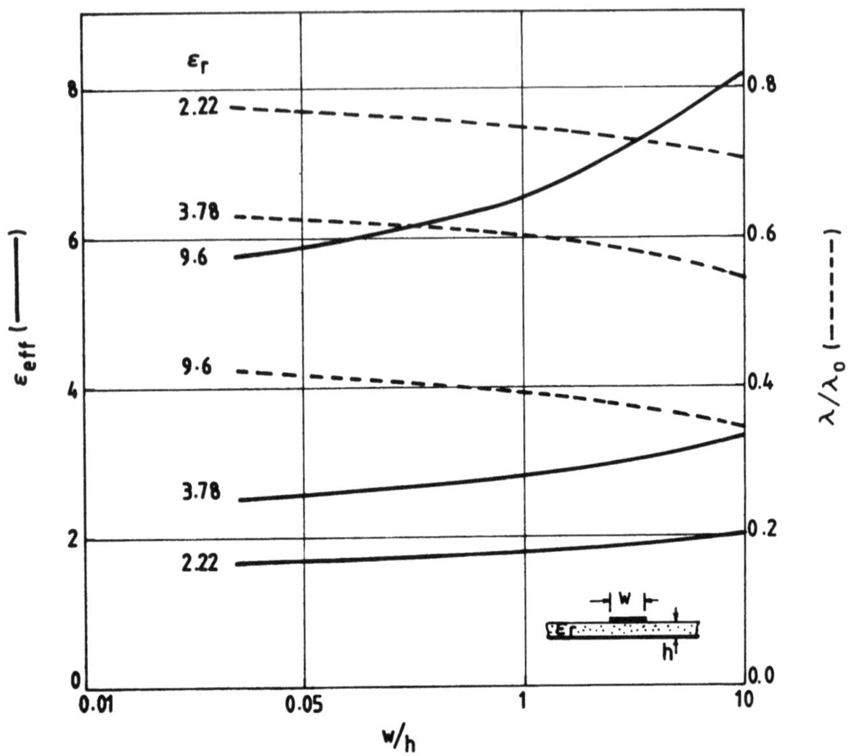


**Fig. 3.** Effective dielectric constant $\varepsilon_{eff}$ and normalized guide wavelength $\lambda/\lambda_0$ of microstrip versus $w/h$ with $\varepsilon_r$ as parameter. $\varepsilon_r = 9.6$ (alumina), 3.78 (fused quartz), 2.22 (RT-duroid).

$\varepsilon_r = 9.6$, the coupling of quasi-TEM mode to the spurious higher order mode occurs at $f_{c,TM} = 53.2$ GHz, thereby ensuring operation only in the dominant mode in the microwave band. The main factor which sets the limit on the highest frequency of operation is the attenuation in the line which increases rather rapidly with an increase in frequency.

From Fig. 5, which illustrates the attenuation versus the height of the substrate in a 50 Ω microstrip, we note that the attenuation is much less with fused quartz and RT-duroid substrates, than with alumina. However, even with these substrates there is a limit on the frequency upto which microstrip geometry can be used since the requirement on making the substrates thinner (for ensuring operation only in the dominant mode) poses problems due to increased conductor loss and critical dimensional tolerances.

The problems of high conductor loss and critical dimensions encountered in microstrips in the millimetre wave range are to some extent circumvented in the suspended substrate configurations (Fig. 1(c)–(e)). As discussed in Sec. 2.3, introducing an air gap between the substrate and the ground plane reduces the effective dielectric constant and consequently, the strip dimensions are nearly two to three times larger than in a microstrip for the same characteristic impedance. Fig. 6 which shows a typical comparison of the impedance characteristics of a

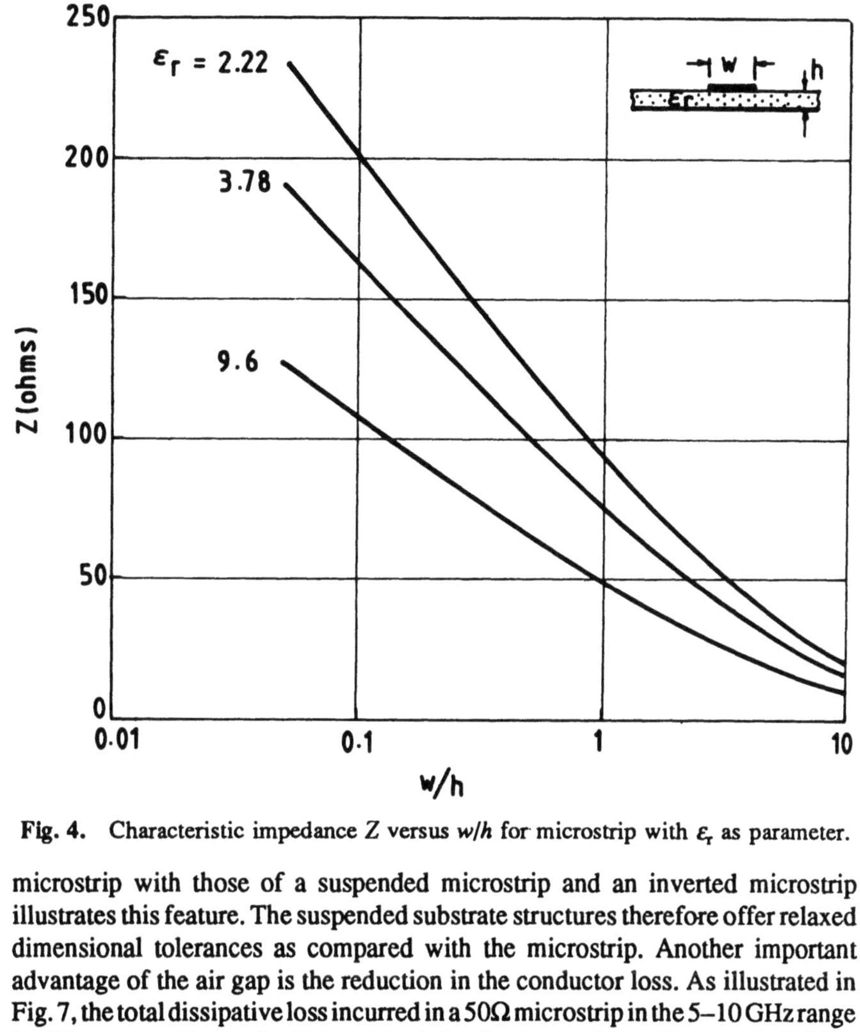


**Fig. 4.** Characteristic impedance $Z$ versus $w/h$ for microstrip with $\varepsilon_r$ as parameter.

microstrip with those of a suspended microstrip and an inverted microstrip illustrates this feature. The suspended substrate structures therefore offer relaxed dimensional tolerances as compared with the microstrip. Another important advantage of the air gap is the reduction in the conductor loss. As illustrated in Fig. 7, the total dissipative loss incurred in a 50Ω microstrip in the 5–10 GHz range is of the same order as that in a suspended or inverted microstrip around 60 GHz. Of the three suspended substrate structures (Fig. 1(c)–(e)), the inverted microstrip is the least utilized, because of the difficulty in incorporating active devices. The suspended microstrip and the suspended stripline which offer easy accessibility to the strip conductor pattern have emerged as popular configurations in the realization of millimetre wave components.

Coplanar structures shown in Fig. 1(f)–(h) tend to be excessively lossy at millimetre wave frequencies. Their utility is limited essentially to microwave frequencies. Circuits using slot-microstrip combination can be easily realized by etching slots on the ground plane side of a microstrip. While the microstrip impedance decreases with an increase in the strip width, the slot impedance increases with an increase in the slot width. The slot-microstrip combination, therefore caters to a wide range of impedance levels, from approximately 20 to 200 Ω and offers considerable flexibility in the circuit design. The coplanar waveguide and coplanar strips are suitable for MMICs built on semi-conductor

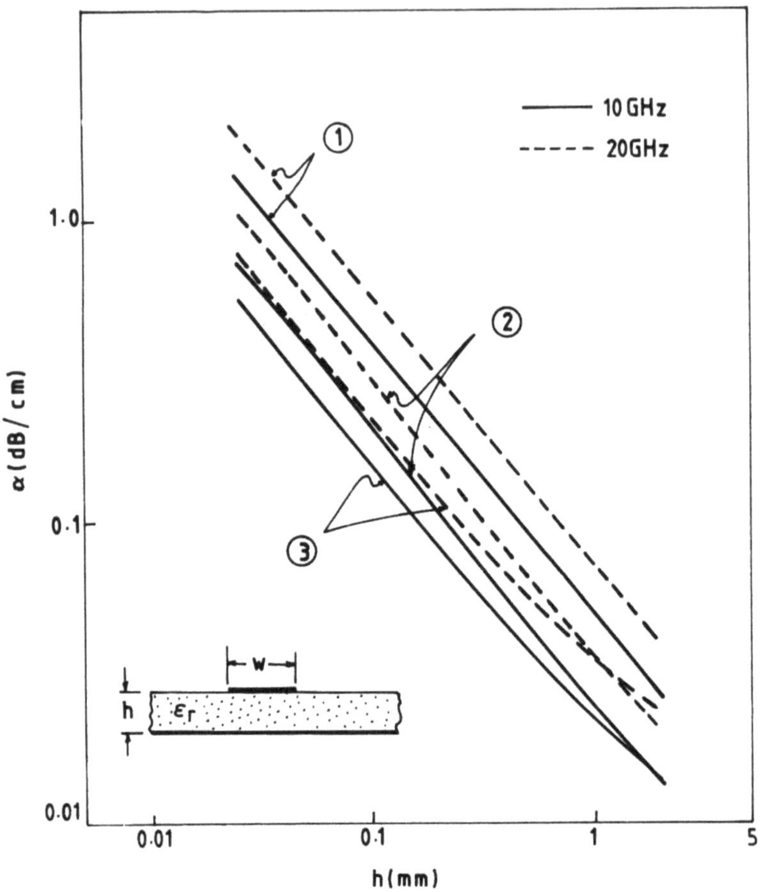


**Fig. 5.** Theoretical attenuation constant $\alpha$ ( $= \alpha_c + \alpha_d$) for a 50 Ω microstrip line as a function of substrate height $h$ with $\varepsilon_r$ as parameter:
1—alumina: $\varepsilon_r = 9.6$, $\tan\delta = 10^{-4}$, $\Delta = 0.05\ \mu m$
2—fused quartz: $\varepsilon_r = 3.78$, $\tan\delta = 10^{-4}$, $\Delta = 0.015\ \mu m$
3—RT-duroid: $\varepsilon_r = 2.22$, $\tan\delta = 5 \times 10^{-4}$, $\Delta = 0.09\ \mu m$

substrates. Both these structures permit mounting of devices in series as well as shunt configurations. The loss in these structures, however, is higher than that in a microstrip. The microstrip is therefore preferred over these transmission lines for MMICs and hence is more frequently used than the coplanar structures.

## 4 BASIC CIRCUIT ELEMENTS FOR MICs

With the planar transmission lines as the basic transmission media, a typical MIC incorporates one or more of the following types of circuit elements-discontinuities, transmission line sections in the form of stubs and resonators, planar lumped elements, dielectric disc resonators and transitions. In the following, we briefly review some of the commonly encountered basic circuit elements in these categories:

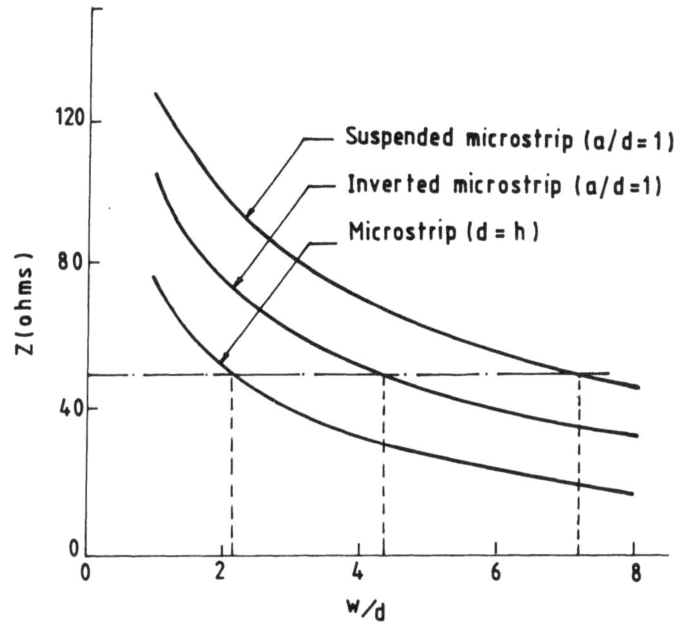


**Fig. 6.** Comparison of impedance characteristics of microstrip (Fig. 1b), suspended microstrip (Fig. 1d) and inverted microstrip (Fig. 1e), $\varepsilon_r = 3.78$.

## 4.1 Discontinuities

Discontinuities in MICs commonly involve a change in the strip conductor width or slot width. Figure 8 shows some of the frequently encountered discontinuities in circuits fabricated using the stripline, microstrip and their variants and Fig. 9 shows the discontinuities encountered in slotlines. Two types of effects may be identified in these discontinuities. One is the generation of fringing electric fields at and in the vicinity of a discontinuity which can be represented in terms of an equivalent capacitance. The second effect is the change in the normal flow of current which can be represented in terms of an equivalent inductance.

*Microstrip Discontinuities* [2, 4, 46, 47]

*Open-end*

Open-end and short circuit terminations are invariably required in any circuit. In a microstrip an open circuit can be easily realized by abruptly terminating a strip conductor whereas a short circuit requires drilling a hole through the substrate and making a connection to the ground plane. Fig. 10a shows a open-ended microstrip and the nature of fringing electric fields at the termination. At lower microwave frequencies, these fringing fields and the resulting increase in the electrostatic energy in the region of the open end may be modeled as a capacitor $C_l$ as shown in Fig. 10b. The capacitance, in turn, is equivalent to extending the microstrip line by a length $\Delta l$ such that the input impedance of this extended line section is equal to the capacitive reactance at the operating frequency.

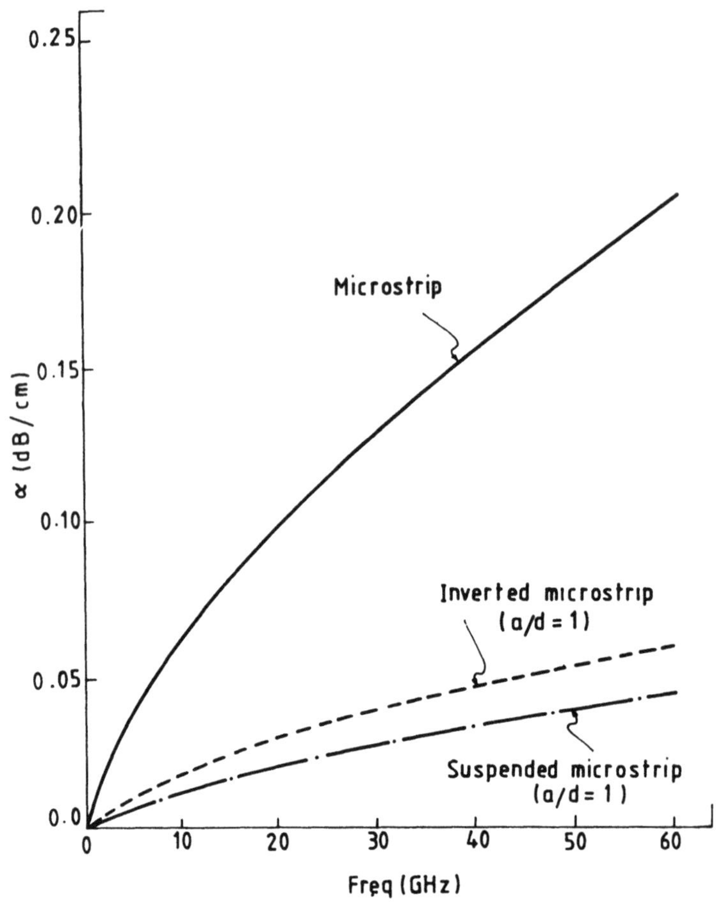


**Fig. 7.** Comparison of attenuation constant $\alpha$ (= $\alpha_c + \alpha_d$) versus frequency of 50 Ω line in microstrip (Fig. 1b), suspended microstrip (Fig. 1d) and inverted microstrip (Fig. 1e)

$$1/\omega C_l = Z \cot (\beta \Delta l) \tag{39}$$

where $Z$ and $\beta$ are the characteristic impedance and propagation constant, respectively, of the microstrip line. Thus, as shown in Fig. 10b, the ideal open circuit lies at a distance $\Delta l$ from the physically terminated open end. Equivalently, in a practical circuit, the physical length of the line must be shorter than the theoretically calculated length of an open circuited line section by the correction factor $\Delta l$.

*Series-gap*

A narrow series gap in a microstrip acts as an open circuit for dc but presents a finite impedance at microwave frequencies. The gap can therefore be used as a dc block in active device circuits and as a RF coupling element between two strip conductors. As illustrated in Fig. 8b, the series gap may be represented as a π-network of capacitors. The shunt capacitors account for the fringing

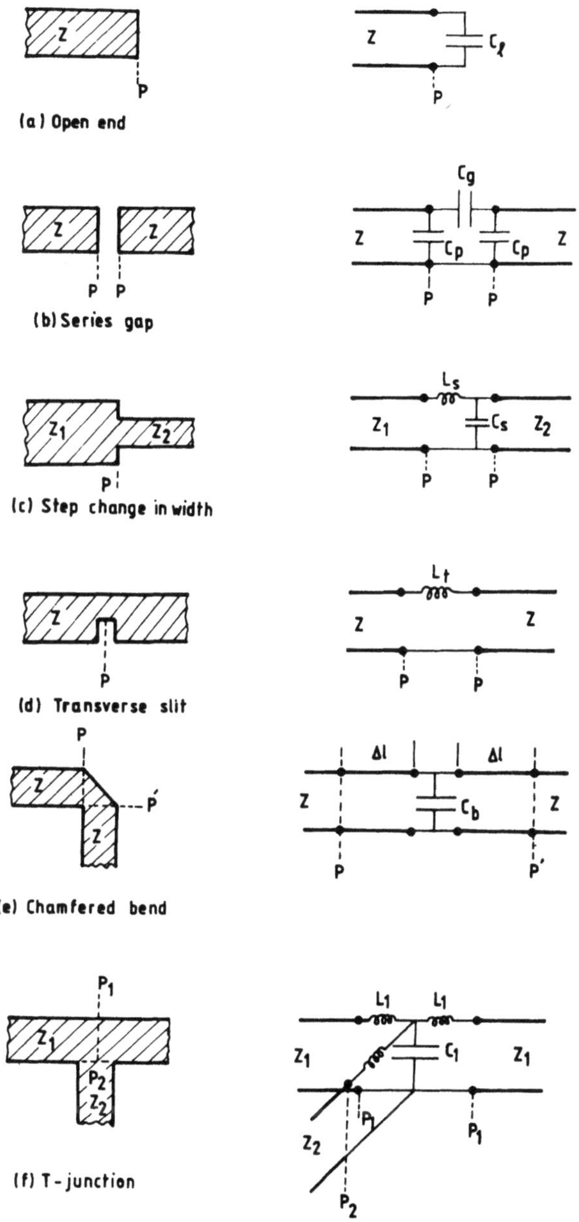


**Fig. 8.** Examples of discontinuities in the strip conductors of planar transmission lines shown in Fig.1(a)–(e) and their equivalent circuits.

electric fields from the open end to the ground plane and the series capacitor accounts for the fringing fields between the two open-ends (see Fig. 11). If the line widths are different, then the two shunt capacitances would have different values. For very large gaps, $C_g$ tends to zero and $C_p$ approaches the end-capacitance of a open-circuited line.

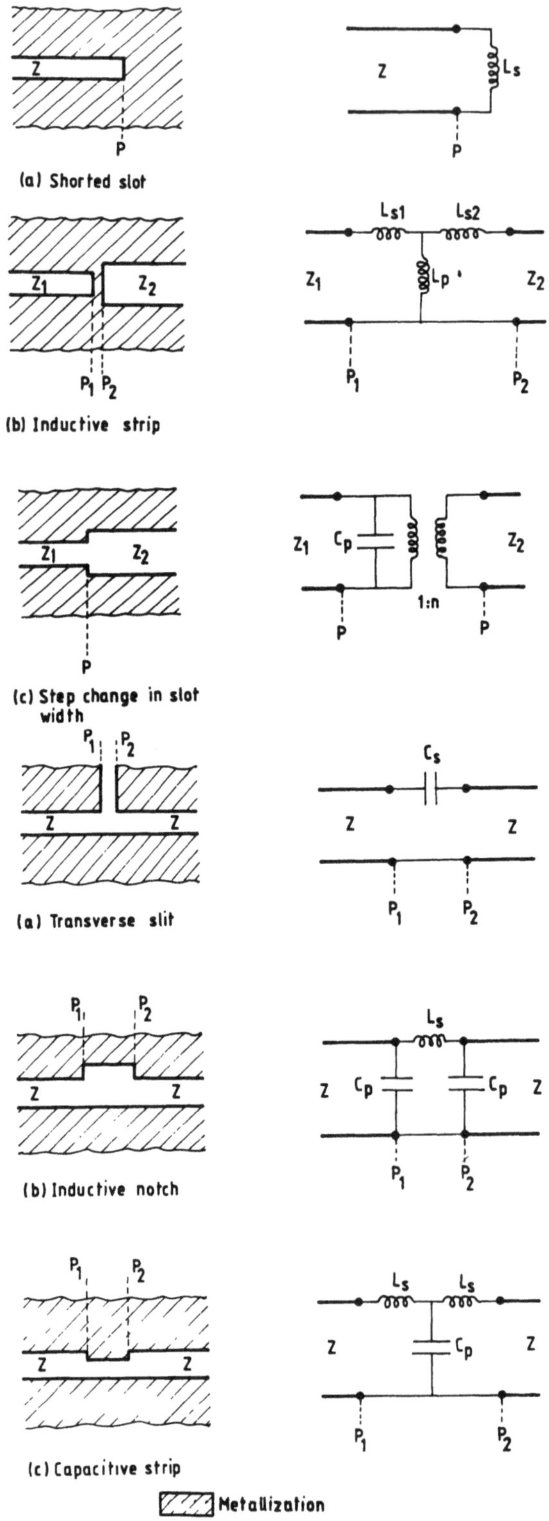


Fig. 9. Examples of slotline discontinuities.

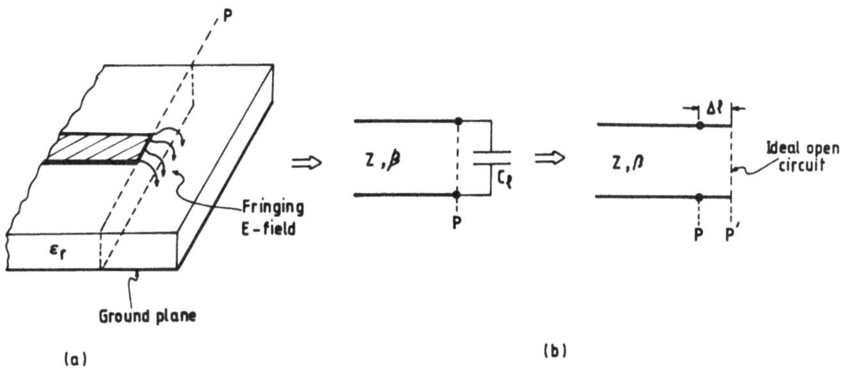


**Fig. 10.** (a) Microstrip with an open end showing fringing electric field and (b) Equivalent end capacitance $C_l$ and end correction $\Delta l$.

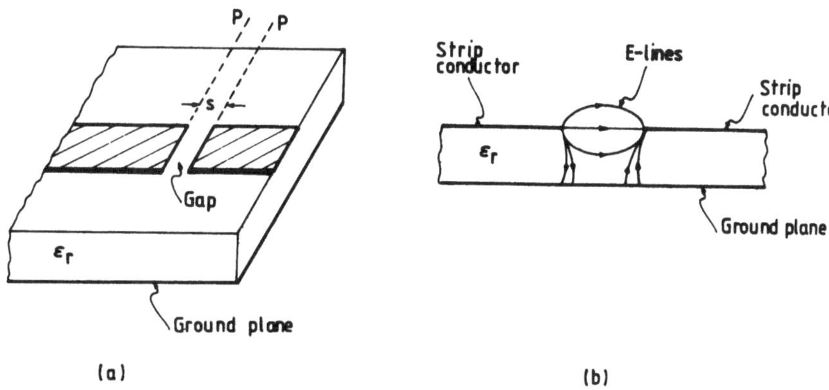


**Fig. 11.** (a) Microstrip with a series gap and (b) Cross-section showing fringing E-field lines.

### *Step Change in Width*

Symmetric and asymmetrical step change in the strip conductor width is common in most microstrip circuits, such as impedance transformers, directional couplers, filters etc. Fig. 12 shows the nature of current flow lines on the strip conductor around the step and the fringing electric field lines due to the excess charge stored at the transition edge. Fig. 8c shows the equivalent circuit. The flexing of current lines can be modeled as a series inductance $L_s$ and the fringing electric fields as a shunt capacitance $C_s$.

### *Transverse Slit in the Microstrip*

A narrow slit in the strip conductor as shown in Fig. 8d can be used to realize a pure series inductance.

### *Chamfered Bend*

Bends in the strip conductor are generally required in order to accommodate the circuit optimally in a given substrate area. A sharp bend in the strip gives

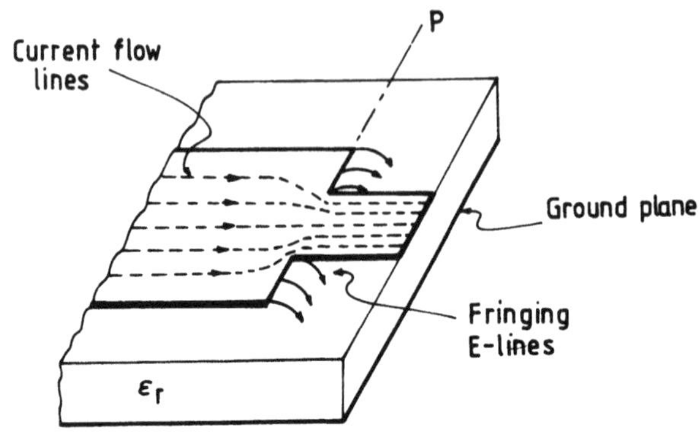


**Fig. 12.** Symmetric microstrip step discontinuity showing current flow lines on the strip conductor and fringing E-field lines at the step.

rise to excess capacitance resulting in poor VSWR. In practice, the effect of the discontinuity reactance is minimized by chamfering the bend as shown in Fig. 8e. The equivalent circuit for the bend section between the planes $P$ and $P'$ can be modeled as a shunt capacitance with transmission line sections of length $\Delta l$ on both sides.

*T-junction*

T-junctions are required in most MICs, such as power dividers, filters, couplers and amplifiers. At the T-junction, two different transmission lines meet at right angles giving rise to fringing electric fields with respect to ground and also distortion in current flow. The junction can therefore be modeled as a three-port network with shunt capacitance and series inductances as shown in Fig. 8f.

*Slotline Discontinuities* [3, 5]

The slotline geometry is complementary to that of the microstrip. While a terminated microstrip represents an approximate open circuit, an abruptly terminated slot represents an approximate short circuit. Fig. 13a shows the geometry of a shorted slot and the resulting current flow around the termination. The additional magnetic energy stored due to the current flow can be modeled by an inductor $L_s$. The inductor, in turn can be modeled by a section of short circuited transmission line of length $\Delta l$ such that the inductive reactance is equal to the input reactance of the shorted transmission line section. That is

$$\omega L_s = Z \tan(\beta \Delta l) \tag{40}$$

where $Z$ and $\beta$ are the characteristic impedance and propagation constant, respectively, of the slotline. Therefore, a physical short at a certain plane $P$ can be considered to produce a perfect electrical short at a plane $P'$ which is $\Delta l$ away from the plane $P$ (see Fig. 13b).

The equivalent circuits of several other slotline discontinuities are shown in Fig. 9. In general, any rerouting of current flow due to the discontinuity giving rise to excess magnetic stored energy is modeled as an inductor and any

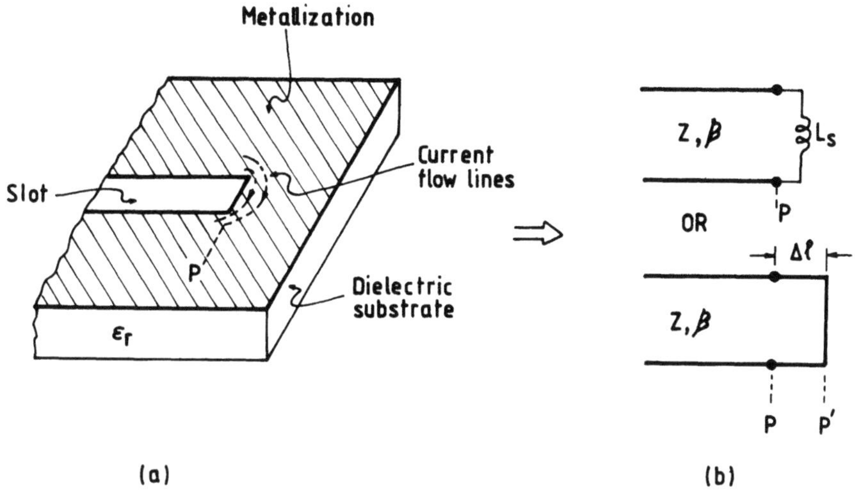


**Fig. 13.** (a) Slotline with a shorted end showing current flow lines around the short and (b) Equivalent end inductance $L_s$ and end correction $\Delta l$.

charge storage giving rise to excess electric stored energy is modeled as a capacitor.

## 4.2 Lumped Constant Elements [48-52]

Lumped constant elements by definition must be very small in size; typically less than $\lambda/20$. Using the planar technique, lumped elements-namely, inductors, capacitors and resistors have been realized up to about 18 GHz [48, 49]. At lower microwave frequencies where the circuits built in microstrip or other planar transmission lines become very large, the use of lumped inductors and capacitors can reduce the circuit size to a considerable extent. The lumped elements also facilitate achieving broadband performance in circuits by virtue of the fact that their values remain fairly independent of frequency. Lumped elements are more prevalent in MMICs than in hybrid MICs since they are easily implementable in miniature from using the monolithic technology.

*Planar Inductors*

Figure 14 illustrates some typical layouts of planar inductors. The simple form of planar inductor is a short section ($l << \lambda/4$) of very high impedance line terminated in very low impedance lines as shown in Fig. 14a. A simple relationship between the length $l$ of the section and its inductance $L$ can be obtained from the expression for the input impedance of a section of short-circuited transmission line. It is given by

$$\beta l = \sin^{-1} (\omega L/Z) \tag{41}$$

where $Z$ and $\beta$ are the characteristic impedance and propagation constant, respectively, of the high impedance line. In microstrip, the characteristic impedance of the inductive line is generally in the range 90 to 110 Ω and that of the low impedance lines is in the range 10 to 15 Ω.

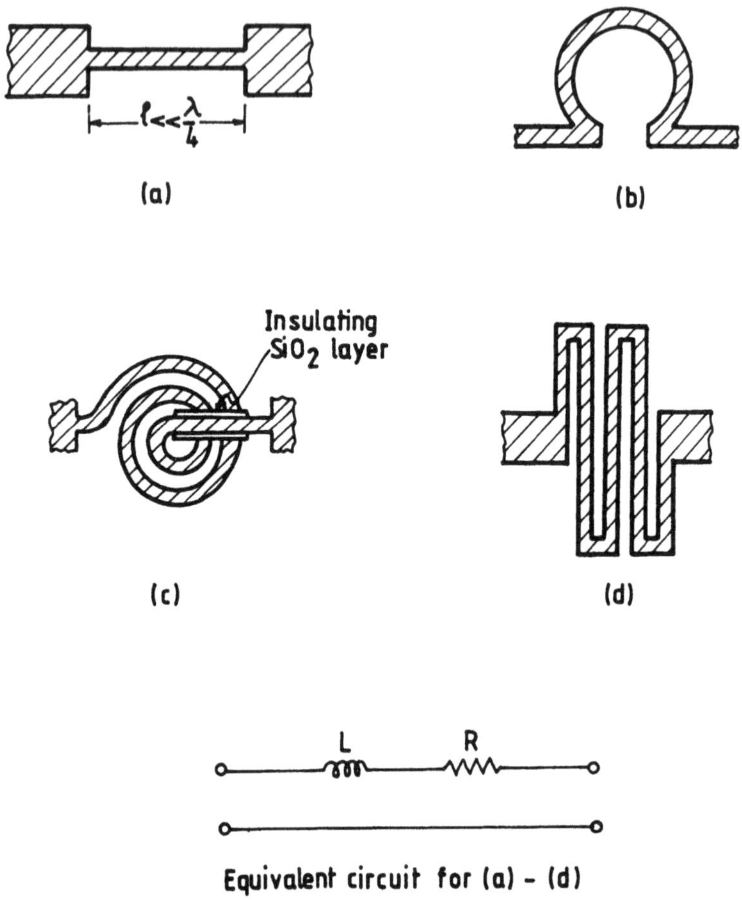


**Fig. 14.** Configurations of planar inductors: (a) Short section of high impedance line; (b) Single turn loop; (c) Spiral; and (d) Meander line.

A thin strip conductor on a dielectric substrate without the ground plane (or with no influence of the ground plane) also acts as an inductor. Single turn loop inductors of the form shown in Fig. 14b can be used for realizing lower inductance values in the range 0.5 nH to about 3 nH. Spiral (Fig. 14c) and meanderline (Fig. 14d) inductors depend on the mutual coupling between the adjacent line segments to achieve higher inductance values upto about 50 nH in a small area. Values higher than this are difficult to achieve in lumped form because of the inter-element fringing capacitance. The $Q$-factor for a spiral is typically 100 which is higher than that for a straight ribbon.

*Planar Capacitors*

A lumped capacitor can be realized as a small length ($l << \lambda/4$) of low impedance line (conductor patch) terminated in very high impedance lines as shown in Fig. 15a. If Z and $\beta$ denote the characteristic impedance and propagation constant,

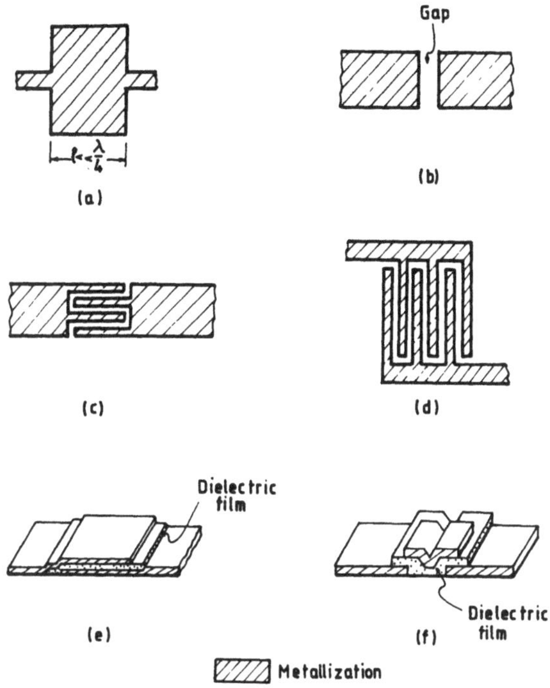


**Fig. 15.** Configurations of planar capacitors: (a) Short length of low impedance line in microstrip; (b) Small gap in a microstrip conductor; (c), (d) Interdigital line and (e), (f) Dielectric film overlay.

respectively, of the line, the length $l$ of the line can be obtained from the expression for the input impedance of a section of open-circuited transmission line. That is,

$$\beta l = \sin^{-1} (\omega CZ) \tag{42}$$

where the value of $Z$ should be low (typically 10–15 Ω in microstrip). Such capacitor patches cascaded alternately with high impedance inductor lines (Fig. 14a) are commonly used as low pass filters in bias networks.

A narrow series gap in a microstrip conductor is useful for realizing very low capacitances of the order of a fraction of a pF. Higher values up to a few picofarads can be achieved in the interdigital capacitors shown in Fig. 15c and d. In these capacitors, the capacitance is formed by the fringing field between the interdigital gaps. The capacitance configurations shown in Fig. 15(a)–(d) require a single metallization layer. Higher values of capacitances can be obtained in overlay structures which use dielectric films such as silicon dioxide (Fig. 15 (e) and (f)). These overlay structures behave as parallel plate capacitors if the film thickness is very small (0.5–1 $\mu$m). Such capacitors offer capacitances as high as 10 to 30 pF over a small area ($\simeq 1$ mm$^2$) and can be easily realized in monolithic technology.

*Planar Resistors*

Planar resistors are used in attenuators, as terminations in directional couplers, power dividers/combiners and circulators and also as isolation resistors in power dividers. Fig. 16 shows typical configurations of a series resistor in a microstrip, terminating resistor for a microstrip and an implanted resistor in MMIC. These resistors are made of high resistivity metal films or cermet films on a dielectric

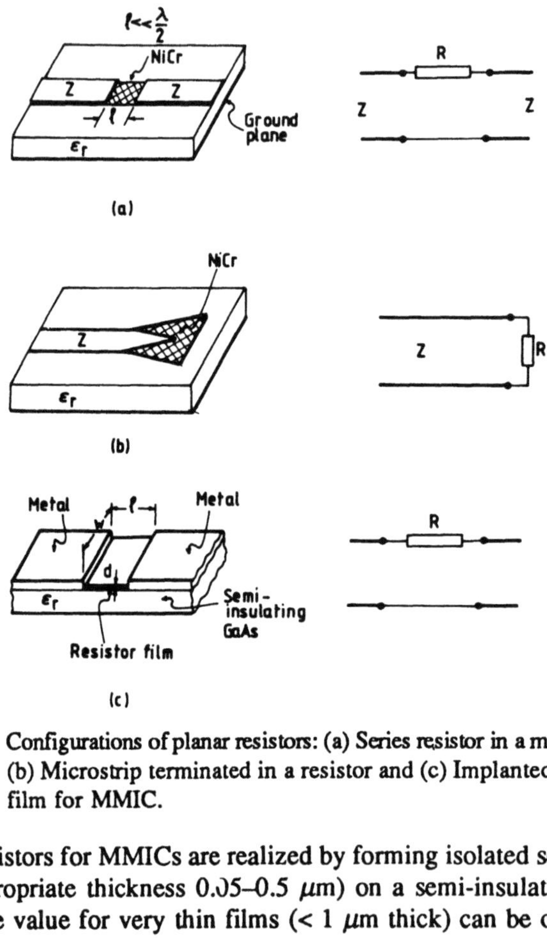


Fig. 16. Configurations of planar resistors: (a) Series resistor in a microstrip; (b) Microstrip terminated in a resistor and (c) Implanted resistor film for MMIC.

substrate. Resistors for MMICs are realized by forming isolated semiconductor films (of appropriate thickness 0.05–0.5 $\mu$m) on a semi-insulating substrate. The resistance value for very thin films (< 1 $\mu$m thick) can be obtained from

$$R = \rho_s l/wd \tag{43}$$

where $\rho_s$ is the specific resistivity ($\Omega$-cm) of the resistor film, and the dimensions $l$, $w$, $d$, which denote the length, width, thickness, respectively of the film are expressed in cm.

## 4.3 Resonators

Resonators for MICs can be classified into the following: lumped element resonators, transmission line resonators and dielectric resonators.

*Lumped Element Resonators*

These are fabricated by combining the planar printed inductors and capacitors described in Sec. 4.2. As examples, Fig. 17a shows a series resonant circuit and Fig. 17b shows a parallel resonant circuit made up of a single turn inductor and an interdigital capacitor. These resonators have rather low $Q$ values typically less than 100 over 4-12 GHz.

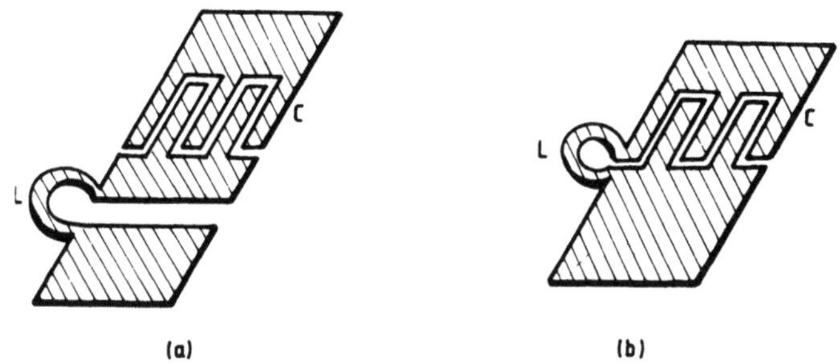


**Fig. 17.** Examples of lumped resonant circuits: (a) Series resonant circuit and (b) Parallel resonant circuit.

*Transmission Line Resonators*

Figure 18a shows a rectangular resonator and Fig. 18b shows a ring resonator in microstrip. Their counterparts in slotline are shown in Fig. 18c and d. The rectangular shaped resonators are commonly used in filter networks. As discussed in Sec. 4.2, the electrical length of these resonators is longer than the physical length owing to the end-effect. For a half-wavelength resonator in microstrip or slotline, resonances occur when

$$l + 2\Delta l = \lambda/2 \tag{44}$$

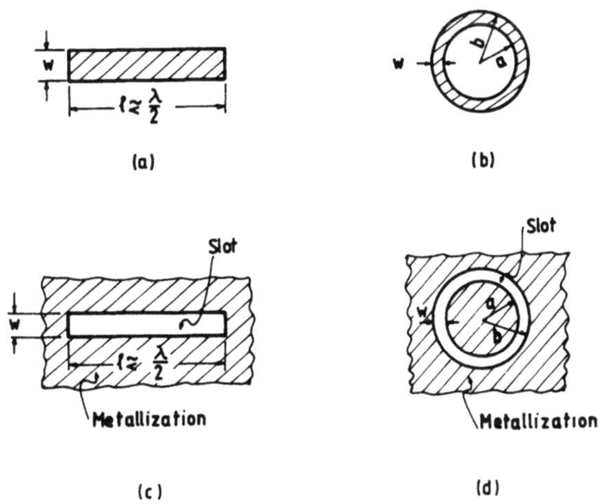


**Fig. 18.** Examples of transmission line resonators: (a) Half-wavelength microstrip resonator; (b) Microstrip ring resonator; (c) Half-wavelength slot resonator and (d) Slot-ring resonator.

where $l$ is the physical length of the resonator, $\Delta l$ is the end correction due to the fringing fields at the termination and $\lambda$ is the guide wavelength in the transmission line.

Unlike the rectangular resonators, the ring resonators do not need any end correction. The resonance condition for the ring resonator is given by

$$\pi(2a + w) = n\lambda \tag{45}$$

where $a$ is the radius of the inner ring, $w$ is the width of the strip/slot and $n$ is an integer indicating the number of azimuthal full-wave variations of the field. Ring resonators are useful in measurement techniques, particularly for measuring the effective dielectric constant of planar transmission lines [53, 54].

*Dielectric Resonators*

Dielectric resonators for MIC application are made of low loss, high dielectric constant ($\varepsilon_r \simeq 30$–100) ceramic mixtures such as the titanates and zirconates (e.g. $Ba_2Ti_9O_{20}$, (Zr-Sn) $TiO_4$). Resonators made of these materials have two important advantages over the lumped and transmission line resonators; viz- very high $Q$-factors in the range 3000–10000 and excellent temperature stability. They are widely used in low loss, narrow band filters and temperature stabilized oscillators [55, 56].

For most applications, resonators are fabricated in the form of a cylinder, although othor shapes such as the ring and rectangle are also used. Figure 19a shows the geometry of an isolated cylindrical resonator. The commonly used resonant mode is the lowest order circular symmetric mode and is denoted as $TE_{01\delta}$. The electric field lines form concentric circles around the z-axis and the magnetic field lines lie in the meridian plane as shown in Fig. 19b. The resonant frequency of a resonator is determined by its dimensions, relative dielectric constant $\varepsilon_r$ and the surrounding medium. For an isolated resonator, the resonant frequency for the $TE_{01\delta}$ mode is given approximately by [56]

$$f_r\,(\text{GHz}) = \frac{34}{a\sqrt{\varepsilon_r}}\,[(a/d) + 3.45];\ a, d \text{ in mm} \tag{45}$$

where $a$ and $d$ are radius and height, respectively of the resonator. This relation is reported to be accurate within about 2% in the range $0.5 < a/d < 2$; $30 < \varepsilon_r < 50$. As an example, typical dimensions of a cylindrical $TE_{01\delta}$ mode resonator having $\varepsilon_r = 38$ and resonating at 10 GHz are $a = d = 2.45$ mm.

Figure 19c shows the arrangement of a dielectric resonator coupled to a microstrip. The $TE_{01\delta}$ mode is easily excited because of the magnetic coupling between the strip conductor and the dielectric resonator. The amount of coupling can be adjusted by varying the lateral distance between the strip conductor and the resonator. It may be noted that the presence of the microstrip alters the resonant frequency of the resonator and also lowers its $Q$-factor [56].

## 4.4 Circuit Examples

The basic circuit elements considered in the preceding sections can be combined in different ways to realize a variety of MIC components. Fig. 20 shows four

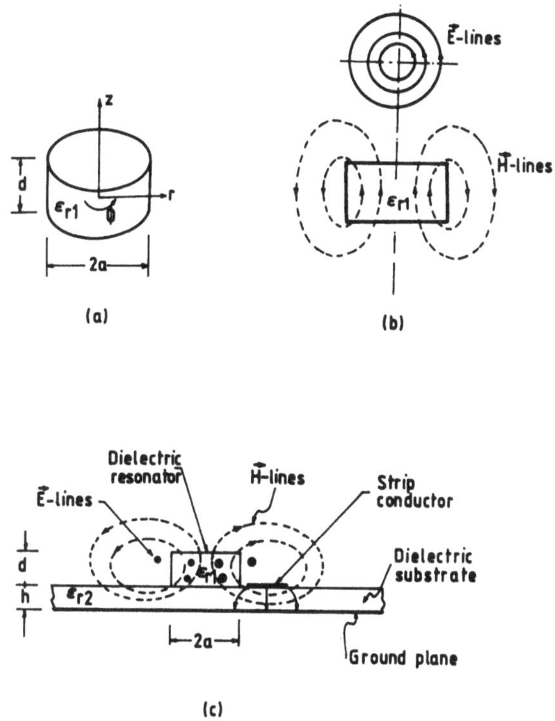


**Fig. 19.** Examples of a dielectric resonator in MIC: (a) Isolated disc resonator; (b) Field lines for the dominant $TE_{01\delta}$ mode and (c) Resonator coupled to a microstrip line.

typical examples. The low pass filter shown in Fig. 20a is made up of series inductors and shunt capacitors cascaded in an alternate fashion. The series inductor is realized as a short section of high impedance line (narrow strip width) and the shunt capacitor is realized as a short section of low impedance line (wide strip). Figure 20b shows a four-port circuit incorporating a ring having a circumference of $3\lambda/2$. This circuit is popularly known as the rat-race hybrid. Ports 1 and 3 are mutually isolated and so are ports 2 and 4. For a signal fed to port 1, equal power division takes place between ports 2 and 4 with signals differing in phase by 180°. For a signal fed to port 3 also, equal power division takes place between ports 2 and 4 but the two signals remain in phase. When two different signals ($V_1$ and $V_3$) are fed to ports 1 and 3, the sum signal appears at port 2 and the difference signal appears at port 4. These properties are used in the design of balanced mixers, *p-i-n* diode phase shifters, and comparator networks.

Figure 20c shows a 3 dB-power division network illustrating the use of a thin film resistor for achieving good isolation between the two output ports (marked 2 and 3). Figure 20d is an example of a directional coupler circuit employing coupling between a microstrip and a slotline to achieve broadband coupling performance. The solid lines indicate the microstrip conductor pattern and the

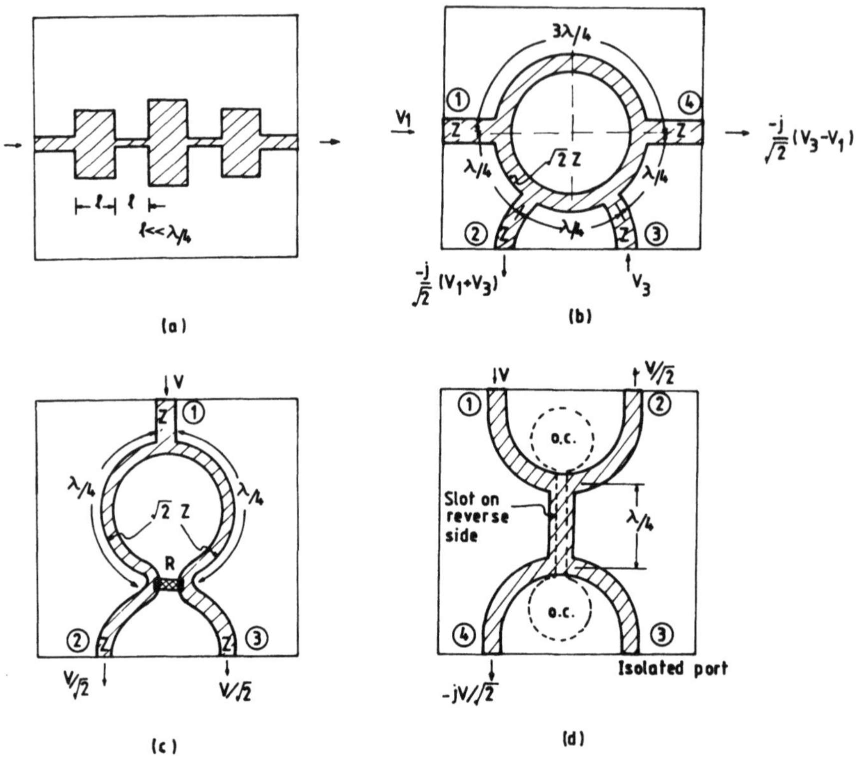


**Fig. 20.** Examples of MIC components: (a) Low-pass filter, (b) Rat-race hybrid, (c) 3 dB power divider, (d) Microstrip-slot directional coupler.

dotted lines represent the slot pattern etched from the ground plane side of the microstrip. The slot line is parallel to the microstrip with its two ends terminated in open circuits (indicated by dotted circles). Power fed to port 1 is equally divided between ports 2 and 4 but the two signals differ in phase by 90°. Port 3 forms an isolated port.

Figure 21 shows a typical microstrip layout of a two-stage FET amplifier which includes in it step junctions, parallel-coupled line (which also serves as a d.c block) and bias lines. The circuit incorporates d.c. block chip capacitors at the microstrip gaps marked $C_1$. Figure 22 shows the photograph of the assembled microstrip amplifier.

## 5. FABRICATION TECHNOLOGY

Two different technologies; *hybrid* and *monolithic* have proved important for the fabrication of MICs. Hybrid MICs permit the use of a wide variety of planar transmission lines described in Sec. 2, whereas for MMICs, the configurations are rather restricted to the microstrip and coplanar structures by virtue of the process technology. The substrate materials appropriate for use in the two technologies are already covered in Sec. 3. In this section, we provide a brief review of the choice of other materials and some key features of the two fabrication technologies [57-59].

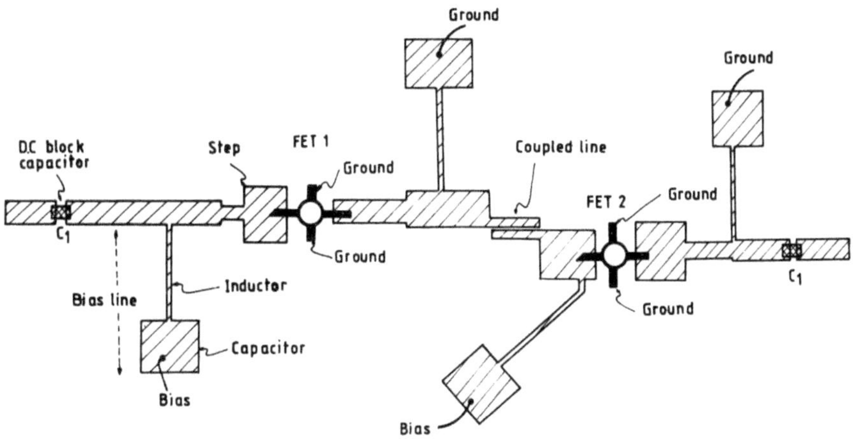


**Fig. 21.** Layout of a two-stage FET amplifier in microstrip configuration.

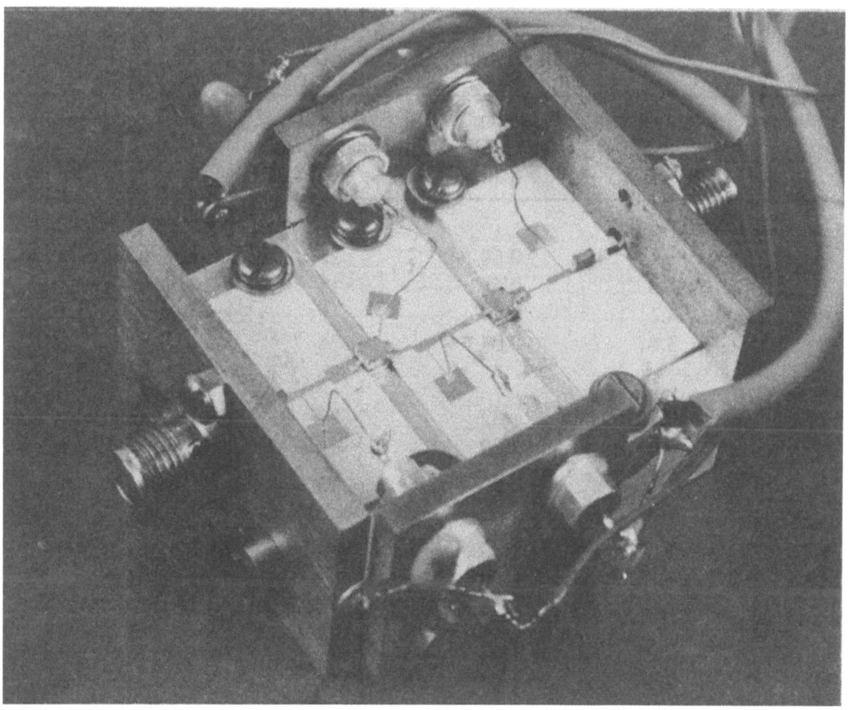

**Fig. 22.** Photograph of the two-stage amplifier.

## 5.1 Materials

*Conductor materials*

Conductor materials for the MICs must have the following desirable properties:

- high conductivity or low RF resistivity so that the ohmic loss is minimized.

- low temperature coefficient of resistance.
- strong adherence to the substrate material.
- good solderability and adaptability to bonding processes (such as thermo-compression and ultrasonic bonding).
- resistant to oxidation.

Table 4 presents a list of conductor materials useful for MICs. The materials listed under category A (Ag, Cu, Au, Al) are good conductors and are used as conducting layers in MICs. For a given operating frequency, the thickness of the deposited layer must be about three to six times the skin depth. These conductors adhere well to the synthetic substrates but on ceramic substrates such as alumina, the adhesion is very poor. The materials listed at category B (Cr, Ta, Ti) have good adherence property but poor conductivity and those at category C(Mo, W) have moderate conductivity and fair adhesion. The materials at category D (Pt, Pd) are barrier metals and are used to separate a good conductor layer (category A) from the adhesive layer (category B).

**Table 4. Conductor materials for MICs [7, 57]**

| Material | Conductivity $\sigma \times 10^{-7}$ (mhos/m) | Skin depth $\delta$ at 10 GHz ($\mu$m) | Coefficient of thermal expansion $\alpha_T$ (/°C × $10^6$) | Remarks |
|---|---|---|---|---|
| *Category A* | | | | |
| Ag | 6.1 | 0.64 | 21 | Good conductivity, poor |
| Cu | 5.8 | 0.66 | 18 | adhesion to ceramic |
| Au | 4.1 | 0.79 | 15 | substrates |
| Al | 3.6 | 0.84 | 26 | |
| *Category B* | | | | |
| Cr | 0.76 | 1.82 | 8.5 | Poor conductivity, good |
| Ta | 0.64 | 1.98 | 6.6 | adhesion to ceramic |
| Ti | 0.21 | 3.5 | 9.0 | substrates |
| *Category C* | | | | |
| W | 1.8 | 1.19 | 6 | Moderate conductivity, |
| Mo | 1.75 | 1.20 | 4.6 | fair adhesion |
| *Category D* | | | | |
| Pt | 0.94 | 1.64 | 9 | Barrier metals as separa- |
| Pd | 0.94 | 1.64 | 11 | tors between categories A and B |

Substrates for MICs are metallized in two-or three-layers. The metal systems commonly employed are, Cr-Au, Cr-Cu-Au, Ta-Au or Pd-Au for hybrid MICs on ceramic substrates and Cr-Au, Ti-Pt-Au and Ti-Pd-Au for MMICs. These materials are commonly deposited using evaporation or RF sputtering. For MMICs, materials such as Ta, Mo and Pt are also deposited using the electron beam evaporation.

*Dielectric Films*

Dielectric films are used in capacitors as overlays, in passive components such as couplers and protective layers for active devices. The properties of commonly used capacitances in the range 0.2 pF–100 pF are required in MICs. Thin-film dielectrics in MICs are listed in Table 5. The desirable properties for these dielectrics are, reproducible dielectric constant, low loss tangent, sufficient break-down voltage and low probability of developing pin holes. The quality factor of silicon monoxide (SiO) is about 50 whereas values higher than 100 are achievable in silicon-dioxide ($SiO_2$), tantalum pentoxide ($Ta_2O_5$) and silicon nitride ($Si_3N_4$). These materials can be deposited using the techniques mentioned in Table 5. For higher power MICs, breakdown voltages in excess of about 200 V are required and these are generally obtained with larger film thicknesses in the range 0.5 to 1 $\mu$m.

**Table 5. Dielectric film materials for MICs [7]**

| Material | $\varepsilon_r$ | Loss tangent (tan δ) | Dielectric strength in $10^6$ V/cm | Deposition technique |
|---|---|---|---|---|
| SiO | 6–8 | 0.01–0.03 | 0.4 | Evaporation |
| $SiO_2$ | 4 | 0.001–0.01 | 10 | Chemical vapour deposition (CVD) |
| $Ta_2O_5$ | 22–25 | 0.01 | 6 | Anodization, evaporation |
| $Al_2O_3$ | 7–10 | 0.2–0.24 | 4 | Anodization, evaporation |
| $Si_3N_4$ | 6.5–7.6 | — | 10 | CVD, sputtering |

*Resistive films*

Examples of planar resistors for hybrid MICs and MMICs are illustrated in Fig. 16. The desirable properties for the resistive films are, good stability, low temperature coefficient of resistance and good heat dissipation capability. The layer thickness is generally less than one skin depth so that the resistance value is independent of frequency. Table 6 lists the properties of important resistive materials. Sheet resistivities in the range 5 to 1000 Ω/square are achievable with these materials. Of the various materials listed, nickel-chrome (NiCr) and tantalum (Ta) are preferred because of their high resilience against environment, heat and aging.

**Table 6. Resistor film materials for MICs [7, 58]**

| Material | Resistivity (ohms/square) | Temp. coeff. of resistance (%/°C) | Stability | Deposition technique |
|---|---|---|---|---|
| Cr | 10–1000 | − 0.1 to 0.1 | Poor | Evaporation |
| NiCr | 40–400 | 0.002 to 0.1 | Good | Evaporation |
| Ta | 5–100 | − 0.01 to 0.01 | Excellent | Sputtering (in A.N) |
| Ti | 5–2000 | − 0.1 to 0.1 | Fair | Evaporation |
| Cr-SiO | 5–600 | − 0.005 to − 0.02 | Fair | Evaporation (or cermet) |

## 5.2 Hybrid Technology

The fabrication of a hybrid MIC basically involves the following three steps: (a) metallization of the substrate, (b) photo-etching the circuit conductor pattern and (c) soldering/bonding the discrete passive and active devices.

Figure 23 illustrates a typical thin film hybrid process involved in realizing a microstrip pattern on alumina substrate.

*Metallization*

The substrate is first chemically cleaned in acetone and sulphuric acid. The metal combination commonly adopted for alumina is Cr-Cu-Au. In order to achieve good adhesion a seed layer of Cr of about 150Å is first deposited followed by a thin conducting layer (0.1 $\mu$m) on the substrate using either simple evaporation or RF sputtering. The high conductivity Cu-layer is built up to about 10 $\mu$m either in the same vacuum run or by electroplating. Finally a flash

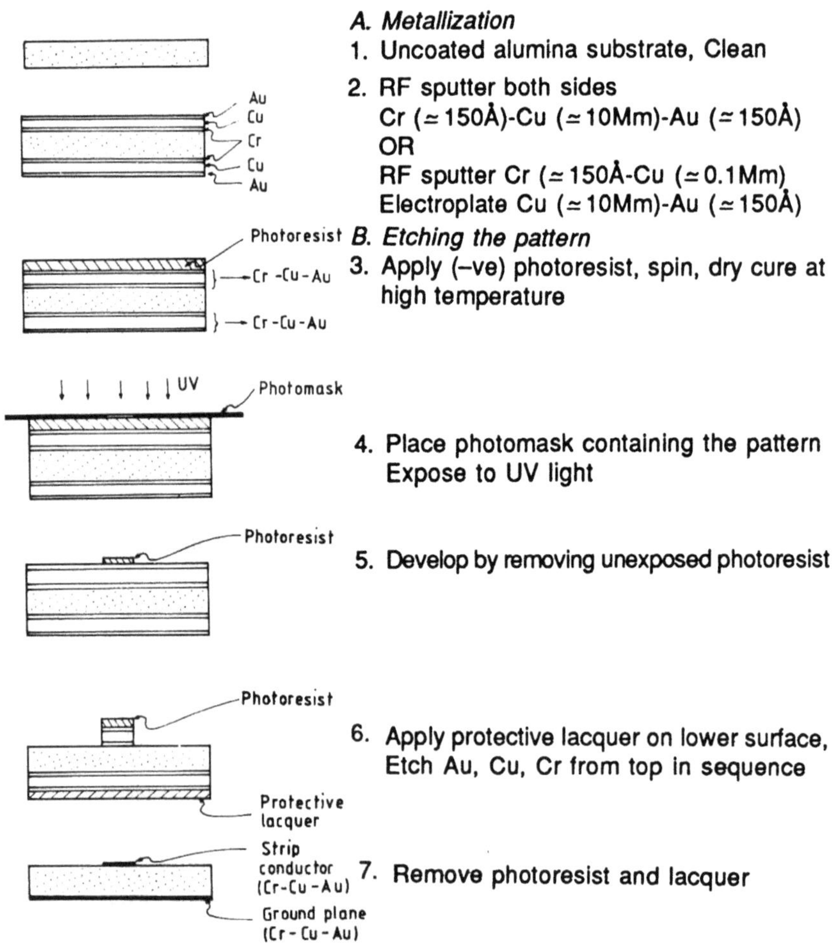


**Fig. 23.** Typical hybrid MIC process for realizing a microstrip.

of Au (150 Å) is given in order to prevent the Cu-surface from oxidation. The Au-top surface is also excellent for ultrasonic bonding whereas a Cu-top surface usually requires soldering for connecting active devices.

*Etching the Pattern*
The etching process for defining the conductor pattern involves different stages owing to the different metallic layers. The substrate is coated with a photo-resist (usually negative photo-resist) using a spinning system. The photo-resist film must be thin for better resolution. The substrate is then dried and cured at high temperature. The photo-mask containing the microwave circuit pattern is placed on the substrate and exposed to ultra-violet light. The substrate is then placed in a photo-resist developer. This will remove the photo-resist from the exposed area and leave the resist on the desired pattern. In the case of a microstrip pattern, the lower metallized surface which is to serve as a ground plane is coated with a protective lacquer. The undesired metal film layers—Au, Cu and Cr on the top surface are then etched away using suitable etchants in sequence (Au etch: potassium iodine mixed with iodine; Cu etch: ferric chloride; Cr etch: sodium hydroxide mixed with ferri-cyanide). Finally, the photo-resist and lacquer are removed leaving the desired conducting pattern. Resistive films if required as part of the circuit can be deposited on the substrate using evaporation or sputtering.

When commercially available Cu-clad plastic laminates such as RT-duroid are used, the first metallization step is eliminated. After cutting the laminate to the desired size, it is cleaned first with a fine abrasive powder or metal polish and then in acetone. The procedure for photo-etching the desired pattern is the same as that described for alumina substrate except that a suitable etchant is used for etching the copper. The pattern can then be electroplated with a thin layer of Au in order to prevent Cu from oxidation.

Attachment of discrete components on plastic substrates is easy since the substrate can be easily drilled to fit diodes and transistors which need one of the terminals to be connected to ground. Several techniques are available for attaching discrete devices-namely soft soldering, epoxy bonding using silver- or gold-based epoxies and wire bonding using either themo-compression or ultrasonic bonding. The first two techniques are commonly used for plastic substrates and bonding is used for gold-coated alumina substrates.

### 5.3 Monolithic Technology

Unlike the hybrid technology which requires a single level metallization process, the monolithic technology involves multi-level process for incorporating all active devices, passive circuit elements and interconnections into the bulk or onto the surface of a semi-insulating semiconductor substrate. Figure 24 illustrates a typical MMIC process on GaAs substrate. The following are the process steps.

*Active Layer Formation*
The first step in the MMIC process is formation of an active layer (*n*-type) on or into a semi-insulating GaAs substrate. Techniques used are either ion-implan-

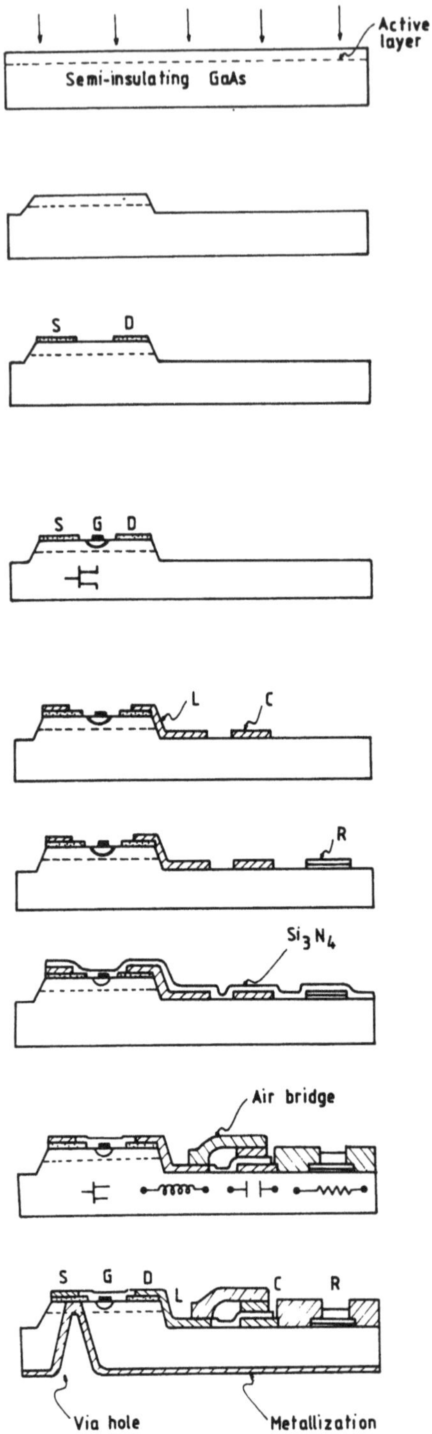


1. Active layer formation
2. Mesa etch
3. Ohmic contacts: Source (S) and drain (D)
   Au-Ge/Ni evaporation or alloying
4. Schottky-gate (G) formation (Ti-Pt-Au)
5. First level metallization (Ti-Pt-Au, Mo-Au)
   Inductor (L), lower plate of capacitor (C), S-D overlay
6. Resistor (R) formation
7. Dielectric film/passivation
8. Second level metallization/air-bridge interconnection
9. Via grounding backside metallization

**Fig. 24.** Typical GaAs MMIC process.

tation or epitaxial growth. In the ion-implantation technique, dopant atoms are bombarded on the GaAs substrate and in the epitaxial technique, an additional *n*-layer film is grown onto the substrate surface. As compared with the epitaxial technique, ion implantation offers better control of doping profiles over a larger substrate area and improved reproducibility.

*Mesa Etch*

Mesa etching is a process by which active device areas are isolated so that current flow is restricted to the desired portions. This is achieved by etching away the surface leaving mesas of active layer at the desired locations. The thickness of the material removed is on the order of 0.8 $\mu m$.

*Ohmic Contact*

Ohmic contacts are formed on the active n-layer by sequential evaporation of Au, Ge and Ni followed by lift-off and alloying in $H_2$ ambient at around 450°C. The total layer thickness is approximately 2000Å.

*Schottky-gate Formation*

The gate material must possess properties of good adhesion to GaAs, good electrical conductivity and thermal stability. The metal system that is commonly used for gate formation in GaAs MMICs is Ti-Pt-Au. Gates are normally defined by contact photolithography, and submicron gates by electron beam lithography.

*First-level Metallization*

This is a overlay metallization which forms inductors, lower plates of MIM (metal-insulator-metal) capacitors, transmission line sections and provides overlays for ohmic contacts. The preferred metal systems are Ti-Pt-Au, Mo-Au, Cr-Pt-Au. For example, sequential layers of Ti, Pt and Au are deposited and patterned by lift-off. The pattern can also be formed by ion-milling through a photoresist mask to remove the unwanted metal. The thickness of this layer is typically 6000Å.

*Resistor Formation*

Resistors are formed by using GaAs material or resistive films such as Ni, Cr, Ti, NiCr and TaN. GaAs resistors are formed between two ohmic pads by using the existing isolation and ohmic contact masks. Thin film resistors made of materials such as Ni, Cr, TaN etc are realized by sputtering.

*Dielectric Film Deposition and Passivation*

Dielectric films are used as dielectric layers for MIM capacitors, cross-over insulation and passivation of active device areas and resistors. The dielectric materials commonly used for this purpose are silicon nitride ($Si_3N_4$) and silicon-dioxide ($SiO_2$). These are normally deposited either by plasma-enhanced chemical vapour deposition or sputtering.

*Second Level Metallization*
The second level metallization is used to form top plates of capacitors, interconnection of components and air-bridges. The metal combination used is Ti-Pt-Au. These materials are evaporated in sequence and patterned by lift-off. Gold electroplating to a thickness of about 10–15 $\mu$m is used for defining air-bridge connections and for realizing transmission lines.

*Via Grounding and Backside Formation*
After completing the process of defining the circuit on the top surface, the GaAs substrate is thinned to about 100-200 $\mu$m. The exact thickness is dictated by the transmission line impedance. Next, via holes are etched through the substrate using ion etching and the entire backside including the holes is metallized to form the RF ground plane.

## REFERENCES


1. H. Howe, *Stripline Circuit Design*, Artech House (MA), 1974.
2. B. Bhat and S.K. Koul, *Stripline-like Transmission Lines for Microwave Integrated Circuits*, Wiley Eastern Ltd. (New Delhi), 1989.
3. K.C. Gupta, R. Garg and I.J. Bahl, *Microstrip Lines and Slotlines*, Artech House (MA), 1979.
4. T.C. Edwards, *Foundations for Microstrip Circuit Design*, John Wiley and Sons (NY), 1981.
5. J. Frey and K. Bhasin (Eds.), *Microwave Integrated Circuits*, (Reprint volume), Artech House (MA), 1985.
6. T. Itoh (Ed.), *Planar Transmission Line Structures*, IEEE Press, NY, 1987.
7. Leo Young (Ed.), *Advances in Microwaves*, Vol. 8, Academic Press, NY, 1974.
8. M.A.R. Gunston, *Transmission Line Impedance Data*, Van Nostrand Reinhold Co., (Lond.), 1972.
9. S.B. Cohn, Problems in strip transmission lines, IRE Trans. Microwave Theory Tech., Vol. MTT-3, pp. 119–126, March 1955.
10. S. Ramo and J.R. Whinnery, *Fields and Waves in Communication Electronics*, John Wiley & Sons, NY, 1965.
11. M.V. Schneider, Microstrip lines for microwave integrated circuits, Bell Syst. Tech, J., Vol. 48, pp. 1421–1444, May-June 1969.
12. H.A. Wheeler, Transmission line properties of parallel strips separated by a dielectric sheet, IEEE Trans. Microwave Theory Tech., Vol. MTT-13, pp. 172–185, March 1965.
13. H.A. Wheeler, Transmission line properties of a strip on a dielectric sheet on a plane, IEEE Trans. Microwave Theory Tech., Vol. MTT-25, pp. 631–647, Aug. 1977.
14. E.O. Hammerstad, Accurate models for microstrip computer-aided design, IEEE MTT-S Int. Microwave symp. Digest, pp. 407–409, 1980.
15. P. Bhartia and I.J. Bahl, *Millimetre Wave Engineering and Applications*, John Wiley & Sons, NY, 1984.
16. E.J. Denlinger, A frequency dependent solution for microstrip transmission lines, IEEE Trans. Microwave Theory Tech., Vol. MTT-19, pp. 30–39, Jan. 1971.

17. E. Yamashita, K. Atsuki and T. Ueda, An approximate dispersion formula of microstrip lines for computer-aided design of microwave integrated circuits, IEEE Trans. Microwave Theory Tech., Vol. MTT-27, pp. 1036–1038, Dec. 1979.
18. R.A. Pucel, D.J. Masse and C.P. Hartwig, Losses in microstrip, IEEE Trans. Microwave Theory Tech., Vol. MTT-16, pp. 342–350, June 1968. Also correction, Vol. MTT-16, pp. 1064, Dec. 1968.
19. E.J. Denlinger, Losses in microstrip lines, IEEE Trans. Microwave Theory Tech., Vol. MTT-28, pp. 513–522, June 1980.
20. B.E. Spielman, Dissipation loss effects in isolated and coupled transmission lines, IEEE Trans. Microwave Theory Tech., Vol. MTT-25. pp. 648–656, Aug. 1977.
21. I.J. Bahl and D.K. Trivedi, A designer's guide to microstrip line, Microwaves, Vol. 16, pp. 174–182, May 1977.
22. M.V. Schneider, Dielectric loss in integrated microwave circuits, Bell Syst. Tech. J., Vol. 48, pp. 2325–2332, Sept. 1969.
23. S.B. Cohn and G.D. Osterhues, A more accurate model of the $TE_{10}$ type waveguide mode in suspended substrate, IEEE Trans. Microwave Theory Tech., Vol. MTT-30, pp. 293–294, March 1982. Also correction in IEEE Trans. Microwave theory tech., Vol. MTT-30, p. 1291, Aug. 1982.
24. F.E. Gardiol, Careful MIC design prevents waveguide modes, Microwaves, Vol. 16, pp. 188, 190–191, May 1977.
25. J.I. Smith, The even-and odd-mode capacitance parameters for coupled lines in suspended substrate, IEEE Trans, Microwave Theory Tech., Vol. MTT-19, pp. 424–431, May 1971.
26. I.J. Bahl and P. Bhartia, Characteristics of inhomogeneous coupled stripline, IEEE Trans. Microwave Theory Tech., Vol. MTT-28, pp. 529–535, June 1980.
27. T.S. Shu, Xiao-xia Qi and Yun-yi Wang, Analysis equations for shielded suspended substrate microstrip and broadside-coupled striplines, IEEE MTT-S Int. Microwave Symp., Digest, pp. 693–696, 1987.
28. S.K. Koul and B. Bhat, Characteristic impedance of microstrip-like transmission lines for millimetre wave applications, Arch. Elek. Ubertragung., Vol. AEU-35, pp. 253–258, June 1981.
29. P. Pramanick and P. Bhartia, CAD models for millimetre-wave fin-line and suspended substrate microstrip lines, IEEE Trans. Microwave Theory Tech., Vol. MTT-33, pp. 1429-1435, Dec. 1985.
30. S.B. Cohn, Slotline on a dielectric substrate, IEEE Trans. Microwave Theory Tech., Vol. MTT-17, pp. 768–778, Oct. 1969.
31. R. Garg and K.C. Gupta, Expressions for wavelength and impedance of slotline, IEEE Trans. Microwave Theory Tech., Vol. MTT-24, p. 532, Aug. 1976.
32. C.P. Wen, Coplanar waveguide: A surface strip transmission line suitable for non-reciprocal gyromagnetic device applications, IEEE Trans. Microwave Theory Tech., Vol. MTT-17, pp. 1087–1090, Dec. 1969.
33. J.B. Knorr and K.D. Kuchler, Analysis of coupled slots and coplanar strips on dielectric substrate, IEEE Trans. Microwave Theory Tech., Vol. MTT-23, pp. 541–548, July 1975.
34. G. Ghone and C. Naldi, Analytical formulas for coplanar lines in hybrid and monolithic MICs, Electron. Lett., Vol. 20, No. 4, pp. 179–181, Feb. 1984.
35. Y. Tokumitsu et al, 50 GHz IC components using alumina substrates, IEEE Trans. Microwave Theory Tech., Vol. MTT-31, pp. 121–128, Feb. 1983.
36. T. Nakagami, H. Yatsuka and M. Ishizaki, Millimetre wave integrated circuits

using alumina substrates, Fujitsu Sci and Tech. J., Vol. 14, No. 4, pp. 33–43, Dec. 1978.

37. P.L. Lowbridge, C. Burnett and B. Prime, W-band high reliability hermetically sealed microstrip receiver, Military Microwave Conf., London, Conf. Digest, pp. 502–511, Oct. 1984.
38. M. Dydyk and B.D. Moore, Shielded microstrip aids V-band receiver designs, Microwaves, Vol. 2, pp. 77–82, March 1982.
39. K. Chang et al, V-band low-noise integrated circuit receiver, IEEE Trans. Microwave Theory Tech., Vol. MTT-31, pp. 146–154, Feb. 1983.
40. R.S. Tahim, G.M. Hayashibara and K. Chang, Design and performance of W-band broadband integrated circuit mixers, IEEE Trans. Microwave Theory Tech., Vol. MTT-31, pp. 277–283, March 1983.
41. L.Q. Bui, N. Ton and D. Ball, A D-band millimetre-wave cross-bar mixer, IEEE MTT-S Int. Microwave Symp. Digest, pp. 555–556, 1984.
42. N.G. Alexopoulos and S.A. Maas, Characteristics of microstrip directional couplers in anisotropic substrates, IEEE Trans. Microwave Theory Tech., Vol. MTT-30, pp. 1267–1270, Aug. 1982.
43. S.K. Koul and B. Bhat, Generalized analysis of microstrip-like transmission lines and coplanar lines with anisotropic substrates for MIC, electro-optic modulator and SAW applications, IEEE Trans. Microwave Theory Tech., Vol. MTT-31, p. 1051–1059, Dec. 1983.
44. S.K. Koul and B. Bhat, *Microwave and Millimetre Wave Phase Shifters, Volume I on Dielectric and Ferrite Phase Shifters*, Artech House, MA, 1991.
45. A.J. Baden Fuller, *Ferrites at Microwave Frequencies*, IEE Electromagnetic Wave Series 23, Peter Peregrinus Ltd, 1987.
46. R. Garg and I.J. Bahl, Microstrip discontinuities, Int. J. Electronics, Vol. 45, pp. 81–87, 1978.
47. R.K. Hoffman, *Handbook of Microwave Integrated Circuits*, Artech House, MA, 1987.
48. M. Caulton, Lumped elements in microwave integrated circuits, in *Advances in Microwaves*, Vol. 8, by L. Young and H. Sobol (Eds.), Academic Press, NY, 1974.
49. D.A. Daly et al, Lumped elements in microwave integrated circuits, IEEE Trans. Microwave Theory Tech., Vol. MTT-15, pp. 713–721, 1967.
50. F.E. Terman, *Radio Engineering Handbook*, McGraw-Hill, NY, 1943.
51. E. Pettenpaul et al, CAD models of lumped elements on GaAs upto 18 GHz, IEEE Trans. Microwave Theory Tech., Vol. MTT-36, pp. 294–304, 1988.
52. R.S. Pengelly and D. Richard, Design, measurement and application of lumped elements up to J-band, Proc. 7th Eur. Microw. Conf., pp. 461–464, 1977.
53. I. Wolff and N. Knoppik, The microstrip ring resonator and dispersion measurements on microstrip lines, Electron. Lett., Vol. 7, pp. 779–781, 1971.
54. K. Kawano and H. Tomimuro, Slot-ring resonator and dispersion measurement on slot line, Electron Lett., Vol. 17, No. 24, pp. 916–917, 1981.
55. J.K. Plourde and C.L. Ren, Applications of dielectric resonators in microwave components, IEEE Trans. Microwave Theory Tech., Vol. 29, No. 8, pp. 754–770, 1981.
56. D. Kajfez and P. Guillon (Eds.), *Dielectric Resonators*, Artech House, MA, 1986.
57. H. Sobol, Application of integrated circuit technology to microwave frequencies, Proc. IEEE, Vol. 59, pp. 1200–1211, 1971.
58. R.W. Bery, P.M. Hall and M.T. Harris, *Thin-Film Technology*, Van Nostrand, 1968.
59. R.A. Pucel (Ed.), *Monolithic Microwave Integrated Circuits*, (A reprint volume), IEEE Press, NJ, 1985.

# 3

# Metallization of Plastics by Electroless Plating

B. Viswanathan
Department of Chemistry
Indian Institute of Technology, Madras, India

Of the available liquid state metallization processes for plastics, electroless process has gained remarkable importance in the last 20 years, even though, the administrator of the invention from NBS Dr. Blum did not anticipate any tremendous commercial use for this process [1]. However, the following statements from P&SF report [2] that "computer memory devices represent a growing market for electroless nickel plating.... The US government has saved the public several million dollars for the past few years by using electroless nickel.... Electroless copper baths without formaldehyde may be the wave of the future in circuit board plating" show that this anticipation was completely proved to be wrong. In fact plating of plastics has become an important segment of the metal finishing industry in US [3] and in many other developed and developing countries. Shipley [4] predicts that commercial usage of electroless plating will grow quite rapidly during the next 10 years (since 1984). Applications will increase in number as design engineers and advanced engineering groups learn more about the capabilities and advantages of electrolessly plated metals and alloys.

Starting from nickel electroless plating by hypophosphite by Brenner and Riddell [5] other metals and reducing agents have been used for the electroless deposition on plastics. The available data given in Table 1, show that only a limited number of metals can be deposited by electroless method, however, this method has gained commercial applications in many sectors like aerospace, automotive, computers, electronics machinery and so on due to the unique properties of this deposition technique. The main reasons for using electroless method in preference to other methods especially electrolytic plating are:

- the feasibility of obtaining uniform deposits over irregular surfaces
- the possibility of direct deposition on nonconductors
- the flexibility of depositing on isolated areas
- the desirable characteristics of the deposits namely, less porous and more corrosion resistant and unique deposit properties
- the possibility of bulk plating and semibulk racking.

**Table 1. Typical metal coatings by electroless deposition—summary of available data**

| Deposit | Year | Medium | Reducing agent | First inventor | Ref. |
|---|---|---|---|---|---|
| Ni-P | 1946 | | Sodium hypophosphite | Brenner & Riddel | [13–17] |
| Co-P | 1947 | alkali | | | [17] |
| Ni | | | Hydrazine hydrosulphides | | [18] |
| Ni-B | 1958 | | Dimethylamineborane borohydrides | | [18, 19] |
| Co-Ni-P | 1973 | | Sodium hypophosphite | | [17, 20, 21] |
| Ni-B-Th | 1981 | | | Matsuoka & Hayashi | [21] |
| Co-B | | acid/alkali | Dimethylamineborane | | [21] |
| Cu | 1947 | alkali alkali alkali | Formaldehyde dimethylamineborane | Narcus | [22, 23, 24–32, 33, 34–36] |
| Pd-P | 1968 | alkali | Sodium hypophosphite | | [22, 37] |
| Pd | 1961 | alkali | Hydrazine ter. amineboranes | Rhoda | [6, 38,39] |
| Au | 1970 | | Amineborane borohydride | Okino | [40, 41, 42] |
| Au-Cu | 1982 | | Formaldehyde | Molenaar | [43] |
| Ag | 1974 | | Dimethylamine borane | Pearlstein | [17, 44] |
| Sn | 1980 | | Titanium trichloride | Warwick & Shirley | [45] |
| Alloy | 1970 | | | Malloy | [46] |
| Polyalloys | 1965 | | | Pearlstein & Weightman | [47] |
| Composites | 1974 | | | Parker | [48] |

The electroless deposit is usually achieved by using a bath containing at least five functional components, namely,

1. A metal salt as the source for the deposited metal.
2. A reducing agent.
3. A complexing agent.
4. A pH adjusting agent.
5. A stabilizing agent and other functional additives.

For each one of these components, one has a variety of substances to choose from. The purpose of this chapter is to examine the science behind this process especially that of electroless copper and to evaluate the relative merits of the choices, especially in the context of metallization of plastics.

## 1. HISTORICAL BACKGROUND OF THIS DEVELOPMENT

As there are more than four important documents already available on this aspect [1, 2, 4, 6] we shall briefly consider the available literature on the developments of the electroless deposition technique. Riddell and Brenner invented this process during their investigations on the electrodeposition of nickel-tungsten and cobalt tungsten alloys on the interior of a liner using an ammoniacal citrate-nickel-tungsten plating bath with an insoluble anode. In order to prevent the oxidation of organic components which affected the quality of the deposit they added various reducing agents to the bath. When they used hypophosphite as the reducing agent, they observed that the current efficiency was higher and at the same time the outside of the liner was also plated. This marked the beginning of the autocatalytic chemical plating which is popularly known as 'electroless' plating today. Though Brenner called it electrodeless process, Blum coined the term 'electroless' to indicate that the process is similar to electro-plating in that a metal coating is produced but the dissimilarity arises in that in the former process no external electric current is involved. It should be remarked that though the effect that sodium hypophosphite would reduce nickel and cobalt salts to the metallic powder was known even 30 years earlier to this invention, the achievement is that by a controlled autocatalytic process one could produce a smooth, coherent metal coating. This was followed by a series of investigations by Gutzeit and his group [7–10] on electroless deposition of Ni-P system, wherein the role of many other additives like chelates, exaltants, buffers, stabilizers were examined and operating conditions for obtaining a desirable rate of deposition were formulated [7–12]. This series of studies has lead to the development of other baths for electroless deposition of other metals and the important inventions in the area of electroless deposition are summarized in Table 1. Electroless nickel using hypophosphite as reducing agent and as it is generally known consists of Ni-P alloy with phosphorus content varying between 3 and 12% depending on both composition and operating parameters. The phosphorus content is usually increased by lowering bath pH and by increasing hypophosphite and orthophosphite concentrations. In many aspects the electroless deposits exhibit different structure, physical and mechanical properties fatigue strength and corrosion resistance as compared to that of electrolytic deposits. A typical compilation of the comparison of the properties of electroless Ni-P alloy and of electrodeposited nickel is given in Table 2 [4].

Electroless copper, has seen remarkable developments starting from Narcus [23] who reported the practical reduction of copper on nonconductors in 1947 and Cahill [49] who described autocatalytic chemical reduction of copper from alkaline copper baths using formaldehyde as the reducing agent. The success story of copper baths is mainly due to the introduction of effective stabilizing agents like mercaptobenzothiazole [50], thiourea [51] cyanide [52] and sulphur compounds [53, 54]. Because of the better corrosion resistance and other favorable properties of electroless copper it is preferred over electroless nickel on plastics.

The interest in electroless cobalt is mainly due to their magnetic properties which can be exploited for recording storage and switching device applications. Commercial usage of electroless cobalt is limited to magnetic applications especially to their magnetic deposits on tapes, discs and wire substrates [55–62].

**Table 2. Comparison of properties of deposits from electroless (Ni with approximately 8% P) and electrolytic nickel [4]**

| Properties | Electrolytic 99+%Ni | Electroless average 92% Ni/8% P |
|---|---|---|
| Composition | 99 + Ni | Average 92% Ni /8% P |
| Appearance | Dull to bright | Semibright |
| Structure | Crystalline | Amorphous |
| Density | 8.9 | Average 7.9 |
| Thickness uniformity | Varies | ± 10% |
| Melting point | 1455 °C | 890°C |
| Hardness as plated ($kg/mm^2$) | 150–400 | 500–600 |
| Heat hardening ($kg/mm^2$) | No effect | 1000 |
| Wear resistance | Fair | Excellent |
| Corrosion resistance | Good (porous) | Very good (few pores) |
| Relative magnetic susceptibility | 36% | 4% |
| Electrical resistivity (micro ohm/cm) | 7 | 60–100 |
| Thermal conductivity (cal/cm/sec/C) | 0.16 | 0.01–0.02 |
| Coefficient of thermal expansion (in/in/F × 10) | 7.5 | 7.8 |
| Modules of elasticity (Ksi) | 30 | 10 |
| Elongation | 6–30% | 2% |
| Internal Stress (ksi) | ± 10 | ± 10 |
| Friction coefficient vs | | |
| steel unlubricated | galling | 0.38 |
| lubricated | 0.2 | 0.2 |

Attempts to produce electroless deposits of polyalloys [63] as well as composites [64] were successful in 1970 in addition to that of Au, Ag and tin.

Electroless gold plating continues to be an area of considerable activity. The studies pursued now involves formulation of baths with long term stability, increased plating rate and improved throwing power. Carbonates of sodium and potassium in combination with amines like mono-, di-, and trialkanol, amines and ethylene di-and tri-amines show higher throwing power in gold plating baths [65]. Combined nickel and gold plating can also be achieved using dimethylamine borane and hydrazine hydrate as reducing agents [66]. A new addition called reduction promoters namely hydrochinone or pyrogallol convert the oxidized thiourea species formed during gold reduction back to thiourea and thus inhibit the precipitation of gold and accelerate gold plating rate [67]. A German patent discloses a weakly acidic or neutral bath containing in addition to KAu $(CN)_2$, 1-hydroxyethane, diphosphoric acid, diethanol-ammonium salt of maleic acid hydrazide and triethanol amminium salt of 1-hydroxylethane-1, 1-diphosphonic acid [68, 69].

## 2. GENERAL SCHEME OF ELECTROLESS PLATING

Normally electroless plating involves a number of steps carried out in sequence as shown in Fig. 1. Normally the sequence of operations involves first sensitization followed by activation and then electroless deposition using chemical reduction reaction with suitable controls either in the form of additives for acceleration or retardation of the deposition rate. However one should remember that there will be a number of rinsing and washing steps which are not explicitly shown in the follow chart shown in Fig.1.

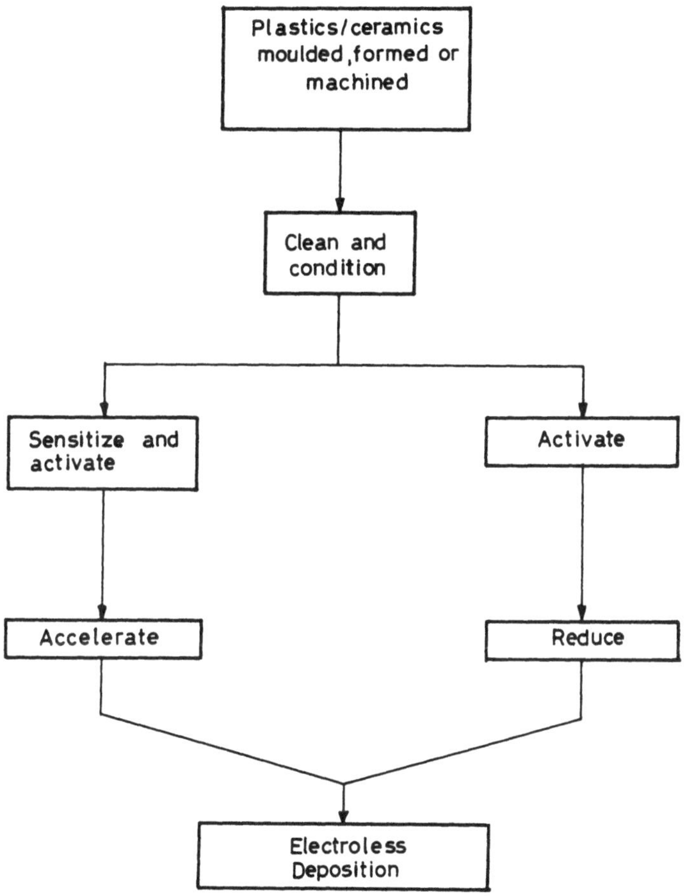


**Fig. 1.** Sequence of general operations in electroless deposition.

## 3. SENSITIZATION AND ACTIVATION

Normal commercial procedures of sensitization for hydrophobic surfaces involves two step immersion sequence in acidic stannous chloride solution followed by immersion in an acidic palladium chloride solution. The first bath is called the sensitizer and the second as activator. The chemistry involved in this process of immersion is dispersion of finely divided palladium (O) which initiates the autocatalytic plating process. The nucleation of palladium (O) is achieved by the redox reaction taking place between Pd(II) ions in the solution as well as tin ions adsorbed on the substrate. This means one should achieve good adsorption of sensitizers. Schlesinger and Kisel [70] have examined the adsorption properties of acidified $SnCl_2$ solution with additives like hydroquinene, triton x-100, thiourea and showed a non-linear relationship between the amount of sensitizer adsorbed and the number of metal islands deposited from electroless nickel bath (citrate bath) per unit area of the substrate. The effectiveness of commercially available sensitizing solutions for electroless copper deposition on hydrophobic substrates has been gauged by the measurement of the contact angles. Feldstein and Weiner [71] have demonstrated that controlled additions of stannic ions with or without sodium chloride especially with aging could yield superior wetting and thus can yield uniform plating of hydrophobic substrates [72–74]. The aging process in the stannic halide solutions probably favors the transformation of an $\alpha$-stannic acid to a $\beta$-stannic acid form. The colloidal nature of $\beta$-stannic acid is consistent with the view that the sensitizers together with the activators form only colloids and not solution complexes [75]. Cohen and West [76–78] postulate that a complex between Sn(II) and Pd(II) is first formed in which the reduction of Pd(II) to Pd(O) occurs subsequently forming a nucleus for the growth of the colloid. The core of the colloid is a metallic alloy of Pd and tin surrounded by stabilizing layers of $SnCl_2$ and Sn $(OH)_2$. According to this model, one should expect the deposition of metallic tin on Pd and the cyclic voltammetric study [79] by Jean Horkans aimed at identifying the chemical state of the catalyst on the activated surface showed that in addition to the stabilizing stannous layer produced by the reduction of Sn(IV) metallic tin also constitutes to the core of the active colloid. The catalytic activity of Pd(O) that are formed seems to be prone to UV irradiation and irradiated Pd(O) species appear to be incapable of becoming catalytically active in the metal plating bath [80, 81].

The structural and composition of the colloidal catalyst particles have been investigated by a variety of techniques like electron diffraction, Mossbauer spectroscopy, Rutherford backscattering and photoelectron spectroscopy [82]. Though there is some convergence on the fcc structure of the active catalyst, there still seems to be some uncertainty regarding the composition of Pd-Sn alloy. It is certain that Pd/Sn ratio is greater than 1, [83, 84]. The various values of the ratio reported in literature could have arisen because the analysis of the data have been performed in the different regions of the active catalyst from the core to the outer layer depending upon the type of technique employed.

In the sensitizing and activating step, certain accelerators like NaOH, EDTA are added which are effective in removing the tin species from the surface and increasing the initial rate of metal deposition. The uniform distribution of the metal deposit is also dependent on the nature of the accelerator used, for example NaOH and EDTA are favorable for uniform deposition of copper [85].

In the sensitization step as stated earlier, various types of tin compounds including tin oxide ($SnO_2$) and stannous and stannic chlorohydroxides, oxyhydroxides and hydrated oxides have been identified. The formation of these compounds and hence the extent of sensitization are dependent on the pH of the solution [86]. The process sequence involves a rinsing step after the activation and sensitization. There seems to be some difference of opinion on the rinse water pH for the activation of plastics, Cohen and Meek [87] propose that acidic solution rinse provides a cleaner, more catalytic surface while Ghorashi [86] claim that a rinse water with pH = 9 provides optimum conditions for obtaining a uniformly activated surface. It is to be remarked that the final rinse with appropriate pH provides the necessary type of species, hydroxide or chloride species on the outer layer of the activator-sensitizer combination which favors the catalytic reduction and deposition of the metal from the solution.

A new concept of the activation can be introduced based on the anchoring the functional groups or ions on the polymer surfaces. For example $Pd^{2+}$ can be exchanged with polymer substrates with functional groups especially carboxylic groups which will avoid the formation of tin chloride shell over the activator and thus promote in a facile manner electroless deposition [87a].

## 4. ELECTROCHEMISTRY OF ELECTROLESS COPPER

Electrochemical nature of the electroless deposition of metals has also received considerable attention [88–91]. Paunovic [88] started it with his mixed potential concept. According to this model, when at an electrode simultaneously two different electrode reactions

$$Ox_1 + n_1e \rightarrow Red_1 \tag{1}$$

$$Ox_2 + n_2e \rightarrow Red_2 \tag{2}$$

take place, both reactions strive to attain the equilibrium state with its own equilibrium potential $E_{eq}$ but in this process attain a steady state mixed potential ($E_{MP}$). The characteristics of this mixed potential are that

(i) Both redox systems are removed from their equilibrium potentials governed by the equations

$$\eta_1 = E_{MP} - E_{1eq} \tag{3}$$

$$\eta_2 = E_{MP} - E_{2eq} \tag{4}$$

(ii) Net electrochemical reaction occurs as they are removed from their equilibrium by the establishment of the mixed potential.

(iii) The condition that the sum of cathodic current densities is equal to the sum of the anodic current densities holds good, since net current cannot flow in the isolated system.

$$\vec{i}_1 + \vec{i}_2 = \overleftarrow{i}_1 + \overleftarrow{i}_2 \tag{5}$$

(iv) Since at the steady state mixed potential, the system is not in equilibrium the change in free energy is not equal to zero.

Extending this mixed potential concept to electroless plating wherein one of the redox couple is the metal electrode and the second one is the reducing agent applied in electroless plating, the corresponding reactions can be written as

$$M^{n+} + n_M e \rightarrow M \quad (6)$$

$$O_k + n_{red} e \rightarrow Red \quad (7)$$

The establishment of mixed potential is shown pictorially in Fig. 2.

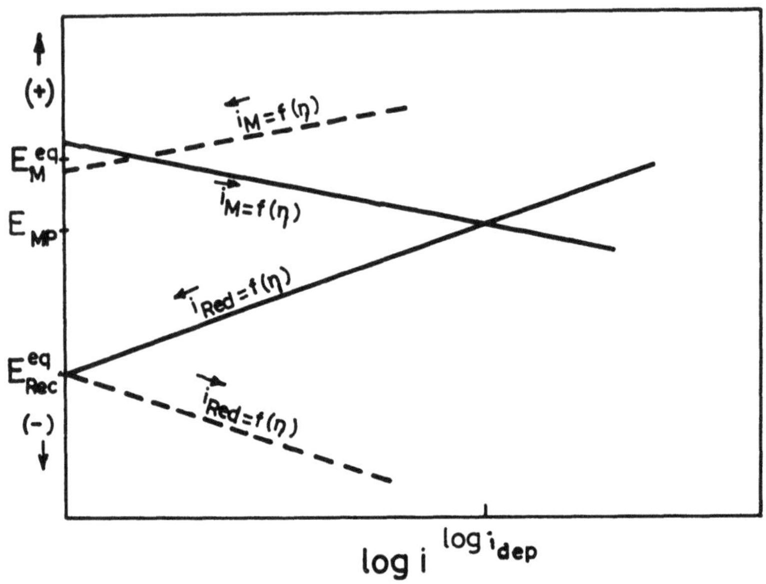


**Fig. 2.** Current-Potential curves for the system with two different simultaneous electrochemical reactions, Kinetic scheme.

Wherein it is seen that the potential of the redox couple is raised from its reversible value $E_{red}^{eq}$ and the potential of the metal electrode is depressed cathodically from its reversible value $E_M^{eq}$ so as to reach the mixed potential $E_{MP}$ value. With the overall reaction at the electrode is represented as

$$M^{n+} + Red \rightarrow M + Ox \quad (8)$$

In essence, the mixed potential can be considered to be the intersection points of $i_M$ and $i_R$. This implies at the steady state mixed potential, the rate of reduction of the metal is equal to the rate of oxidation of the reducing agent or

$$i_{\text{electroless deposition}} = \overrightarrow{i}_M = \overleftarrow{i}_{red} \quad (9)$$

This implies that one can a priori calculate the values of $E_{Mp}$ and deposition rate $i_{dep}$ if one knows the current potential functions $i = f(\eta)$ of the individual electrode processes. The parameters which determine the current potential curves for the partial electrode reactions (6) and (7) also determine $E_{Mp}$ and $i_{deposition}$. Parameters like the variation of the exchange current density (Fig. 3), concentration of reactants (since $i_0 = \eta F \overrightarrow{k} C_{OX} = \eta F \overleftarrow{k} C_{red}$, where $k$s are rate constants and $C$s are concentrations), temperature and the type of overvoltage (charge transfer, mass transport or reaction overvoltage) that control the values of $E_{MP}$

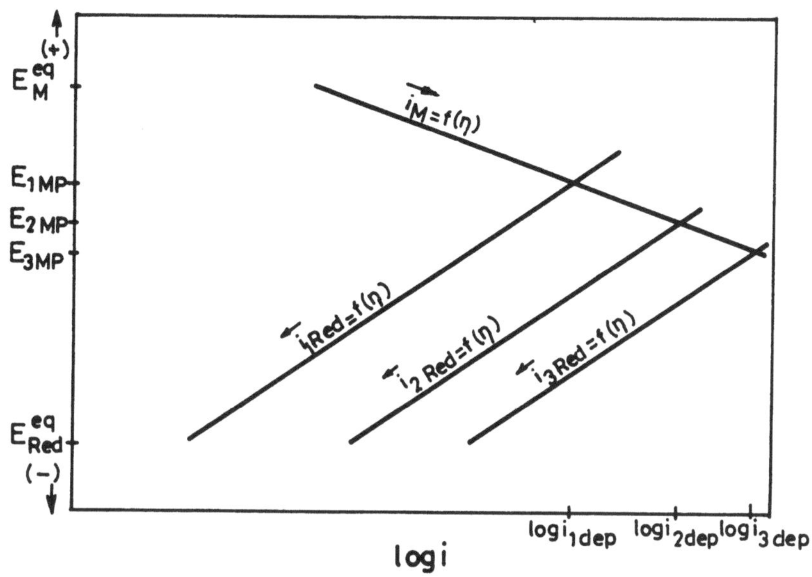


**Fig. 3.** Variation of the rate of electroless deposition and the mixed potential with the exchange current density.

as well as $i_{deposition}$. Another aspect that has been investigated is to account why certain metals with negative free energy of adsorption of hydrogen can be deposited using a specific reducing agent like formaldehyde [92, 93].

The solution chemistry of electroless copper bath has been probed by a variety of electrochemical techniques. Using cyclic voltammetry technique below the mixed potential value (– 0.7 V) Tam [94] has identified that methylene glycolate ($CH_2O_2H$) and its copper (11) complex are the electroactive species in solution. The methleneglycolate anion is adsorbed on the copper sites with the C–H bonds adjacent to the copper surface which then undergoes electro-dehydrogenation to give rise to formate anion as well as molecular hydrogen evolution [95]. As seen from the data in Table 1 only a limited variety of reductants are used in electroless process namely hypophosphite, formaldehyde, borohydride, dialkylamine borane and hydrazine. Normally electroless plating is accompanied by hydrogen evolution though it is not directly related to metal deposition. Van der Meeraker [96] proposed a universal mechanism for electroless deposition which involves the dehydrogenation of the reductant as the first step.

$$RH \rightarrow R + H \quad (10)$$

$$R + OH^- \rightarrow ROH + e^- \quad (11)$$

$$H + H \rightarrow H_2 \quad (12)$$

$$H + OH \rightarrow H_2O + e^- \quad (13)$$

$$M^{n+} + ne \rightarrow M \quad (14)$$

where RH represents the reductant. It is seen from this mechanism that hydrogen atom combination step (12) indicates that metals with dehydrogenation-hydrogenation catalytic activity can promote this reaction step favorably. Extending this proposal Ohno et al [97] propose that the oxidation of the reductants proceeds mainly along with hydrogen evolution step on copper, silver and gold while hydrogen ionization mechanism predominates on Co, Ni, Pd and Pt. The catalytic activity series of the metals for anodic oxidation of reductants does not coincide with that for hydrogen electrode reaction.

Electroless copper obtained using hypophosphite as reductant is supposed to proceed through the following steps

$$H_2PO_2^- \rightarrow HPO_2^- + H$$

$$Cu^{2+} + 2H_2PO_2^- + 2OH^- \xrightarrow{\text{catalyst surface}} Cu^0 + 2H_2PO_3^- + H_2 \quad (15)$$

when complexing agents like EDTA are used octahedral complexes are formed while with ligands like oxalate, glycine square planar or square pyramidal type complexes are proposed. Multinuclear (usually dimeric) complexes are reported with citrate, tartrate, malate which did not show any ESR signal indicating strong antiferromagnetic coupling, and in the presence of these ligands significant plating rates were observed [98, 99]. Eventhough complexation is one of the essential steps in the electroless plating no definite relationship could be established between the plating rate and the stability constant of the possible complexes that are formed. However ligands which can form dimeric copper species exhibit higher plating rates (Refer to the data assembled in Table 3 [100]).

The mixed potential theory proposed for electroless deposition does not seem to predict correctly the experimentally obtained mixed potential and the current potential ($i/V$) curve is not a simple sum of the half cell $i/V$ responses in the case of copper electroless deposition by formaldehyde in presence of EDTA. When the two half cell reactions are

$$Cu\,(EDTA)^{2-} + 2e^- \rightarrow Cu\,(metal) + EDTA^{4-} \quad (16)$$

$$2H_2CO + 4OH^- \rightarrow 2HCOO^- + H_2\,(gas) + 2H_2O + 2e^- \quad (17)$$

strongly coupled then the assumption that the reduction of Cu $(EDTA)^{2-}$ is not perturbed by the presence of HCHO or vice versa does not hold good. It is therefore necessary to invoke in the reaction mechanism of formaldehyde induced reduction of copper, the formation of a Cu (EDTA)/HCHO complex on the electrode followed by the reduction of the complex. This is one of the results reported in literature which does not conform to the mixed potential model [101]. However, recent studies by Hung and Ohno [102, 103] have shown that reactions involved in electroless copper deposition reduced by hypophosphite though complicated still follows the mixed potential theory. The stability of HCHO bath for the electroless copper deposit is a problem that has received attention. It is generally believed that thick copper deposit can be obtained if the additives that have propensity for adsorption on copper surface are excluded. Among the additives NaCN was found to leave the chemical purity of the deposit in tact [104].

**Table 3. Plating rates and stability constants with copper (II) and nickel (II) ions for various complexing agents (M = metal ion and L = ligand; *K* value will change with pH and ionic strength of the medium)**

| Complexing agent | Stability constant with $Cu^{2+}$ log *K* | Stability constant with $Ni^{2+}$ log *K* | Plating rate pH = 9.2, 65°C, $mg/cm^2$. h | Plating rate pH = 5.3 $mg/cm^2$. h |
|---|---|---|---|---|
| Sodium citrate | ML (5.9) | | | |
| | $M_2L_2$ (13.2) | ML (5.4) | 7.6 | 0 |
| Sodium potassium | ML (3.39) | | | |
| tartrate | $M_2L_2$ (8.24) | ML (2.06) | 6.6 | 0.8 |
| Oxalic acid | $ML_2$ (9.21) | | | |
| | ML (4.84) | ML (5.16) | 1 | 0 |
| Glycine | ML (8.15) | ML (5.78) | 0.19 | |
| | $ML_2$ (15.03) | $ML_2$ (10.58) | | 0.3 |
| | | $ML_3$ (14.0) | | |
| Malic acid | ML (3.42) | ML (3.17) | 4.2 | 2.9 |
| | $M_2L_2$ (8.0) | | | |
| EDTA (sodium | ML (18.7) | ML (18.52) | 0.7 | 0.8 |
| salt) | $ML_2$ (11.7) | | | |
| Phthalic acid | ML (3.15) | ML (2.17) | 0 | 0.9 |
| Succinic acid | ML (2.6) | ML (1.6) | 0.2 | 3.7 |
| Lactic acid | ML (2.45) | ML (1.64) | 0.2 | 1 |
| | $ML_2$ (4.08) | $ML_2$ (2.76) | | |
| | $ML_3$ (4.3) | $ML_3$ (3.1) | | |
| Ethylenediamine | ML (10.6) | ML (7.35) | 0.8 | 0.6 |
| | $ML_2$ (19.6) | $ML_2$ (13.54) | | |
| | | $ML_3$ (17.71) | | |
| Cyanide | $ML_4$ (25) | | | |
| Diphosphate | $ML_2$ (11) | | | |
| Ammonia | $ML_4$ (14) | | | |
| Pyridine | $ML_4$ (7) | | | |
| Dipyridyl (22′) | $ML_3$ (7) | | | |
| 2, 9-Dimethylphenanth- | | | | |
| roline-(1, 10) | $ML_2$ (11.7) | | | |

Data reproduced from [27] and [35].

## 5. CHEMISTRY OF ELECTROLESS COPPER BATHS

Compositions of a typical electroless copper bath is given in Table 4. Though free cupric ions can exist in acid medium, its presence in + 2 state requires a complexing agent in alkaline medium used in the electroless plating baths. Typical reducing agents like formaldehyde releases the necessary electrons for the reduction of the metal ions in alkaline medium while in acid solutions it is directly converted to carbondioxide according to the reaction

$$HCHO + H_2O \rightarrow CO_2 + 2H_2 \tag{18}$$

**Table 4. Typical bath compositions normally used in electroless plating**

| Deposit | Bath specifications (concentration) | End uses |
|---|---|---|
| Copper | Copper sulphate (0.04 M)<br>Ethylene diaminetetraacetic acid (0.10 M)<br>Rochelle salt (0.14 M)<br>Formaldehyde (0.20 M)<br>pH = 12.5<br>Temperature 65°C | PCBs, MICs EMI |
| Nickel | Nickel sulphate or chloride (0.10 M)<br>$Na_3C_6H_5O_7$ (0.10 M)<br>$NaH_3PO_2$ (0.20 M)<br>Triethanolamine (0.15 M)<br>pH = 5.7<br>Temperature 75°C | Aerospace, automotive, computers, electronics and machinery |
| Gold | KAu $(CN)_2$ (0.005 M)<br>Dimethylamine borane (DMB) (0.05 M)<br>KOH (.80 M)<br>KCN (0.035 M)<br>lead acetate 15 ppm<br>$K_2CO_3$ (0.5–1.0 M)<br>pH = 5–7.5<br>Temperature 80°C | PCB, Connector taps, chips and selective plating |
| Palladium | Palladium chloride (0.06 M)<br>Rochelle salt (0.09 M)<br>Ethylenediamine (0.04 M)<br>Sodium Hypophosphite (0.04 M)<br>pH = 8.5<br>Temperature 70°C | PCBs and connectors |
| Silver | Sodium silver cyanide (0.01 M)<br>Sodium cyanide (0.02 M)<br>Sodium hydroxide (0.02 M)<br>DMAB (0.05 M) | Contact switches, wave guides, reflectors |

However, the effective concentration of HCHO could be reduced in alkaline medium due to a number of side reactions like the disproportionation reaction (Cannizaro reaction)

$$2HCHO + OH^- \rightarrow CH_3OH + HCOO \qquad (19)$$

which is responsible for the aging of the bath and its instability.

Addition of methanol may shift the reaction (19) to the left and thus make available the necessary amount of formaldehyde for the reduction reaction. The precipitating seeds in the bath could promote the reduction of cupric ions to cuprous ions according to the reaction

$$2Cu^{2+} + HCHO + 5OH^- \rightarrow Cu_2O + HCOO^- + 3H_2O \qquad (20)$$

Since the bath does not contain complexing agents for cuprous ions and because of low solubility of cuprous oxide the following reactions may also take place in the bath

$$Cu_2O + H_2O \rightarrow Cu^0 + Cu^{2+} + 2OH^- \quad (21)$$

$$2Cu^+ \rightarrow Cu^{2+} + Cu^0 \quad (22)$$

These reactions result in fine particles of copper dispersed randomly in the solution. Since cuprous ions are not easily reduced by formaldehyde the bath instability also occurs. There are various ways one can avoid the formation of cuprous oxide. They are: (1) agitation or aeration of the bath, (2) use of low concentration of the constituents, (3) employing polyelectrolytes like cellulose, ether, polyvinyl pyrolidine, polyvinyl alcohol, gelatin peptone polyamides etc to cover the cuprous oxide to prevent the catalytic effect, (4) Decreasing the free concentration of cupric ions by using suitable complexing agents and (5) removing the nucleating impurities in the materials used especially in the Rochelle salt.

## 6. NUCLEATION AND GROWTH OF ELECTROLESS DEPOSITS

The general description of conventional deposition processes cannot be applied to electroless deposition obtained by the chemical reduction reaction taking place at the interface. The chemical reaction is normally initiated at the catalytic sites on the surface by a direct three dimensional nucleation step. The copper nuclei thus formed are about 25 Å in diameter, and form aggregates (nearly a few hundreds angstroms thick) and the ultimate size of the aggregate is decided by the energy considerations and at the limit of energetically stable condition crystallization process sets in wherein unstable aggregates are transformed into a relatively large grains. Repeated nucleation and recrystallization give rise to twin faults between two adjacent nuclei with twin orientation having a common {111} boundary [105]. In the case of mixed electroless deposits like Ni-P, Co-P, Ni-Co-P though nucleation takes place at the catalytically active sites, the aggregates are in a liquid like solid solution of the species which undergoes crystallization in the subsequent heat treatment steps [106].

The nucleation, growth and crystallization processes in electroless deposition are dependent on the chemical reduction reaction and depending upon the conditions employed the structure of the crystallized phase of the metal deposit can be different from what one obtains from simple metal deposits formed by other methods.

## 7. KINETIC STUDY OF ELECTROLESS COPPER PLATING

The rate law [107] applicable to electroless copper plating from HCHO-Rochelle salt baths has been empirically formulated as

$$\text{Rate} = 18.5\ [Cu^{2+}]^{0.47}\ [OH^-]^{0.18}\ [HCHO]^{0.07} \exp\ [18.7\ (T-313)/T]$$

which has been shown to be applicable to a limited extent. This type of rate laws [108] indicate a complex mechanism involving interaction among hydroxide ions, HCHO, and cupric ion in the reduction process. Similar rate law has also been found applicable when EDTA is used as complexing agent and HCHO as the reducing agent [109]. Donahue et al [110] have deduced the relationship

$$1/r_{X_s} = 1/r_0 + KC_{X_s}/r_0$$

where $r_{X_s}$ and $r_0$ are the plating rates in excess of ligand (N, N, N′, N′-tetrakis (2-hydroxy-propyl) ethylenediamine (quadrol) and in the absence of excess ligand respectively, $K$ is the adsorption coefficient and $C_{X_s}$ is the concentration of the ligand when it is present in excess. The applicability of the experimental data to this equation was considered to imply that adsorption-inhibition effect is operating in the electroless plating of copper by formaldehyde [110]. From a coulostatic method, Sato et al [111, 112] showed that the plating current density of copper on glass substrates can be expressed as

$$i_{ELP} = K/R_P$$

where $K$ is a constant and $R_P$ is the polarization resistance. The deposition rate decreases with aging of the bath while the proportionality constant $K$ ( = 27.3 mV) is independent of the substrate used as well as aging of the bath. It is therefore obvious that the kinetics of electroless plating though amenable to various kinds of measurements is complicated and only empirical kinetic laws can be generated on them which depend on a number of process parameters like substrate used, additives added, aging of the bath and the nature of reducing agent used. There is possibly a correlation between the rate of dissociation of the complex (when the ligands like tartrate, EDTA, Quadrol, 2-diamine N, N, N′, N′-tetraacetic acid, triethanolamine [113] are used) and deposition rate [114] and the hydroxyl concentration depenndence on the deposition rate depends on the nature of the ligand used [115]. But for these few generalizations the kinetics of electroless deposition still appears to be grossly empirical.

## 8. THERMAL AND MECHANICAL PROPERTIES OF ELECTROLESS COPPER

The electroless deposit for manufacturing printed circuit boards differs from the conventional subtractive process in that (i) a permanent, non-strippable resist is an integral part of the board, (ii) copper is not etched from the board, (iii) conductor resolution is better (iv) uniform plating over the entire panel is possible. However, the additive method has its own limitations. One of them is that hair line cracks occurring at hole corners during thermal testing especially for specific military specifications. The bath specifications were usually monitored in bulk preparation by automatic controllers, for example concentration of formaldehyde (bisulphate titration), copper (colorimetric analysis) and cyanide (ion selective electrode) and pH, solution density and mixing potential of plating solution. Typical mechanical data on two different electroless copper deposits are assembled in Tables 5 and 6 [116, 117]. Deposits from electroless baths containing EDTA and sufficient amount of glycine (upto 1 : 1 mole ratio so that condensation product is formed and thus controls the release of formaldehyde

**Table 5. Properties of electroless copper deposits**

| Property | Standard bath[a] | High quality bath[a] | Crack free deposit[b] | |
|---|---|---|---|---|
| | | | Class 1 | Class 2 |
| Elongation % | 2.5–3.0 | 6–12 | 4.3 | 7.1 |
| Tensile strength ($kg/mm^2$) | 25 | 30–60 | 59 | 21.8 |
| Grain size (dia.) parallel to substrate (m) | — | 0. 3–0. 7 | 0.3 | 0.7 |
| Grain size (dia.) perpendicular to substrate (m) | — | 6–7 | 7 | 6 |
| Cracks in through-hole after thermal stress | Yes | None | No | No |

[a]Data taken from [120].
[b]Data reproduced from [116].

or the free formaldehyde concentration) with HCHO have nearly equal elongation values and greater tensile strength than electrolytic copper from acid sulphate solutions containing no additives. The mechanical properties have a direct correlation with the deposition rate. The internal stress in the electroless copper is attributed to (a) entrapped hydrogen resulting from the plating process and (b) fine particles of copper obtained from bath decomposition occluded in the deposit [119].

**Table 6. Typical properties of copper foils obtained by electroless plating (data reported in [118])**

| Plating temp. °C | 38 | | | 49 | | |
|---|---|---|---|---|---|---|
| Plating rate m/hr | 1.9 | 2.82 | 3.43 | 3.45 | 4.32 | 5.69 |
| mil/hr | 0.073 | 0.111 | 0.135 | 0.14 | 0.170 | 0.224 |
| Tensile strength (thousand PSI) | 52.8 | 37.5 | 35.8 | 45.4 | 49.2 | 31.0 |
| Tensile strength after annealing (thousand PSI) | — | 45.1 | — | — | — | 48.9 |

In electroless copper baths certain additives like cyanides and 1, 10-phenanthroline are added as bath stabilizers. Though there are claims that addition of cyanide improves the mechanical strength of the deposit Junginger et al [120, 121] have shown that bivalent copper complex is adsorbed on the extraneous germs which are reduced to metallic copper by formaldehyde and the bath is affected in the presence of oxygen and complexing agents. The texture of deposit is dependent on the presence of surfactant (100 textured) while in the absence of surfactant the growth is (111) textured. There are two forms of hydrogen [122] (diffusible and residual) that exists in electroless copper deposits. The diffusible molecular hydrogen in the microvoids impairs the ductility of the metal deposit through its effect on lattice strain induced by high pressure and through the effect of voids on the structural integrity of the deposit.

## 9. APPLICATIONS OF ELECTROLESS COATINGS

### 9.1 Printed Circuit Boards [123]

Printed circuit boards is one of the important elements of electronics industry. Size, volume and circuit density are the factors wherein printed circuit boards (PCB) technology scores over other processes. Printed circuitry is an accepted technique by which interconnection of active components is made on solids or flexible polymer substrates. These contacts are made by either substractive (etch down) or additive process. Additive process is gaining importance due to cost saving and other pollution problems associated with etch down operations. In the additive printed circuitry either semi-additive or true additive procedures are employed wherein the sequence of steps are slightly altered, in the semi-additive process electroless plating precedes the masking while it is in the reverse order in the additive process. In the additive printed circuitry technique additives like monionic surfactants or cyanides are incorporated so as to obtain copper deposits of unusual physical properties like good surface conductivity, good adhesion, and durability. One can also use the photochemical activity of tin (II) compounds to create catalytically active sites including Pd (O) sites. These developments in addition to the simplicity of the process are expected to make this process more attractive and acceptable for printed circuitry manufacture.

### 9.2 Semiconductor Devices

The contact to silicon transistors are made using soft solders using intermediate metallization. The intermediate metallization should have the following characteristics:

(1) Good contact to both n- and p-type silicon.
(2) Good adhesion to the surface.
(3) The metal must be solderable.
(4) The metal-solder interface should be inert and should not form brittle intermetallics.

In the silicon transistor chips the contacts are usually made using solder. Direct soldering is not possible and intermediate metallization is necessary. This metallization layer should have certain desirable characteristics as stated above. Electroless nickel seems to meet the requirements. This technique has also been extended to III-V compounds as well.

In addition production of a variety of miscellaneous electronic components like capacitors, resistors, connectors, diodes, heat sinks, relays lead frames employ electroless deposition to a great extent.

### 9.3 Microwave Integrated Circuits (MIC)

The electroless technique has now been extensively used to produce a variety of passive and active microwave components on hard substrates like alumina, titanate ceramics, ferrites as well as soft substrates like PTFE (polytetrafluoroethylene). The replacement of Cr-Au metallization by electroless copper can result in 15–25% decrease in microwave attenuation. Recently microwave strip lines have also been designed by electroless copper plating [124]. The usefulness of this technique as compared to copper clad PTFE is demonstrated by the data given in Tables 7 and 8.

**Table 7. Microwave Properties of commercially available copperclad PTFE materials manufactured by different firms [124]**

| Material | Designed resonance frequency (GHz) | Observed resonance frequency (GHz) | Dielectric constant | Loss tangent (tan δ) |
|---|---|---|---|---|
| Rogers corporation RT Duroid 5880 | 8 | 7.67 | 2.41 | 0.0031 |
| Taconic | 8 | 7.96 | 2.24 | 0.0036 |
| Rogers corporation RT/Duroid 5880 | 10 | 10.03 | 2.2 | 0.00279 |
| Hindusthan Fluorocarbons | 10 | 10.8 | 1.89 | 0.0022 |
| 3M company | 10 | 9.7 | 2.35 | 0.0035 |
| Taconic | 10 | 9.75 | 2.325 | 0.0035 |
| Rogers corporation RT/Duroid 5880 | 16 | 15.9 | 2.21 | 0.0024 |
| Taconic | 16 | 15.7 | 2.27 | 0.0036 |
| Rogers corporation RT/Duroid 5880 | 18 | 18.94 | 2.01 | 0.01 |

**Table 8. Microwave properties of electroless copper coated PTFE substrates [124]**

| Material | Designed resonance frequency (GHz) | Observed resonance frequency (GHz) | Dielectric constant | Loss tangent (tan δ) |
|---|---|---|---|---|
| RT/Duroid | 8 | 7.555 | 2.48 | 0.0081 |
| RT/Duroid | 10 | 9.878 | 2.26 | 0.0030 |
| RT/Duroid | 16 | 15.63 | 2.29 | 0.0115 |

## 9.4 Electromagnetic Interference Shielding

Communication and control circuitry are highly vulnerable to interference by electromagnetic or radiofrequency waves. Electromagnetic interference can paralyse effectively HF communications, Ballisatic missile guidance and light aircraft navigation. Electroless copper, or Ni-P or Ni-B plated ABS/polypropylene oxide/polyester have been tested for shielding and copper coated with nickel and phosphorus seems to provide maximum shielding.

## 9.5 Thin Magnetic Film Devices

Cobalt-phosphorus films deposited on plastic substrates have been used in magnetic recording tapes. In addition high density thin film magnetic discs for storage applications in computers are produced using electroless cobalt deposit.

The Ni-P or cobalt deposits should have a high polish and uniformity, otherwise the pits and non-uniformity will lead to loss of information and make the disc scratchy and noisy.

### 9.6 Automobile Exterior Parts [125]

The production and use of plastics has grown dramatically during the last three decades. Most of the electroplated exterior automotive plastic trim is fabricated from acrylonitrile-butadiene-styrene (ABS) because of its low cost and good physical properties. Some plated exterior automotive trim is molded from modified polyphenylene oxide (MPPO) because of higher heat distortion temperature and lower coefficient of thermal expansion than ABS type of plastics. The direction of studies in this application is the development of improved stabilizers and rate controllers like cyanides and other organic derivatives that will enable rack processing.

## 10. CONCLUSION

Electroless plating has thus been receiving considerable attention in recent times with respect to new bath formulations with improved features like long term stability, increased plating rate, and increased throwing power from which purer deposits or deposits with consistent composition and quality can be obtained. Non-conventional methods like electroless gold plating on superconducting $YBa_2Cu_3O_{7-x}$ ceramic oxides from non-aqueous baths are also attempted for specific applications [126]. Plastics with thin uniformly metallized coatings is used for various electronic and machinery applications and the need in this sector is increasing more than what is expected. Structural studies of the deposits still seem to interest scientists [127]. On the whole as expected, electroless deposition appears to have a challenging future.

## REFERENCES

1. A. Brenner. Plating and Surface Finishing, p. 24, July 1984.
2. P and SF report, Plating and Surface Finishing, p. 36 (1984).
3. D.A. Arcilesi, Plating and Surface Finishing, p. 90, June 1981.
4. C.R. Shipley, Jr., Plating and Surface Finishing, p. 92 (1984).
5. A. Brenner and G.E. Riddell, Proc. AES., **33**, 23 (1946); **34**, 156 (1947); US Patent, 2, 532, 283; 2, 532, 284 (1950).
6. W. Goldie, 'Metallic Coatings on Plastics' Electrochemical Publications Ltd., Middlesex, England (1968).
7. G. Gutzeit and E.J. Ramirez, US Patent, 2, 658, 842 (1953).
8. G. Gutzeit and A. Krieg, US Patent, 2, 658, 841 (1953).
9. P. Talmey and G. Gutzeit, US Patent, 2, 762, 723 (1956).
10. G. Gutzeit, Plating, **46**, 1158, 1275, 1317 (1959); **47**, 63(1960).
11. E.B. Saubestre, Metal Finish., **60**, 67 (June 1962); **6049** (July 1962); **60**, 45 (August 1962); **60**, 59 (September, 1962).
12. F. Pearlstein, 'Electroless Plating' in F.A. Lowenheim Ed., Modern Electroplating, 3rd Ed., John Wiley & Sons, NY 1974 Ch. 31.

13. K.S. Rajam and S.R. Rajagopalan, Technical Memorandum, MT-TM-8601 (1986); MT-TM-8801 (1988) NAL, Bangalore, India.
14. K.S. Rajam, S.R. Rajagopalan, M.S. Hegde and B. Viswanathan, Materials Chemistry and Physics, **27**, 141 (1991).
15. M. Papini and F. Papini, Thin Solid Films, **115**, 1 (1984).
16. M. Schlesinger and J. Kisel, J. Electrochem. Soc., **136**, 1658 (1989).
17. S.L. Chow, N.E. Hedgecock, M. Schlesinger and J. Rezek, J. Electrochem. Soc., **119**, 1614 (1972).
18. D.J. Levy, Electrochem. Tech., **1**, 38 (1963).
19. J.W. Dini and P.R. Coronado, Plating, **54**, 385 (1967). Deutsches Aulegeschrift (DAS) 1, 137, 918 (1958).
20. M. Gulla, US Patent, 3764, 352 (1973); 3, 832, 168 (1974).
21. M. Matsuoka and T. Hayashi, Plating and Surace Finishing, 68, 66 (July 1981).
22. A. Sergienko, US Patent, 3, 418, 143 (1968).
23. H. Narcus, Met. Finish., **45**, 64 (sep 1947).
24. M. Paunovic and R. Zeblisky, Plating and Metal Finishing, **72**, 52 (Feb. 1982).
25. K.F. Blurton, Plat. Met. Finishing, 52 (Jan, 1986).
26. S. Mizumoto, H. Nawafune, M. Kawasaki, A. Kinoshita and K. Araki, ibid., 48 (Dec, 1986).
27. R. Junginger, J. Electrochem. Soc., **135**, 2299 (1988).
28. C.I. Courduvelis and G. Sutcliffe, plating **67**, 71 (Sep, 1980).
29. Y. Okinaka and H.K. Straschil, J. Electrochem. Soc., **133**, 2608 (1986).
30. S. Nakahara, Y. Okinaka and H.K. Straschil, J. Electrochem. Soc., 136, 1120 (1989).
31. C.E. Baumgartner, Plating and Surface Finishing, 53 (Dec, 1989).
32. K. Kondo, J. Ishikawa, O. Takenaka, T. Matsubara and M. Irie, J. Electrochem. Soc., **137**, 1859 (1990).
33. F. Pearlstein and R.F. Weightman, Plating, 474 (1972).
34. Aina Hung, Plating and Surface Finishing, 62, (Jan, 1988); 74 (April, 1988).
35. Aina Hung, Ken-Ming Chen, J. Electrochem. Soc., **136**, 72 (1989).
36. Aina Hung, and Izumi Ohno, J. Electrochem Soc., **137**, 918 (1990); Plating and Surface Finishing, 54 (March, 1990).
37. F. Pearlstein and R.F. Weightman, Plating, **56**, 158 (1969).
38. R.N. Rhoda, J. Electrochem. Soc., **108**, 707 (1961).
39. W.V. Hough, J.L. Little and K.E. Warheit, US Patent, 4, 255, 194 (1981).
40. Y. Okinaka, Plating, **57**, 914 (1970).
41. J.F. McCormack, US Patent, 3, 589, 916 (1971).
42. M.F. El-Shazly and K.D. Baker, US Patent, 4, 337, 091 (1982).
43. A. Molenaar, J. Electrochem, Soc., **129**, 1917 (1982).
44. F. Pearlstein and R.F. Weightmen, Plating, **61**, 154 (1974).
45. M.E. Warwick and B.J. Shirley, Trans, Inst. Met. Finish., **58**, 9 (1980).
46. G.O. Mallory, US Patent, 4, 019, 910 (1977). G.O. Mallory and T.R. Horhn, Plat. Sur. Finis., **66**, 40 (April 1979). G.O. Mallory, Plating and Surface Finishing, **63**, 34 (June 1976).
47. F. Pearlstein and R.F. Weightman, Electrochem Technol., **6**, 427 (1965).
48. K. Baker, Plating, **61**, 834 (1970).
49. A.E. Cahill, AES Proc., **44**, 130 (1957).
50. F. Pearlstein US Patent 3, 222, 195 (1965).
51. M. Saito, Prod. Finish., **31**, 7, 57 (1967).
52. R.J. Zeblisky, J.F. McCormack, J.D. Williamson and F.W. Schneble, US Patent, 3, 095, 309 (1963).
53. F.W. Schneble, R.J. Zeblisky, J.F. McCormack and J.D. Williamson, US Patent, 3, 361, 580 (1968).

54. E.B. Saubestre, Plating, **59**, 563 (1972).
55. R.D. Fisher and W.H. Chilton, J. Electrochem. Soc., **109**, 485 (1962).
56. L.D. Ransom, and V. Zentner, J. Electrochem Soc., **111**, 1423 (1964).
57. Y. Moradzadeh, J. Electrochem, Soc., **112**, 891 (1965).
58. M.G. Miksic, R. Travieso, A. Arcus and R.H. Wright, J. Electrochem. Soc., **113**, 360 (1966).
59. J.S. Sallo, Plating, **54**, 257 (1967).
60. J.S. Judge, J.R. Morrison, D.E. Speliotis and J.R. DePaw, Plating, **54**, 533 (1967).
61. R.D. Fisher and W.H. Chilton, J. Electrochem. Soc., **121**, 1023 (1974).
62. F. Pearlstein and R.F. Weightman, J. Electrochem. Soc., **121**, 1023 (1974).
63. D.W. Baudrand, Plat. Surf. Finish., **66**, 18 (Nov. 1979).
64. N. Feldstein, Met. Finish., **81**, 35 (Aug 1983).
65. US Patent, 4978559 (1990).
66. US Patent, 4985076 (1991).
67. US Patent, 4979988 (1990).
68. European Pataent, 418715 (1991).
69. German Patent, 284253 (1990).
70. M. Schiesinger and J. Kisel, J. Electrochem. Soc., **136**, 1658 (1989).
71. N. Feldstein and J.A. Weiner, Plating, **59**, 140 (1972).
72. N. Feldstein and J.A. Weiner, J. Electrochem. Soc., **119**, 668 (1972).
73. N. Feldstein, J.A. Weiner and G.L. Schnable, J. Electrochem. Soc., **119**, 1486 (1972).
74. N. Feldstein, S.L. Chow, and M.Schlesinger, J. Electrochem. Soc., **120**, 875 (1973).
75. E. Matijevic, A.M. Poskanzer and P. Zuman., Plating, 62, 958 (1975).
76. R.L. Cohen and K.W. West. Chem. Phys. Lett., **16**, 128 (1982).
77. R.L. Cohen and K.W. West, J. Electrochem. Soc., **120**, 502 (1973).
78. R.L. Cohen and R.L. Meek, J. Colloid and Interface Sci., **55**, 156 (1976).
79. Jean Horkans, J. Electrochem. Soc., **130**, 311 (1983).
80. B.W. Baulis, C.C. Huang and M. Schlesinger, J. Electrochem. Soc., 126, 394 (1979).
81. S.L. Chow, N.E. Hedgecock, M. Schlesinger and J. Rezek., J.Electrochem. Soc., **119**, 1013 (1972).
82. T. Osaka, H. Nagasaka and F. Goto., J. Electrochem. Soc., **127**, 2343 (1980).
83. N. Feldstein and J.A. Weiner, J. Electrochem. Soc., **120**, 475 (1973).
84. J.J. Grunwald, S. Gottesfeld and D. Laser, Plating and Surf. Finish., 71 (Oct. 1981).
85. J. Horkans, J. Kim, C. McGrath and L.T. Romankiw, J. Electrochem. Soc., **134**, 300 (1987).
86. H.M. Ghorashi, Plating and Surf. Finish., 694 (1975).
87. R.L. Cohen and R.L. Meek, Plating and Surf. Finish., 47 (1976).
87a. R.L. Jackson, J. Electrochem. Soc., **135**, 3172 (1988).
88. M. Paunovic, Plating, **51**, 1161 (1968).
89. S.M. El-Raghy and A.A. Abo-salama, J. Electrochem. Soc., **126**, 171 (1979).
90. F.M. Donahue, J. Electrochem. Soc., **119**, 72 (1972); **121**, 887 (1974).
91. P. Bindra and J. Tweedie, J. Electrochem. Soc., **130**, 1112 (1983).
92. P. Bindra and J. Rolden, J. Electrochem. Soc., **132**, 2581 (1985).
93. D. Vitkavage and M. Paunovic, Plating and Surf. Finish., 48 (April 1983).
94. T.M. Tam, J. Electrochem. Soc., **132**, 806 (1983).
95. S. Gottesfeld, J. Beery, M. Paffett, M. Hollander and C. Maggiore, J. Electrochem. Soc., **133**, 1344 (1986).
96. J.E.A.M. Van Den Meeraker, J. Appl. Electrochem., **81**, 395 (1981).

97. I. Ohno, O. Wakabayashi and S. Haruyama, J. Electrochem. Soc., **132**, 2323 (1985).
98. A. Hung. Plating and Surf. Finish., 62 (Jan, 1988).
99. A. Hung, Plating and Surf. Finish., 74 (April 1988).
100. A. Hung and K.M. Chen., J. Electrochem. Soc., **136**, 72 (1989).
101. B.J. Feldman and O.R. Melroy, J. Electrochem. Soc., **136**, 640 (1989).
102. A. Hung and I. Ohno, J. Electrochem. Soc., **137**, 918 (1990).
103. A. Hung and I. Ohno, Plating and Surf. Finish., 54 (March 1990).
104. C.E. Baumgartner, Plating and Surf. Finish., 53 (Dec. 1989).
105. R. Sard, J. Electrochem. Soc., **117**, 864 (1970).
106. S.L. Chow, N.E. Hedgecock, M. Schlesinger and J. Razek, J. Electrochem. Soc., **119**, 1614 (1972).
107. F. L. Shippey and F.M. Donahue, Plating, 43 (Jan, 1973).
108. F.M. Donahue and F.L. Shippey, Plating, 135 (Feb., 1973).
109. A. Molenaar, M.F.E. Holdrinet and L.K.H. Van Beek., Plating, 238 (March, 1974).
110. F.M. Donahue, D.J. Sajkowski, A.C. Bosio, L.L. Schafer, J. Electrochem. Soc., **129**, 717 (1982).
111. N. Sato, M. Suzuke and Y. Sato, J. Electrochem. Soc., 130 1485 (1983).
112. M. Suzuki, N. Sato, K. Kanno and Y. Sato., J. Electrochem. Soc., **129**, 2183 (1982).
113. K. Kondo, J. Ishikawa, O. Takenaka, T. Matsubara and M. Irie, J. Electrochem. Soc., **137**, 1859 (1990).
114. M. Paunovic, J. Electrochem. Soc., **124**, 349 (1977).
115. S. Ohno, Plating, 58 (1971).
116. K.F. Blurton, Plating and surf. Finish., 52 (Jan., 1986).
117. M. Paunovic and R. Zeblisky, Plating and Surf. Finish., 52 (Feb., (1985).
118. S. Mizumoto, H. Nawafune, M. Kawasaki, A. Kinoshita and K. Araki, Plating and Surf. Finish., 48 (Dec., 1986).
119. T. El-Mallah, S.M. El-Raghy, E.A. Abou-Saif, M.Y. Saada and A.E. El-Mehairy, Plating and Surf. Finish., 61 (Oct., 1980).
120. K. Junginger, J. Electrochem. Soc., **135**, 2299 (1988).
121. R. Junginger and G. Elsner, J. Electrochem. Soc., **135**, 2304 (1988).
122. Y. Okinaka and H.K. Straschil, J. Electrochem. Soc., **133**, 2608 (1986).
123. N. Feldstein, Plating, 146 (Feb., 1974).
124. V. Sundarajan, V. Sivasubramanian, K.C. James Raju, B. Viswanathan and V.R.K. Murthy, Indian J. Tech., **28**, 609 (1990).
125. D.A. Arcilesi, Plating and Surf. Finish., 90 (Jan., 1981).
126. US Pataent 4971944 (1990).
127. K.S. Rajam, Indira Rajagopal, S.R. Rajagopalan and B. Viswanathan, Materials Chemistry and Physics (in Press).

# 4

# Methods of Measurement of Dielectric Constant and Loss in the Microwave Frequency Region

V.R.K. Murthy
Department of Physics
Indian Institute of Technology, Madras, India

In the recent years, there has been considerable interest in the microwave properties of dielectric and magnetic materials usually termed as microwave materials. Every cost effective microwave device is a success story in coping with the limitations of the materials.

The measurements in microwave frequency region may be made purely for scientific purposes for e.g., in connection with the investigation of relaxation phenomena. However, due to the increasing importance of the microwave materials in telecommunications, in the design of microwave circuit components and quasi-optical components such as substrates, dielectric wave guides, radiation absorbing materials (RAM), one needs high precision in measurements and in some cases, the dielectric loss should be critically measured.

In this chapter, various methods are discussed to measure the dielectric constant, loss, complex magnetic susceptibility, microwave conductivity on powders and single crystals. These methods can also be used to study the microwave characteristics of thin films. All these methods have been verified on standard samples and found agreeable.

## 1. CAVITY RESONANT METHOD

### 1.1 Using a Rectangular Waveguide

Cavity perturbation method has been widely used to study the dielectric and magnetic parameters in the microwave frequency region. The dispersive and dissipative terms of the materials are directly related to the change in the resonant frequency and quality factor of the cavity from the respective empty cavity values.

In order to understand the overall properties of a material, the solutions of Maxwell's equations in the particular configuration must be understood.

The standard perturbation analysis of this problem is found elsewhere [1, 2]. As the resonant cavity is analogous to an isolated lumped resonant circuit, the standard expressions for calculating the values of capacitance and inductance of a rectangular cavity excited in its fundamental mode have been arrived [3].

From the Maxwell's electromagnetic equations, for a rectangular waveguide with broad dimension $a$, narrow dimension $b$ and length $l$ along $X, Y, Z$ directions respectively, the orientations of respective electric and magnetic fields can be deduced as $E_y$, $H_x$ and $H_z$ (where $E_z$, $E_x$ and $H_y$ are zero). In this case, the assumption is that the cavity is parallel resonant circuit with a capacitance $C_y$ corresponding to an average value of $E_y$ and the resultant inductance $L_{xz}$ from the parallel combinations of two inductances $L_x$ and $L_z$ corresponding to the average values of the respective magnetic fields $H_x$ and $H_z$. From the low frequency lumped circuit elements analogy

$$C_y = \frac{(\varepsilon_0 al)}{b} \tag{1}$$

$$L_{xz} = \frac{\mu_0 N^2 al}{(a^2 + l^2)} \tag{2}$$

Where $l = P\lambda_g/2$ for $TE_{10p}$ mode and $N$ has usual meaning similar to the number of turns in a solenoid. If $N$ is replaced by $1/\pi$, the above equations result to

$$f_0 = \frac{1}{2\pi\sqrt{L_{xz} C_y}} \tag{3}$$

Where $f_0$ is the frequency of the cavity.

### *1.1.1 Dielectric constant $\varepsilon'$ and loss $\varepsilon''$*

A sample of complex dielectric susceptibility $X_e = X_e' - jX_e''$ (equals to $\varepsilon' - 1 - j\varepsilon''$) is kept at the maximum electric field location of the cavity. A thin sample (for example a maximum of 3 mm diameter or side) is taken such that the electric field is uniform through out the volume of the sample. In order to simplify the problem, the sample is assumed to be cylindrical or rectangular with uniform cross-sectional area $S$ and length is greater than or equal to $b$ so that it will occupy the narrow section of the cavity.

After the introduction of the sample, the empty cavity's resonant frequency and $Q$ factor alter due to change in the overall capacitance and conductance of the cavity without perturbing the inductance.

The resonant frequencies with and without the samples are $f_1$ and $f_0$, correspondingly the dielectric constant $\varepsilon'$ is given by

$$\varepsilon' = 1 + \frac{f_0^2 - f_1^2}{f_1^2}\,\frac{V_c}{4V_s} \tag{4}$$

and the dielectric loss $\varepsilon''$ is given by

$$\varepsilon'' = \frac{V_c}{4V_s}\left[\frac{f_0^2}{f_1^2}\left(\frac{1}{Q_1} - \frac{1}{Q_0}\right)\right] \tag{5}$$

where $V_s$ is the volume of the sample, $V_c$ is the volume of the cavity, $Q_1$, $Q_0$ are the respective quality factors of cavity with and without the sample.

### 1.1.2 *Magnetic permeability $\mu'$ and loss $\mu''$*

In this case, the sample is introduced at maximum magnetic field location. The sample is aligned along $H_x$ for which $H_z$ and $E_y$ are least perturbed. The permeability $\mu'$ is given by

$$\mu' = 1 + \frac{1}{K}\frac{V_c}{V_s}\left(\frac{f_0 - f_1}{f_1}\right) \tag{6}$$

$$\mu'' = \frac{V_c}{KV_s}\left(\frac{1}{2Q_1} - \frac{1}{2Q_0}\right) \tag{7}$$

where

$$K = \frac{2a^2}{a^2 + l^2} \tag{8}$$

The details of the experimental arrangement used in this type of measurement is given in [3].

## 1.2 Measurement of Dielectric Constant and Loss on Small Single Crystals Using a Circular Cavity

The principle is based on the perturbation theory developed by Buranov and Shchegolev [4]. This method is proposed for measuring the dielectric constant and loss of cylindrical samples whose length is smaller compared to wavelength, while its transverse dimensions are comparable or less than the skinlayer thickness.

The cavity is a circular cavity as shown in Fig. 1 operating in $TE_{011}$ mode with its $Q \approx 10{,}000$. The appropriate equations for the perturbation caused by insertion of a small sample at an electric field antinode are

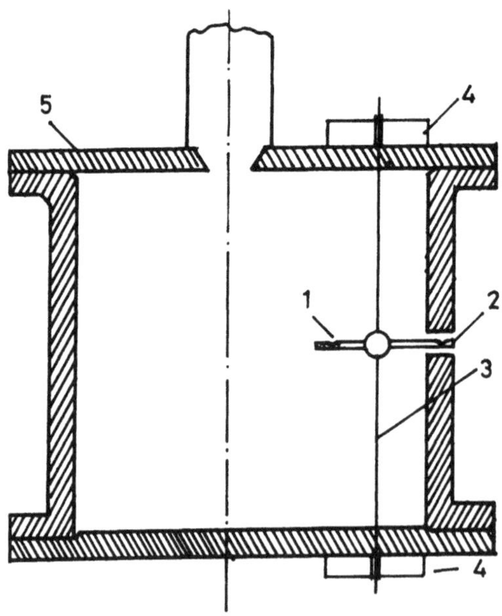


**Fig. 1.** Device for mechanical movement of the sample. 1—Sample, 2-Yokes, 3—Capron filament, 4—Pinions and 5—Top cover of the cavity.

$$\frac{\Delta_0}{2} = \frac{\Delta}{2f_0} = \frac{\alpha\varepsilon''}{[1 + n(\varepsilon' - 1)]^2 + (n\varepsilon'')^2} \tag{9}$$

$$\Delta f_0 = \frac{\Delta f}{f_0} = \frac{\alpha[(\varepsilon' - 1)(1 + n(\varepsilon' - 1) + (n\varepsilon'')]}{[1 + n(\varepsilon' - 1)]^2 + (n\varepsilon'')^2} \tag{10}$$

where $f_0$ is the centre frequency of the cavity, $\Delta$ is the change in the half power bandwidth of the cavity with sample and without, $\Delta f$ is the change in the centre frequency of the cavity with and without. For a cylindrical cavity operating in $TE_{011}$ mode, the filling factor $\alpha = 2.1$ (volume of the sample/volume of the cavity). The depolarization factor $n$ is evaluated for (a) long cylindrical sample ($L >> d$) as

$$n = \frac{3}{2}\left(\frac{d}{L}\right)^2\left[\ln\frac{4L}{d} - \frac{7}{3}\right] \tag{11}$$

and (b) for an ellipsoid of revolution as

$$n = \frac{bc}{a^2}\left[\ln\frac{4a}{b + c} - 1\right] \tag{12}$$

where $a$, $b$, $c$, are ellipsoid semiaxes. Consequently, the dielectric constant $\varepsilon'$ and loss $\varepsilon''$ are evaluated by

$$\varepsilon' - 1 = \frac{1}{n}\Delta f_0 \frac{\left(\frac{\alpha}{n} - \Delta f_0\right) - \left(\frac{1}{2}\Delta_0\right)^2}{\left(\frac{1}{2}\Delta_0\right)^2 + \left(\frac{\alpha}{n} - \Delta f_0\right)^2} \tag{13}$$

$$\varepsilon'' = \frac{\alpha}{n^2} \frac{\frac{1}{2}\Delta_0}{\left(\frac{1}{2}\Delta_0\right)^2 + \left[\left(\frac{\alpha}{n}\right) - \Delta f_0\right]^2} \tag{14}$$

These results assume that the sample is completely permeated by the microwave field i.e the skin depth $\delta > t$, where $t$ is the thickness of the sample perpendicular to the field. There are several difficulties with this technique, probably the most serious is that many poor conductors exhibit much higher apparent conductivities (often 5 or 6 orders of magnitude) than at dc. Thus any high conductivity microwave results should be checked and confirmed by dc or low frequency conductivity measurements. Nevertheless, this technique is useful as an excellent screening method.

The other drawbacks have to do with geometrical factors, especially the depolarization factor. However, this method can be comfortably used on anisotropic crystals also.

## 1.3 Measurement of Dielectric Constant and Loss on Powder Samples Using Rectangular Cavity

Many samples are available in powder form and some of them are prepared in small quantities. The measuring technique and theory [5] in the foregoing part are in general and may be applied to research on such powders.

The real and imaginary parts of the dielectric constant are given by

$$\varepsilon' + \frac{1}{N} - 1 = \frac{\frac{1}{N}\left(\frac{\delta\omega_0}{\omega_0}\right)\left(\frac{NV_c}{2V_s}\right) + 1}{\left[\left(\frac{\delta\omega_0}{\omega_0}\right)\left(\frac{NV_c}{2V_s}\right) + 1\right]^2 + \left[\delta\left(\frac{1}{2Q}\right)\left(\frac{NV_c}{2V_s}\right)\right]^2} \tag{15}$$

$$\varepsilon'' = \frac{1}{N}\frac{\delta\left(\frac{1}{2Q}\right)\left(\frac{NV_c}{2V_s}\right)}{\left[\left(\frac{\delta\omega_0}{\omega_0}\right)\left(\frac{NV_c}{2V_s}\right) + 1\right]^2 + \left[\delta\left(\frac{1}{2Q}\right)\left(\frac{NV_c}{2V_s}\right)\right]^2} \tag{16}$$

where the parameter $\delta$ refers to the change in frequency or $Q$ factor, $V_c$, $V_s$ are the volumes of the cavity and sample respectively. It can be seen that if we take $N = 1/3$, the equations are valid for a spherical sample and $N = 0$, they are valid for rod of infinite length.

However, it is explicitly agreed that one should analyze the depolarization factor for the evaluation of dielectric parameters. However, by taking the powder sample of known weight into a polyfoam cylindrical tube of known diameter that can be used as a sample holder for cavity measurement, the depolarization factor $N$ can be calculated using Wurschchmid's method [6]. The corresponding factors such as volume of the sample and cavity can be easily determined.

Further, the author has theoretically calculated $\varepsilon'$ and $\varepsilon''$ for various values of $N$ taking known weights of the sample into the polyfoam sample holder and computed graph between various values of depolarization factor and corresponding values of $\varepsilon'$ for each weight of the sample. Thus for a known value $\varepsilon' = 14$ sample the experimentally measured values are fitted into the above equation for the evaluation of $N$ that has agreed with theoretial value of $N$. However, for higher values of $\varepsilon'$, a slightly inevitable disagreement between theoretical and experimental values of $N$ could be seen which can be interpreted due to slight changes in cylindrical shape of the polyfoam sample holder, inhomogeneity of the sample packing and so on.

## 1.4 Microwave Conductivity Measurements on High $T_c$ Superconductors

Since the discovery of superconductivity with transition temperature ($T_c$) above 90K in $YBa_2Cu_3O_{7-x}$ system, extensive studies have been carried out on superconducting ceramic oxides. Even in superconducting state, the ac resistivity does not vanish and it is increasing with frequency. The reason may be the thermal dissociation of the superconducting fluid into normal electrons in between these $T_c$s obtained by direct current (dc) and microwave frequency (ac) and also due to the effect of thermally excited quasi particles.

The method of microwave conductivity measurement is described [7] using a rectangular cavity.

In this study, the change in $Q$ (quality factor) values were measured using a rectangular cavity in frequency modulation mode. The value of $Q$ for empty cavity is of the order of $10^3$ at room temperature.

Figure 2 shows the flow chart of the microwave conductivity measuring system employed. A rectangular cavity of $TE_{106}$ mode with resonant frequency at 9 GHz (X-band) was used. The cavity was coupled to the system and evacuated

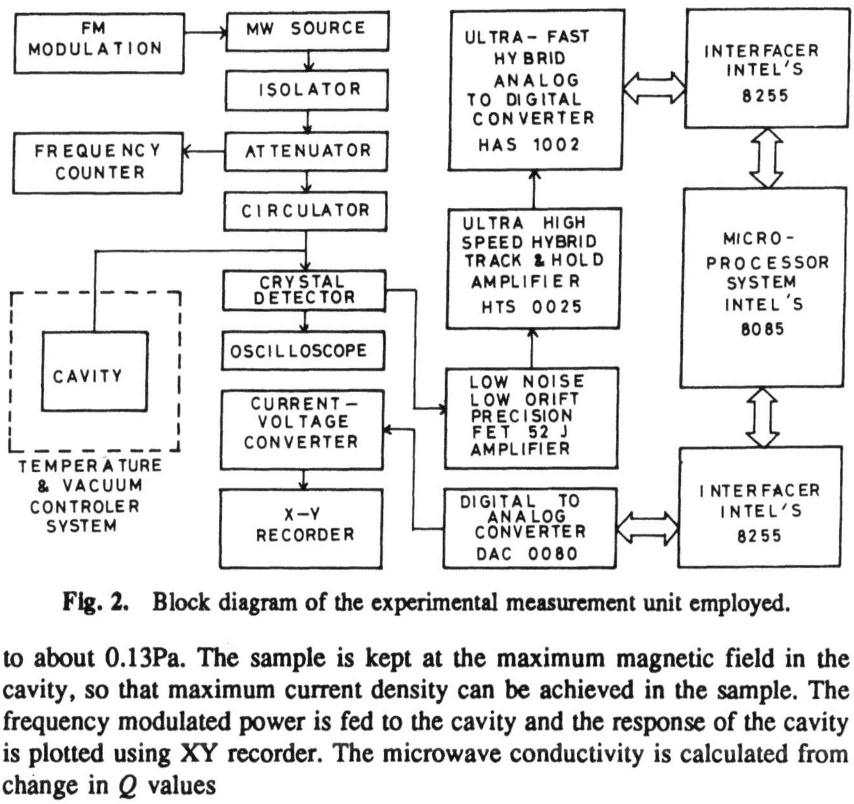


**Fig. 2.** Block diagram of the experimental measurement unit employed.

to about 0.13Pa. The sample is kept at the maximum magnetic field in the cavity, so that maximum current density can be achieved in the sample. The frequency modulated power is fed to the cavity and the response of the cavity is plotted using XY recorder. The microwave conductivity is calculated from change in $Q$ values

$$\Delta\left(\frac{1}{Q}\right) = \frac{1}{Q_1} - \frac{1}{Q_2} = P_0\,(\omega_0 W) \tag{17}$$

where $Q_1$ and $Q_2$ are the quality factors of the cavity with and without the sample, $P_0$ is Joule loss due to sample, $\omega_0 = 2\pi f_0$, $f_0$ being the resonance frequency of the empty cavity and $W$ is the power stored in the cavity. The $W$ and $P_0$ are calculated using the relations

$$W = \varepsilon_0 E_0^2\, \frac{abd}{8} \tag{18}$$

$$P_0 = \frac{R_s P^2 \pi^2 E_0^2 a_s b_s}{4d^2 \omega_0^2 \mu_0^2} \tag{19}$$

where $E_0$ is the maximum value of the electric field, $a$, $b$, $d$ are the dimensions of the cavity along breadth, width and length respectively, $R_s = (\sigma\delta)^{-1}$ is the surface resistivity of the sample and $\delta$ is the skin depth at the frequency of operation = $(2/\omega_0 \mu_0 \sigma)^{1/2}$ $a_s b_s$ is the total surface area of the sample, $\varepsilon_0$, $\mu_0$ are the permittivity and permeability of vacuum. From the above relations, the conductivity is given by

$$\sigma = \frac{2P^4 \pi^4 (a_s b_s)^2}{(\omega_0^2 \varepsilon_0^2 d^6 \mu_0^2)\,(ab)^2 \left(\Delta \frac{1}{Q}\right)^2} \tag{20}$$

where $P$ has usual meaning in $TE_{mnp}$ mode.

From the above relation one can see that conductivity is proportional to square of the microwave loss $(\Delta(1/Q))^2$.

## 2. DIELECTRIC MEASUREMENTS USING SWEPT FREQUENCY TECHNIQUES

In this approach, a swept frequency dielectric measurement [8] is presented which is shown to have advantages over conventional technique. This method shows a basic accuracy of better than 5% depending on the material under test. The most conventional methods [9, 10, 11] are simplified considerably if the sample length is an odd or even number of quarter wavelengths in the material. However this condition is usually difficult to achieve in practice since knowledge of the wavelength inside the sample implies knowledge of the complex dielectric constant one is trying to measure.

It becomes quiet obvious that for measuring dielectric constant, a very desirable condition is to have a sample length of one quarter wavelength or multiple thereof. It is rather awkward to change the physical dimensions of the sample in a continuous manner, instead the electrical length of the sample can be varied by changing the frequency. If now, the assumption is made that the dielectric properties of the material under test do not change appreciably with frequency over a relatively short range, then changing the electrical length by sweeping over some band, has the same effect as changing the physical length of the sample. In this way, it is possible to create conditions such as a quarter wavelength which makes it relatively easy to calculate the complex dielectric constant of the material. This approach forms the basis for the swept frequency method. The experimental measurement is basically a reflection measurement which relies on interface reflections and the attenuation of the sample to give information about the complex dielectric constant.

The basic experimental setup consists of a levelled microwave sweep generator that sends the incident wave to the sample, which will reflect and transmit some of the energy. If the reflected signal is detected and displayed versus frequency, a superimposed standing wave pattern results.

The dielectric constant $\varepsilon'$ and loss tangent ($\tan\delta = \varepsilon''/\varepsilon'$) are given by the following relations

$$\varepsilon' = \frac{1}{2} E \left[ 1 + \sqrt{1 - \frac{4P}{E}} \right] \tag{21}$$

where $E = (c/21\Delta f)^2$ and

$$\tan\delta = \frac{c}{(2\pi f_1 l\varepsilon')} \sqrt{\varepsilon' - P \ln(\Delta)} \tag{22}$$

$c$ = velocity of electromagnetic wave (light); $l$ = length of the sample; $P = (f_2P_2 + f_1P_1)/(P_1 + P_2)$; $f_2, f_1$ = frequencies of adjacent maxima in the region of swept frequency. $P_1, P_2$ = approximately the ratios of cut-off frequencies to the actual frequencies at the adjacent maxima.

$$\Delta f = f_2 - f_1$$

$$\Delta = \frac{(1 - R|r_1|)}{(R - |r_1|)} \tag{23}$$

$R$ = amplitude of reflection maximum.

$$|r_1| = \frac{\sqrt{1-P} - \sqrt{\varepsilon_1 - P}}{\sqrt{1-P} + \sqrt{\varepsilon_1 + P}} \tag{24}$$

Thus from three experimental quantities, viz, the sample length, the frequency difference between two extremes and the amplitudes of maxima, one can evaluate $\varepsilon'$ and $\tan\delta$. This method can also help one to decide whether or not there are any inhomogenities present in the sample such as large voids and cracks.

## 3. MEASUREMENT OF DIELECTRIC CONSTANT AND LOSS ON HIGH PERMITTIVITY MATERIALS

### 3.1 Ferroelectrics

Lanagan et al [12] have described this method for measuring on high permittivity materials. This method can be essentially used on ferroelectrics and is based upon S-parameters measurement using microwave network analyzer.

The specimen is a discontinuity for travelling waves and generally the incident wave incidents at an angle on the air/dielectric boundary as shown in Fig. 3. Since the incident wave has a reflection coefficient equal to unity and the reflected wave can be described as complex quantity given by

$$\rho = \frac{1 - (\varepsilon^x)^{1/2}}{1 + (\varepsilon^x)^{1/2}} \tag{25}$$

where $\varepsilon^x = \varepsilon' - j\varepsilon''$.

The remainder of the incident wave is transmitted through the dielectric material until the next dielectric discontinuity is reached. The magnitude and

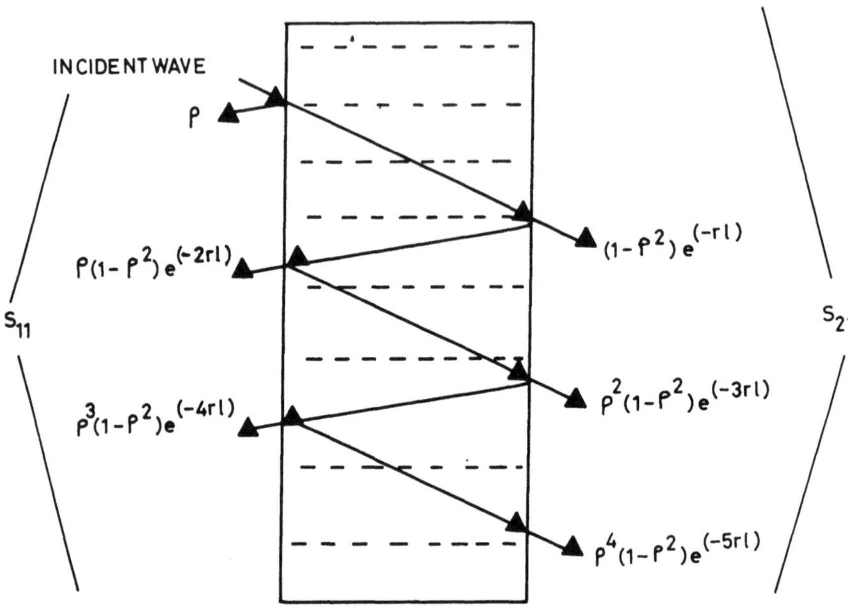


**Fig. 3.** Plane wave incident to the interface of two different dielectric media.

phase of the transmitted wave as a function of the propagation constant $\gamma$ is expressed as

$$\gamma = \alpha + j\beta - j\,(\mu^{x}\varepsilon^{x})^{1/2} \tag{26}$$

$$\beta = \frac{2\pi\varepsilon'}{\lambda_0} \tag{27}$$

where $\tan\delta = \varepsilon''/\varepsilon'$; $\lambda_0$ is free space wavelength and $\mu^{x}$ is complex permeability which is unity in case of nonmagnetic materials.

The sum of the reflected and transmitted waves can be expressed in terms of the scattering parameters $S_{11}$ and $S_{21}$ which characterize two port device such as a waveguide connected on both ends. The scattering parameters are usually taken as complex quantities in terms of magnitude and phase given by

$$S_{21} = |S_{21}|\, e^{i\theta} \tag{28}$$

The internal reflections shown in figure are accounted for by an infinite geometric series and relations for $S_{11}$ and $S_{21}$ are given by

$$S_{11} = P\,\frac{1 - \exp(-2\gamma l)}{1 - \rho^2 \exp(-2\gamma l)} \tag{29}$$

$$S_{21} = \frac{(1 - \rho^2)\exp(-2\gamma l)}{1 - \rho^2 \exp(-2\gamma l)} \tag{30}$$

$l$ is the sample thickness.

In order to minimize the complication to solve for $\varepsilon'$ from the above equation, an approximation is made in the above equation which requires $S_{21}$ measurement at several frequencies. For high dielectric constant and high loss materials, $\rho^2 \exp(-2\gamma l)$ can be assumed to be less than unity. The above equation is simplified to the following equation $S_{21} = (1 - \rho^2) \exp(-2\gamma l)$. A plot of $\log S_{21}$ vs frequency is linear as per the above equation. The dielectric constant and loss interms of $S_{21}$ phase and magnitude are given by

$$\varepsilon' = \left[\left(\frac{\Delta\theta}{\Delta v}\right)\frac{C}{2\pi d}\right]^2 \tag{31}$$

$$\tan\delta = \left(\frac{\Delta |S_{21}|}{\Delta v}\right)\left[\frac{8.686\, C}{\pi d\sqrt{\varepsilon'}}\right] \tag{32}$$

The parameter $c$ is the speed of light in free space. $(\Delta | S_{21} | / \Delta v)$ and $(\Delta\theta / \Delta v)$ are the slopes of magnitude and phase with respect to frequency $v$. The phase difference is expressed interms of radians and magnitude difference in dB.

### 3.2 Dielectric Resonator Ceramics

The most important characteristics of a dielectric resonator are high dielectric constant ($\varepsilon'$ of the order 100) and high $Q$ (> 10,000) and temperature coefficient of resonant frequency should be minimum. The operating frequency of a particular mode of dielectric resonator can be controlled by the physical dimensions of the sample and the dielectric constant. The dielectric constant can be measured by the method suggested by Hakki and Colemon [13] and Courtney [14].

The measurement method involves placing the dielectric resonator (usually taken in the form of a disc) between two perfectly conducting plates. When the metal plates make contact with dielectric resonator, the field equations can be expressed as a transcendental equation relating resonant frequency, dielectric constant and size of the resonator. In general the resonator can support various TE, TM and hybrid HEM modes. Any mode can be used to measure the dielectric constant. However, $TE_{0nl}$ is most commonly used mode for the evaluation of $\varepsilon'$ due to its insensitiveness to the presence of air gap between the metallic plates and the dielectric resonator, the zero index in $TE_{0nl}$ refers to the field variation along the azimuthal direction, $n$ and $l$ refer to the field along the radial and axial direction respectively. However, $TE_{011}$ mode can be identified from all the existing modes both by mode chart and experimental arrangement that can be further used for $\varepsilon'$ evaluation [13].

The experimental arrangement is shown in Fig. 4. The short circuited resonator is operated as a transmission resonator with small coupling antennas used to couple in and bring out the power. The antennas are connected to the network analyzer which can gauge the resonance frequencies of various modes. Among these modes, $TE_{011}$ can be easily identified.

The characteristic equation for $TE_{0nl}$ mode is

$$\alpha\frac{J_0(\alpha)}{J_1(\alpha)} = -\beta\frac{K_0(\beta)}{K_1(\beta)} \tag{33}$$

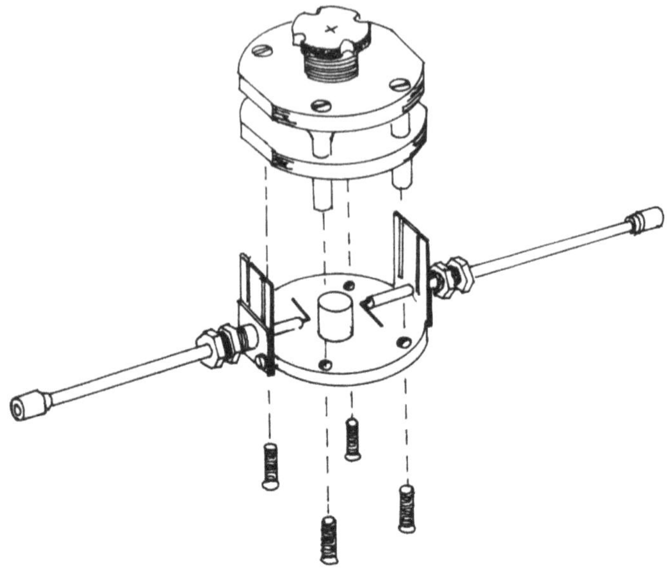

**Fig. 4.** The Courtney holder.

where $J_0(\alpha)$ and $J_1(\alpha)$ are two Bessel functions of the first kind of the orders zero and one respectively; $K_0(\beta)$ and $K_1(\beta)$ are the modified Bessel functions of the second kind of orders zero and one respectively; $\beta$ is the radial wave number outside the dielectric resonator and $\alpha$ is the radial wave number inside the dielectric resonator. The values of $\alpha$ and $\beta$ are given by

$$\beta = \frac{\pi D}{\lambda_0}\left[\left(\frac{l\lambda_0}{2L}\right)^2 - 1\right]^{1/2} \tag{34}$$

$$\alpha = \frac{\pi D}{\lambda_0}\left[\varepsilon' - \left(\frac{l\lambda_0}{2L}\right)^2\right]^{1/2} \tag{35}$$

$\lambda_0$ is the free space wavelength, $D$ is the diameter of the resonator, $L$ is the length of the resonator, $l = 1, 2, 3\ldots$ corresponds to the multiple half wavelengths along the axial directions of the resonator.

The $Q$ of the dielectric resonator is measured using a test holder given in Fig. 5. The same can be used for temperature coefficient of resonant frequency ($\tau_f$). The measurement can be done with network analyzer. The $Q$ can be calculated by measuring the magnitude and phase of the reflection coefficient as a function of frequency as described by Kaifez and Hwan [15]. The temperature coefficient of resonant frequency is measured using the equation

$$\tau_f = \frac{1}{f_0}\frac{\Delta f}{\Delta T} \tag{36}$$

where $f_0$ is the resonant frequency at the starting temperature.

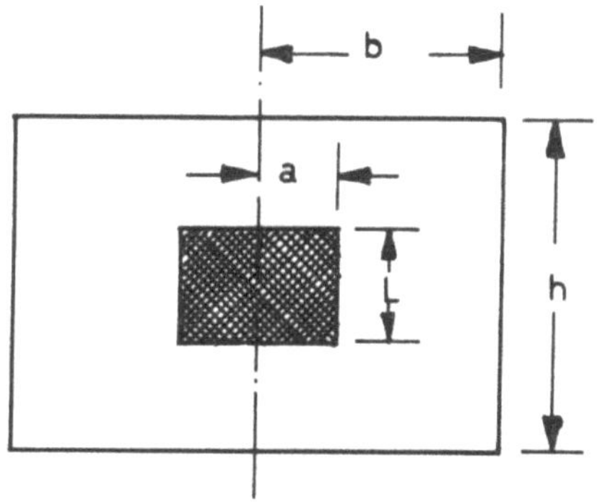


**Fig. 5.** Dielectric resonator in a cylindrical cavity.

Thus, in this chapter, emphasis was given to describe the various reliable experimental methods to measure the dielectric constant, loss and conductivity in the microwave frequency region. The materials include polymers, ceramic oxides such as ferrites, titanates, dielectric resonators and high $T_c$ superconductors.

## REFERENCES

1. J.S. Artman and T.E. Tammelwald, *J. Appl. Phys.*, 26, 1124, 1955.
2. B. Lax and K.J. Button, *Microwave Ferrites and Ferrimagnetics*, McGraw-Hill, New York, 1962.
3. V.R.K. Murthy and R. Raman, *Solid State Commn.*, 70, No. 8, 847, 1989.
4. L.T. Buranov and I.F. Shchegoler, Translated from *Pribory i Tekimilon Eksperiments*, No. 2, 171, March-April, 1971.
5. Iroshi Kobayashi, *Jap. Jl. Appl. Physics*, 10, No. 3, 345, 1971.
6. H. Zijlstra, *Experimental methods in magnetism. 2* (North Holland), Ch. 2, p.70, 1967.
7. R. Pragasam, N.S. Raman, V.R.K. Murthy and B. Viswanathan, *Physica Status Solidi*, Vol (a), 125, 583, 1991.
8. W.R. Tinga and E.M. Edwards, *Jl. Microwave Power*, 3 (3), 112, 1968.
9. S. Roberts and A. Von Hipple, *Jl. Appl. Phys.*, 17, 610, 1946.
10. W.H. Surber Jr. *Jl. Appl. Phys.*, 19, 514, 1948.
11. C.G. Montgomery, *Techniques of microwave measurements*, (McGraw-Hill Book Co. Inc, New York), Ch. 10, 1947.
12. M.T. Lanagan, J.H. Kim, D.C. Dube, S.J. Jang and R.E. Newnham, *Ferroelectrics*, 82, 91, 1988.
13. B.W. Hakki and P.D. Coleman, *IRE Transactions on Microwave Theory and* Techniques, Vol MTT-8, pp. 401–410, 1960.
14. William E. Courtney. *IEEE Transactions on Microwave Theory and Techniques*, Vol MTT-8, pp. 476–485, 1973.
15. D. Kaifez and E.J. Hwan, *IEEE Trans. on Microwave Theory and Techniques*, Vol. MTT-32, pp. 666–670, July 1984.

# 5

# Microwave Ferrites

G.P. Srivastava
Department of Electronic Science
South Campus, Benito Juraz Road, New Delhi, India

Bijoy K. Kuanr
Zakir Husain College
Jawaharlal Nehru Marg, New Delhi, India

## 1. INTRODUCTION

Emergence of ferrites arises due to the search for ferromagnetic materials with low eddy current losses. The high resistivity of these materials is the primary factor for controlling the eddy current losses, which is useful as cores for transformers and inductors. The present status is that ferrites have been established as materials of immense industrial use and ferrite devices find numerous applications in entire frequency range. At high frequencies, as in the communication field, the advantage of the ferrites become more pronounced specially at microwave levels. Great bulk of the microwave applications will be quite impossible without the assessment of ferrite and garnet materials.

The foundation of modern interest in ferrites was laid by Snoek [1] in 1946, thereafter, the basic theory of magnetism developed by Neel [2] in 1948 was the starting point for the rapid expansion of research and development activities in this field of materials. The importance of ferrites is due to the fact that they possess the combined properties of a magnetic material and an electric insulator [3]. As a results, a high frequency electromagnetic wave can be propagated through ferrites with very low attenuation. The work of Gorter [4], Goodenough [5] and Blasse [6] showed that it is possible to make ferrites with different substitution having a very wide range of saturation magnetisation and Neel temperature values. These materials are also important from the point of view of theoretical investigation because of their different type of spin arrangements and exchange interactions.

Because of the large application of ferrites in microwave frequency range, a designer of the ferrite devices has to know the detailed characteristics of the material. Important parameters needed to characterise the ferrite material at high frequencies are the resonance line-width ($\Delta H$) and spin-wave line-width ($\Delta H_k$). $\Delta H_k$ is a measure of the power handling capability of the material.

There has been steady interest in developing new or improved latching type high power ferrite phase shifters for use in phased array scanning antenna. The device performance of phase shifter is limited by the ferrite material parameters mentioned above. Inadequacies in currently available commercial ferrite and garnet for use at high power levels has prompted the need for further material development with improvement in power-handling capability as well as hysteresis-loop properties. In addition, to the requirement for suitable hysteresis-loop properties, materials for microwave latching phase shifters should have low dielectric losses.

Lithium ferrite with chemical formula $Li_{0.5}Fe_{2.5}O_4$ is an inverse spinel [7] having a wide range of device applications. Several research programs have been undertaken on this series of material to study its fundamental properties and especially to develop high-power microwave materials from it. The utility of Lithium ferrite and its substituted compositions as a class of materials for application in microwave components and memory devices is due to the following characteristics of the materials [8, 9].

1. Lithium ferrite can be prepared with a low value of $\Delta H$, with proper substitutions of non-magnetic materials of the order of 50 oersted. This is an extremely useful parameter for use in microwave devices.
2. Lithium ferrite can be used in high-power microwave devices. The power-handling capability can be raised with appropriate non-magnetic substitutions (like relaxing impurities) in its composition. Hence lithium ferrite have come up as a low cost substitution for expensive garnets.

The unique properties of lithium ferrites which render them invaluable in microwave applications is the employment of appropriate chemical substitution in the basic formula unit. Although a wide range of substituted lithium ferrites are available commercially, those in general used at microwave frequencies include Ni, Zn, Mg, Al, Co, Ti etc.

The proper doping level and amount of dopants leads to material with high Curie temperature, good density, low microwave dielectric losses, high hysteresis loop squareness, cover a wide range of saturation magnetization values, low stress sensitivity, high resistivity, high permeability and above all improved microwave power handling capability.

## 2. FERROMAGNETIC RESONANCE (FMR) AT LOW POWER

When a ferrimagnetic material is subjected to the simultaneous action of a rf field perpendicular to a dc magnetic field, under suitable conditions the sample is found to absorb energy from the rf field. This phenomenon is referred to as ferromagnetic resonance. The absorption of microwave power by the sample occurs when the precessional frequency of the magnetization vector about the direction of the externally applied dc field equals the frequency of rf field. This resonant absorption of the electromagnetic radiation by the ferromagnetic material is called FMR. In a usual FMR experiment the applied field has two components—a large time independent field $H_{app}$ and a small perpendicular sinusoidally varying field, 'sin $\omega t$'. The first ferromagnetic resonance experiment was performed by Griffithis [11] in 1946. Until today a lot of experiments have been made by various workers for the complete understanding of relaxation processes in ferromagnetic insulators. In a typical FMR experiment a sample is placed in a uniform magnetic field large enough to magnetize it parallel to the field direction. The absorption of rf power from the sinusoidally varying rf signal is a function of $H_{app}$ for fixed frequency. A resonance peak is obtained if the absorption is plotted as a function of $H_{app}$. The full width, in field units, at half maximum absorption point is designated as the resonance line-width $\Delta H$. This simple description of FMR contains the three basic ingredients common to resonance phenomena are precession, resonant response and relaxation. The understanding of FMR was advanced considerably by Kittel [12] when he pointed out the role of demagnetizing fields in determining the resonance conditions. The usual resonance condition with $\gamma$ (gyromagnetic ratio) is satisfied for most materials when demagnetizing [13, 14, 15] field are taken into account. A phenomenological description of the FMR absorption was given by Van Vleck [15]. He discussed the observed line-width and various possible sources for line broadening. Microwave phenomena have further expanded with discovery of YIG [16].

Sparkes [18] has made a detailed study of FMR in YIG and pointed out the important role played by spin-waves. Kittel's theory was modified [19] for the ferrimagnetic material which can be described by two sub-lattice systems of oppositely directed magnetization. Further Brown and Park [20] had shown that there were two types of responces, one in which the two sublattice magnetizations precesses about the applied field direction out of phase i.e. the exchange resonance and the other when they precess in phase, which is the ferromagnetic case. The size effect of FMR in ferrites was observed by Beljers and Polder [21]; the resonance field was found to be a function of the diameter of the ferrite sphere.

Yager et al [22] have obtained the linewidth of single crystal nickel ferrite. It is 70 oersteds. Extensive studies were made on the variation of $\Delta H$ with frequency [23], temperature [24], and ferrite composition [25]. The behaviour of $\Delta H$ with temperature could be attributed to the cation distribution [26] and the valence state of the iron ion [27], in the ferrite. Lecraw and Spencer [28] had shown that the linewidth in pure YIG was of the order of few millioersteds.

## 2.1 Spin Waves

In the usual FMR experiments the rf field is applied perpendicular to the dc field and in resonance the uniform mode (i.e. the spins are parallel to one another) alone is excited. This is a normal mode of the spin system. Spin wave analysis has been used many times for the description of the resonance behaviour of ferromagnetic spinel as well as garnet systems. The wavelength of the spinwave is, under most conditions, much smaller than the wavelength of an electromagnetic wave of the same frequency. Therefore, a spin wave can be excited by virtue of its dipole moment and this is small because the positive and negative contributions arising from different parts of the wave tend to cancel out to a high degree. Other modes in which the spins do not precess in phase with each other also exist. Such disturbances in the spin system are called spin waves and is characterised by a wave vector $\hat{k}$ whose magnitude is $|k| = 2\pi/\lambda$, where $\lambda$ is the spinwave wavelength.

For the uniform mode $k = 0$. The relaxation process assumed that energy is scattered from the uniform precessional mode into higher order modes [29]. But Clogston et al [30] provided a clear picture for the understanding of the exact mechanism of the energy transfer by taking into account the effect of the dipole-dipole interaction. The dipole-dipole interactions broadens the spinwave dispersion relation into a band, which was named as spinwave manifold. The spinwave manifold may be defined as that region of the dispersion relation where spin wave analysis is valid. In this manifold $k = 0$ spinwaves also exist, which are degenerate with the uniform precession.

The relaxation process assumes that energy is scattered from the uniform precessional mode into degenerate spinwave modes. The analysis of this scattering process depends upon the knowledge of the relationship between wavelength and resonant frequency of such modes. According to the theory of Clogston [30] the relaxation process can be visualized as follows. In a resonance experiment the rf field puts its energy into the uniform precession. Thus the system has an excess of uniform precession magnons ($k = 0$). The $k = 0$ magnon can relax to the lattice through three channels as shown in Fig. 1. In the first process a $k = 0$ magnon can be annihilated and a $k = 0$ degenerate magnon can be created in a two magnon process, with the $k = 0$ magnon eventually relaxing to the lattice through processes involving magnons and phonons. Secondly a $k = 0$ magnon can relax to the lattice via interactions with magnons other than the degenerate magnons. In the third process, a $k = 0$ magnon can directly relax to the lattice through processes not involving other magnons. The two magnon process [31] has the dominant mechanism for relaxation in polycrystalline ferrites.

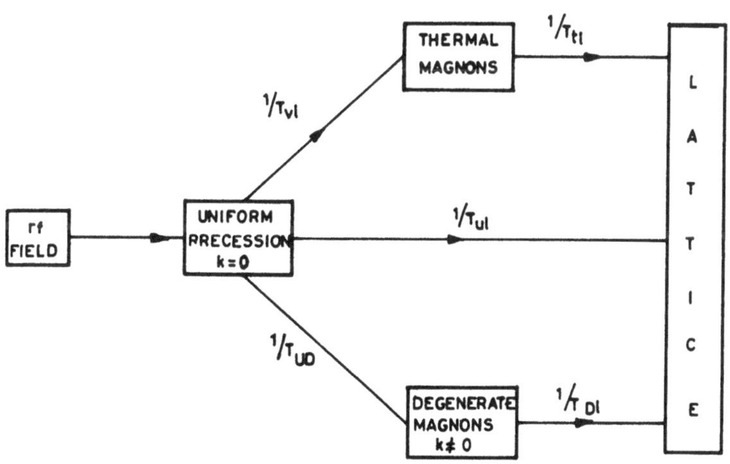


**Fig. 1.** General scheme for relaxation processes.

The derivation of the spinwave manifold hold as long as the wavelength of the spinwave is much smaller than the size of the sample so that the surface demagnetization fields of such magnons can be neglected. For large wavelengths these fields must be taken into account. Walker [32] solved the problem by including these surface demagnetization fields and the resulting modes are called the magnetostatic modes. The uniform mode has $k = 0$ and therefore is a magnetostatic mode. Walker's calculations showed that the magnetostatic modes has a spread wider than the spinwave manifold i.e. the upper limit for the Walker modes does not correspond to an extension of the upper limit of the spinwave manifold. The bottom limits however, do coincide.

The spinwave manifold shown in Fig. 2 can be divided into two parts. A very flat region where the exchange interaction plays no part is the first region and the second is a curved region where the exchange effect is appreciable.

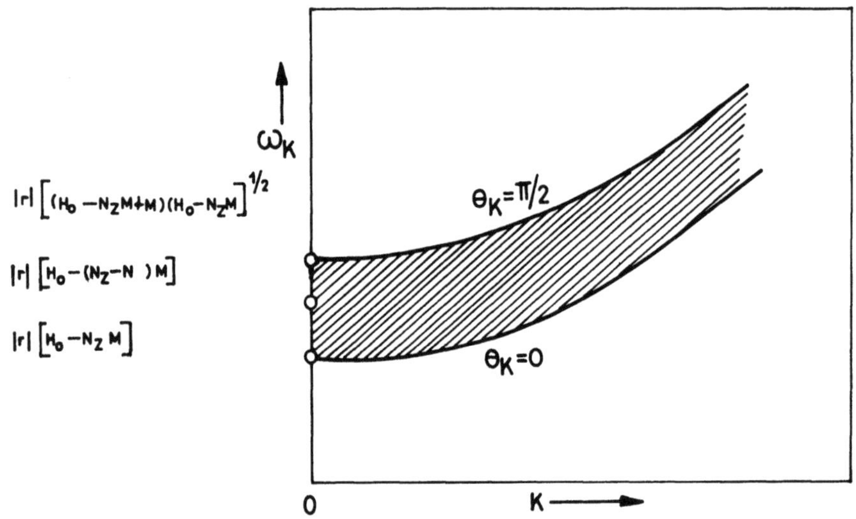


**Fig. 2.** The spin wave spectrum.

Geschwind and Clogston [33] and Schlomann [34] have explained linewidths in polycrystalline materials by calculating the coupling of the uniform precession with degenerate spinwaves of medium $k$ values (i.e. in the no exhange region). They have neglected the effect of spinwave with high $k$ values, where the exchange effects are appreciable. In spinel ferrites the scattering arises from a distribution of the magnetic ion on the octahedral sites. The coupling between the uniform precession and the high $k$-spin waves is thought to be stronger in the spinel than in the garnet. So the spinels have greater line width as compared to those of garnets.

## 3. HIGH POWER FMR

Now it is a well known fact that ferrimagnetic materials display nonlinear loss characteristic at high levels of peak microwave power. It has been known since the work of Damon [35] and of Bloembergen and Wang [36] that at high power the phenomena of ferromagnetic resonance are quite different from those observed at low power levels. Two unexpected effects come to light:

1. The main resonance line appeared to saturate and broaden steadily as the signal power was increased beyond a threshold value.
2. Secondly, at a similar signal power, an additional rather broad absorption peak appeared, a few hundred oersted below the dc magnetic field required for the main resonance.

These two effects are shown here in Fig. 3.

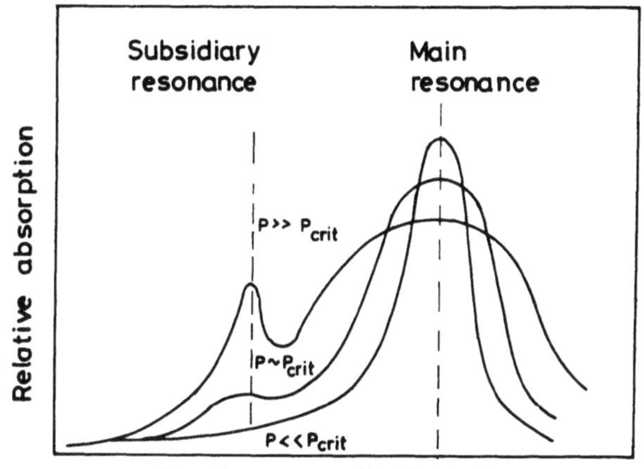


**Fig. 3.** Subsidiary absorption and premature saturation.

The presence of these nonlinear effects can seriously interfere with the performance of microwave devices such as isolator, circulator and phase shifter.

The reduction of main resonance can impare the performance of resonance isolators. The subsidiary resonance which appeared at low dc field cause serious deterioration in the high power performance of phase shifter and circulators. So it is important to understand the cause and cure of these nonlinearity.

Suhl [37] has explained these nonlinearities theoretically. Both the above mentioned effects arise from power dependent coupling between the so-called uniform mode of magnetic precession which is driven by the rf field and certain spinwaves which become excited if the applied microwave field exceeds a critical value. The spin waves which have the same frequency as that of the applied field are responsible for the saturation of the main resonance and the spin waves having the frequency one half of the signal frequency causes the subsidiary absorption peak. The threshold value required for these nonlinearities differ for the two types, but, both threshold ($h_{crit}$) depend upon the linewidth, the saturation magnetization, the geometry of the sample and the operating frequency. In some cases the subsidiary absorption peak coincides with the main resonance, this happens when threshold field are low.

Suhl [38, 40] presented a theoretical investigation on the behaviour of ferrites at high microwave signal levels and showed that the nonlinearities were connected with two kinds of instability [39] of the uniform precessional motion of the total magnetization against certain spinwave disturbances. And these disturbances grow exponentially when the signal level exceed a certain threshold value.

Schlomann [41] had developed a theory of ferromagnetic resonance in which dipolar interaction was taken into account by means of the spinwave formalism. His theory predicts a very strong frequency and shape dependence of the line width for the case in which the homogeneous mode of precession was approximately degenerate with long-wavelength spinwaves propagating in direction perpendicular to the dc field. The influence of inhomogeneities on the saturation of the ferromagnetic resonance was investigated. Schlomann et al [42] observed the nonlinear effect arising from spinwave instability in a microwave magnetic field applied parallel to the dc field which termed as parallel pumping spinwave instability. They have investigated on cobalt and zinc substituted nickel ferrite and single and polycrystals of rare earth substituted garnets. Their results indicated the increase of spinwave linewidth with increasing wave number and decreasing angle between propagation direction and dc magnetic field. LeCraw et al [43] studied FMR at 9.3 and 3 GHz in single crystal YIG and observed that $h_{crit}$ was the lowest at room temperature, but as the temperature was increased $h_{crit}$ rises quite abruptly. More works [44] on parallel pumping technique have been made over a wide range of power levels and applied dc fields. A confluence process was given to explain the susceptibility curves, obtained on single-crystal YIG. Sounders and Green [45] measured subsidiary absorption on polycrystalline aluminium substituted YIG and nickel ferrites, and shown that spinwave linewidth was independent of aluminium content. Schlomann et al [46] had analysed the power-handling capability of substituted garnets used in circulators and phase shifters. They have defined the "high power figure of merit" for the measure of the suitability of microwave ferrites, for high power applications. They observed for ferrites, containing appreciable substitutions of strong relaxers, a high figure of merit ($F_{hp}$) by reducing the saturation magnetization. Single crystal nickle zinc and lithium zinc ferrites [47] were studied at room temperature and 77°K and at 9.0 GHz. His experimental data showed that spinwave linewidth reduced by a large amount when measured at 77°K.

For oblique pumping, the threshold microwave field amplitude ($h_{crit}$) required for the unstable growth of spinwave was investigated [48] with a linearly polarized microwave field applied at an arbitrary angle with respect to the static magnetic field. Green et al [48] performed their experiment on YIG spheres at 9.2 GHz and used a pulsed magnetron as source with rectangular $TE_{102}$ cavity. They have derived theoretical expressions for the oblique pumping $h_{crit}$, and have obtained a fair agreement of their experimental data with the theory by assuming a transverse demagnetizing factor $N_t$, greater than value of 1/3. Extensive data on the oblique-pumping threshold as a function of pumping angle have been obtained for external fields and the theory of parallel and perpendicular pumping in saturated ferromagnetic insulator was extended to include oblique pumping. Patton [49] has given a quite versatile theory which was applicable to a large variety of microwave field configuration which were not accessible on the basis of earlier theories [37, 41, 42]. Patton [49] has developed equations for first order instability threshold and showed that sample geometry and the pump configuration have a strong influence on the instability threshold. The first order spinwave instability threshold has also been measured [50] as a function of sample shape (sphere, rod and thin disk), sample orientation, pump configuration and magnetization state. The effect of sample shape and pumping configuration on the threshold field for saturation was compared with the theory. They have observed a consistent correlations between the data and the theory which provided that domain shape plays the same role in determining the threshold for partially magnetized samples that sample shape does for saturated material.

Spinwave linewidth on polycrystalline YIG observed as a function of porosity [51] pump configuration [52] and grain size [53, 60, 63] provides a clear picture for the understanding of threshold mechanism. A spin wave transit time model [55, 61] was given to explain the experimental data. Effect of nonmagnetic inclusions on the spinwave linewidth in polycrystalline YIG was observed by Schotter [57]. The threshold field for nonlinearity for first order processes on the various configuration of static field angles were studied [54] on single crystal YIG. Schotter [57] has studied the effect of rare-earth doped YIG on spinwave linewidth and showed that the high power capability of ferrites ($\Delta H_{k\to 0}$) was increased with rare-earth doping. Parallel pump [58] and subsidering absorption [59] spinwave linewidth have been studied in porous polycrystalline YIG (as a function of temperature) and single crystal YIG respectively. The effect of crystalline anistrophy for any arbitrary microwave pumping configuration on spinwave instability theory [60] of single crystal YIG [62] provided a good agreement with experimental data. Spinwave linewidth in polycrystalline YIG [63] was compared to the theory of Sawado [62] for spinwave scattering by nonmagnetic pores.

### 3.1 High Power Effects

Great technical interest lies in high power FMR. One can selectively excite spinwave of a particular wave number through proper choice of dc field and can couple energy into that spinwave. The basic thing is to pump microwave energy into the spin system and observe the change in susceptibility with increasing amplitude of microwave power. This process results in a greater excitation of the $k = 0$, i.e. uniform precession spinwaves, which in turn results in an greater excitation of $k = 0$ spinwaves to which the uniform precession is coupled. If the rate of energy transfer to a particular $k = 0$ spinwave exceed the rate at which energy can be transferred from the spinwave, the amplitude of the spinwave increases exponentially, hence there is a abrupt change in susceptibility. The instability condition depends on the rate at which energy can be dissipated from the particular $k = 0$ spinwave. This rate is characterized by a linewidth parameter called as spinwave linewidth, $\Delta H_k$, which has to do with resonant excitation at high power. $\Delta H_k$ is a linewidth only in the frequency swept linewidth sense. The threshold microwave field amplitude $h_{crit}$ at the break point of the susceptibility curve can be related to the relaxation rate for spin-wave mode with the lowest threshold for prescribed experimental conditions. $\Delta H_k$ is defined as $2Dk/\gamma$, where $\gamma$ is the gyromagnetic ratio. It can also be defined in terms of lifetime of a magnon ($k$) of certain $k$ as $1/\gamma k$. So, the measurement of nonlinearity threshold provides information about the relation of a particular $k = 0$ spinwaves. The excitation of specific modes provide a detailed investigation of the spinwave spectrum and the energy transfer mechanism with the spin system. At high power, coupling between the microwave field and certain spinwave modes leads to a non-linear behaviour on the response curve. The discovery of non-linear phenomena provide a new insight into microwave relaxation process.

The instability threshold can be determined by two techniques those are perpendicular or transverse pumping and the other is parallel pumping technique. The configuration in which the microwave field is perpendicular to the dc magnetic field, as in case of low power experiments, is the transverse pumping and in case of parallel pumping, the rf field is applied parallel to the dc magnetic field. Only the later technique is used in this investigation for determination of instability threshold.

### *3.1.1 Transverse Pumping Instability*

After the works of Damon [35] and Bloenbergen and Wang [36] two principle non-linear phenomena were established in the usual FMR experiment (rf field perpendicular to dc magnitude field) at high power. First one is the premature saturation of the main resonance, i.e. the value of susceptibility, $\chi''$, at resonance declines at a power level far fellow that for saturation. Second effect is often labelled as subsidiary absorption i.e. the appearance of a subsidiary peak at values of dc magnetic field below that required for main resonance. Both effects are shown in Fig. 3. The origin of these peculiar high power effect was first explained by Suhl [37]. In his theory, Suhl proposed that the parametric excitation of spin-waves by the uniform mode was the cause of the observed behaviour which is due to dipole-dipole interaction. At low power levels the spinwave amplitudes stay essentially at their thermal values because of the small amplitude of uniform precession. But as the power level increases, the uniform precession amplitude grows and more and more energy is pumped into the spinwave modes.

#### I. SUBSIDIARY ABSORPTION

Suhl [37] proposed a mechanism for the high power effect which is bound on the interaction of the uniform mode with spinwaves of short wavelengths. In "first-order Suhl process", where there is an appearance of a subsidiary peak, energy from a single $k = 0$ magnon is transferred to and creates two magnon, having wave vector, $k$ and $-k$ respectively. This is shown is Fig. 4a.

Due to the conservation of energy principle these magnons must have one-half the frequency of the $k = 0$ magnon. One particular pair, of magnon, among many pairs with $\omega_k = \omega/2$ and with equal and opposite $k$, will be strongly couple to the $k = 0$ mode and consequently lead to a nonlinearity in the absorption.

For a ellipsoid the critical field derived by Suhl [37] can be written as

$$\gamma h_{crit} = \frac{2\omega_k\{[\omega_k^2 + (\omega/2 - \omega_k)^2]\,[\eta_0^2 + (\omega_r - \omega)^2]\}^{1/2}}{\omega_M \sin\theta_k \cos\theta_k\,(\omega_k + \omega_0\,\omega_M N_z + \omega_k\,\eta_k k^2)} \tag{1}$$

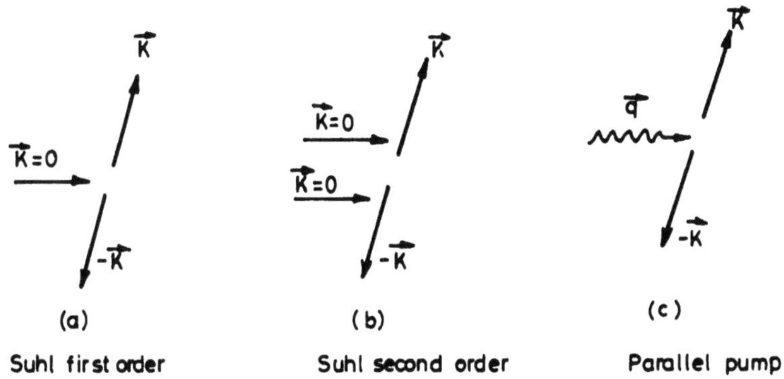


**Fig. 4.** Schematic representation of processes involved in the instability thresholds.

where $\eta_k = \gamma \Delta H_k \eta_0 = \gamma \Delta H/2$ and $\omega_r = \omega_0 (N_z - N_t) \omega_m$.

It can be seen from the numerator of Eq. 1 that unless $\omega_k = \omega/2$ or $\omega_r = \omega$ or both, $h_{crit}$ is as large as $\omega^2/\omega_M$. The condition for minimum of the above equation can be obtained with $\omega_k = \omega/2$, because of momentum conservation, to explain the nonlinear effects. Two cases can be considered, one in which the subsidiary absorption occur at a low dc magnetic field than that required for main resonance and the other when the main resonance and subsidiary absorption coincide.

**Case 1**

This corresponds to the experimental observations when $\omega_k = \omega/2$ but $\omega_r$ not close to $\omega$, the subsidiary absorption occurs at an applied field less than that required for main resonance. The threshold for this case can be written as [37]

$$\gamma h_{crit} = \frac{2\omega \eta_k (\omega_0 - \omega_r)}{\omega_M \sin\theta_k \cos\theta_k (\omega/2 + \omega_0 - \omega_M N_z)} \tag{2}$$

The minimization of $h_{crit}$ be done with $\theta_k$ other than zero or $\pi/2$. This condition simply tells that the eligible spinwave must propagate on a cone with an intermediate value about the direction of propagation i.e. $\theta_k = 45°$. Thus the subsidiary absorption is confined to a range of fields like

$$\omega_0 < (\omega/2 + \omega_M N_z) \tag{3}$$

which is a necessary condition for experiments.

**Case 2**

Merging of the subsidiary absorption with the main resonance occur for $\omega_k = \omega/2 = \omega_r$. This condition occurs provided

$$N_t \omega_M > \omega/2 \tag{4}$$

Here the final result for threshold is

$$h_{crit} = \frac{2\omega \eta_k \eta_0}{\omega_M(\omega/2 + \omega - N_t \omega_M) \sin\theta_k \cos\theta_k} \tag{5}$$

This subsidiary absorption refers to a reduction in the susceptibility at the main resonance.

II. PREMATURE SATURATION OF THE MAIN RESONANCE

The premature saturation of the main resonance is often called the "second order Suhl process", which results from a 4 magnon process. Two uniform precession magnons, $k = 0$, are annihilated and a pair of magnon created having the same energy and wave vector $k$ and $-k$, as shown in Fig. 4b.

Under practical conditions $\omega = \omega_k = \omega_r$ is the only situation, for which $h_{crit}$ is small enough. Though in principle this instability can occur when the driving field differs from the resonance frequency. The uniform precession is most strongly coupled with spinwave of $\theta_k = 0$ i.e. the z-directed spinwave propagating along the direction of the magnetization, have minimum threshold. The final expression for $h_{crit}$ for a sphere, becomes

$$h_{crit} = \Delta H/2 \cdot (\Delta H_k/4\pi M_s)^{1/2} \tag{6}$$

The approximate value of the ratio of susceptibility is given by

$$\pi''/\chi_0'' = \sqrt{P_{crit}/P_{inc}} = h_{crit}/h \tag{7}$$

The general instability threshold for subsidiary absorption, to a good approximation, is given by

$$h = (\omega/\omega_M)\,(\Delta H_k/\sin\theta_k \cos\theta_k) \tag{8}$$

The conventional interpretation of subsidiary absorption involves the excitation of spinwaves with wave numbers $k = 0$ propagating at an angle $\theta_k$ with respect to the static field, for fields below the butterfly curve minimum. In this field range $h_{crit}$ and $k$ both are taken to increase with decreasing field. Whereas, above the butterfly curve minimum $k$ is about zero and $\theta_k$ decreases [68]. Subsidiary absorption results on singal crystal YIG [67] showed a broad flat minimum below the butterfly curve minimum, rather than the smooth increase predicted theoretically.

III. PARALLEL PUMPING

Schlomann et al [42] pointed out that the parametric excitation of spinwaves and associated instability effects could also result when the microwave field is parallel to the static field. In many respects the parallel pump threshold is easier to understand physically than subsidiary absorption. The uniform mode can also be driven by a longitudinal rf field at twice the uniform precession frequency, provided the rf amplitude exceeds a critical level and the precession path is elliptical. Some spinwaves precess on ellipsoidal cones, rather than a circular cone due to their dipolar interaction, and pair of these modes with equal and opposite wave number can be excited by a sufficiently strong rf field parallel to the dc field. In this case, the sample will absorb power only if the rf field exceeds the instability threshold. Direct coupling between the parallel pump microwave field and certain spinwave modes is possible, because of the spinwave ellipticity. Figure 5 shows an elliptical cone.

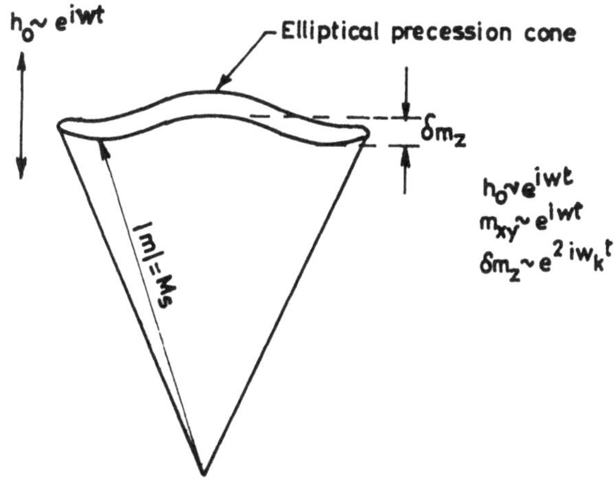


**Fig. 5.** Schematic representation of elliptical precession cone for spin-waves which leads to the parallel pump instability.

If the magnitude of $M$ is constant the elliptical orbit results in a wobble in the $z$-component denoted by $\delta m_z$. A spin wave mode at $\omega_k$ can couple to the parallel pump microwave field at $\omega = 2\omega_k$. The amplitude of $\delta m_z$ is dependent on the polar angle $\theta_k$; at $\theta_k = 0$, the amplitude of $\delta m_z = 0$ i.e., the ellipticity is zero and at $\theta_k = \pi/2$, $\delta m_z$ and hence ellipticity is maximum. The amplitude of threshold field for instability, $h_c$, in terms of spinwave linewidth $\Delta H_k = 2\eta_k/\gamma$ is given by

$$h_c = (\omega/\omega_m)(\Delta H_k/\sin^2\theta_k)_{min} \tag{9}$$

where $\omega$ is the operating frequency. From Eq. 9 it is evident that the threshold diverges at $\theta_k = 0$, because the ellipticity is then zero. The above expression has to be minimized for $\omega_k = \omega/2$. At small dc fields the line $\omega_k = \omega/2$ intersects the spinwave spectrum for all values of $\omega_k$, the lowest threshold is for $\theta_k = \pi/2$. Hence, for the most susceptible magnons Eq. 9 becomes

$$h_c = (\omega/\omega_m)\Delta H_k \tag{10}$$

From standard microwave theory the threshold field, $h_c$, incident on the sample is given by the relation [37, 38]

$$h_c = \frac{25Q_1(1 - \Gamma_s)}{f_0 g V_c} \tag{11}$$

where $h_c$ is in Oe, $P_c$ is the critical incident power in watts, $f_0$ is the magnetron operating frequency in GHz, $V_c$ is the volume of the cavity in cm$^3$, $Q_1$ is the loaded $Q$ of the cavity, $\Gamma_s$ is the reflection coefficient related to return loss by

$$\Gamma_s = 10^{-(\Gamma_{dB}/20)} \tag{12}$$

$g$ is the cavity geometrical parameter given by

$$g = (1/8)[1 + (d/a)^2] \tag{13}$$

where $d$ and $a$ are respectively the length and the width of the cavity.

As $\Delta H_k$ is dependent on the wave number $k$, the increase in dc field decreases the wave number and hence decreases $h_c$ to a minimum. The wave number becomes zero at a characteristic dc internal field strength given by

$$H_{int} = -2\pi M_s + [(2\pi m_s)^2 + (\omega/2\gamma)^2]^{1/2} \tag{14}$$

For higher fields there are no $\theta_k = \pi/2$ magnons satisfying the condition $\omega_k = \omega/2$. Threshold field, $h_c$, increases rapidly as $\sin^2 \theta_k$ decreases, because of the decrease of the magnitude of $\delta m_z$ and hence the ellipticity of precession cone.

### 3.2 High Power Microwave Measurements

The power handling capability of the ferrimagnetic materials can be determined by finding the value of threshold microwave field, $h_{crit}$, at which the loss characteristics of the materials become nonlinear. Two methods are available for the determination of the critical microwave field for non-linearity threshold. They are

1. Observation of deterioration of the trailing edge of the transmitted pulse.
2. Measurement of insertion loss versus incident power level at a given biasing dc field $H_{app}$.

The onset of nonlinearity can be observed by noting the distortion of the trailing edge of the pulse. This method requires a perfect square pulse at low power level. The CRO trace of the pulses transmitted at various power levels through $TE_{102}$ reflection type microwave cavity containing the sample shows that the nonlinearity becomes first noticeable at the trailing edge of the pulse. With the increase of microwave power a large region of the pulse was affected. The nonlinearity sets quite abruptly, so the experimental determination of the threshold field from the CRO traces was ambiguous. Here the threshold is measured by the second method. The insertion loss of the test section was plotted against the rf power and the transition from linear to nonlinear state was observed. The pumping configuration used for this study is the parallel pumping.

#### *3.2.1 X-Band High Power Microwave Bench*

The experiment can be performed around a reflection cavity high power spectrometer. A pulsed magnetron (like BEL 4J 52A-1, India), can be used as a high power source operating at a fixed frequency (like 9.4 GHz). The instrumentation material research is shown in Fig. 6.

The pulse width and the pulse repetition frequency (PRF) should be chosen after some important consideration. These are

(1) Pulse width should be larger than the time taken for the instability to set in the material after the power is incident on it. This is dependent on the relaxation time of the material which is of the order of few nanoseconds.

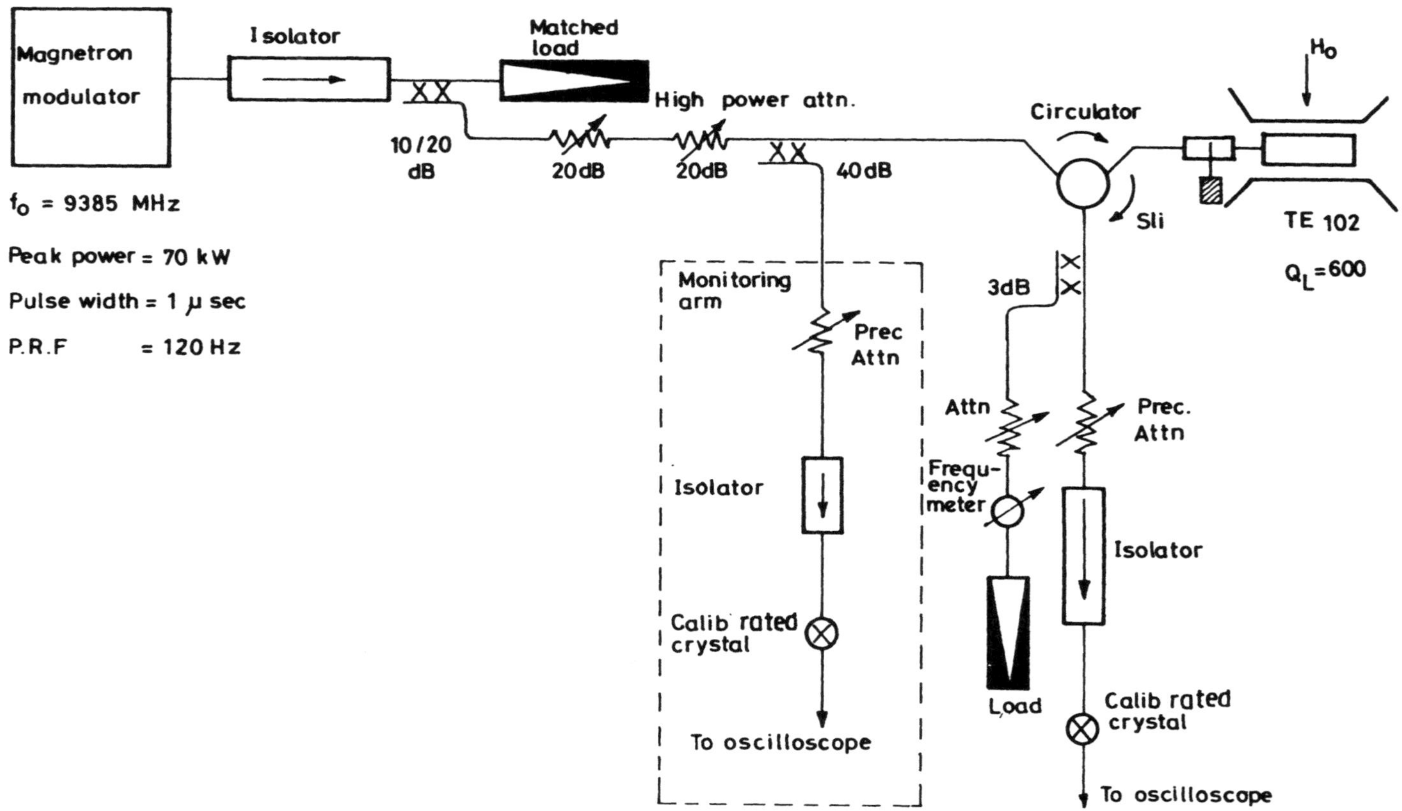


**Fig. 6.** X-Band high power bench: $h_{crit}$ measurement in reflection mode.

(ii) Pulse repetition frequency should be chosen in such a way that the duty cycle should be as low as possible i.e. of the order of $10^{-5}$, to avoid sample heating.
(iii) Pulse width should be larger than the response time of the cavity.
(iv) The leading and the trailing edges of the magnetron pulses should be avoided.

A pulse width of 1 $\mu s$ and PRF of 50 or 100 Hz has been found suitable for all requirements.

A $TE_{102}$ rectangular microwave cavity of a low $Q$, should be used so that a slight detuning caused by the insertion of the ferrite sample could not change the reactance of the cavity.

The RF power provided by the magnetron operating in the pulsed mode, passes through a high power ferrite isolater into a 10 dB directional coupler. The unused power of the main arm of the directional coupler is terminated in a high power load in order to minimize frequency pulling effect due to mismatch. The desired RF signal is incident on the cavity through a circulator and a shorting switch, which preceed the cavity. The shorting switch helps to measure the incident power as well as the reflected power from the cavity which is monitered with calibrated diode detector connected to a CRO. The precession variable attenuator which precede the detector is kept at a constant level such that the crystal output is constant. This ensured that the deviation of crystal behaviour from ideal square law does not affect the reading. The power absorbed by the sample as a function of increasing input power level is measured at a fixed biasing field $H_{app}$. From the graph of the return loss versus input power level the sharp increase in the cavity reflected power gives the onset of instability. This procedure is repeated for various values of the applied dc field. The threshold field for non-linearity ($h_c$) was calculated using the standard relation given in Eq. 11. These values of $h_c$ was plotted as a function of $H_{app}$ and the curve obtained is usually called as "Butterfly curve". Spin wave line-width $\Delta H_k$ is calculated from the $h_c$ (minimum) value using the relation in Eq. 10. The crystal IN21 with mount and the isolator gives an output of 200 mV on the CRO when the power incident is 200 mW. All measurements are obtained with spherical samples of 2 to 3 mm diameter to permit the application of perturbation technique. The sample is placed in a region of minimum electric field and maximum magnetic field, by a teflon rod sample holder inside the cavity. Such placement insured small dielectric losses and nearly uniform microwave magnetic field throughout the sample. The cavity is continuously tuned to the magnetron frequency at each biasing field by a tuning screw.

## 4 MICROWAVE STUDIES

### 4.1 High Power Microwave Studies

#### *4.1.1 Parallel Pumping Instability Threshold*

The results obtained on parallel pumping technique at high power microwave levels on $Gd^{3+}$ substituted YIG and $Co^{2+}$ and $Al^{3+}$ substituted LiTi ferrites are given. The microwave field is applied parallel to the static field. The method of measurements is described in section 3.2.

Figure 7 shows the variation of critical fields, $h_{crit}$ as a function of the static applied field for some specimen of garnet and ferrites respectively. These curves are called the "Butterfly Curves". Here the angle ($\psi$) between the pump field and the static field is zero degree. The well known butterfly shape of the curves are apparent for all these samples. It is evident from the figure that the $h_{crit}$ decrease to a minimum. In this field range, the instability corresponds to $\theta_k = \pi/2$ spinwaves with $k$ values which are gradually decreases to zero at the minimum, this is the one with lowest relaxation rate. The applied static field at which the wave number decreases to zero and the butterfly curve has a minimum for a isotropic material is given in Eq. (14) [42],

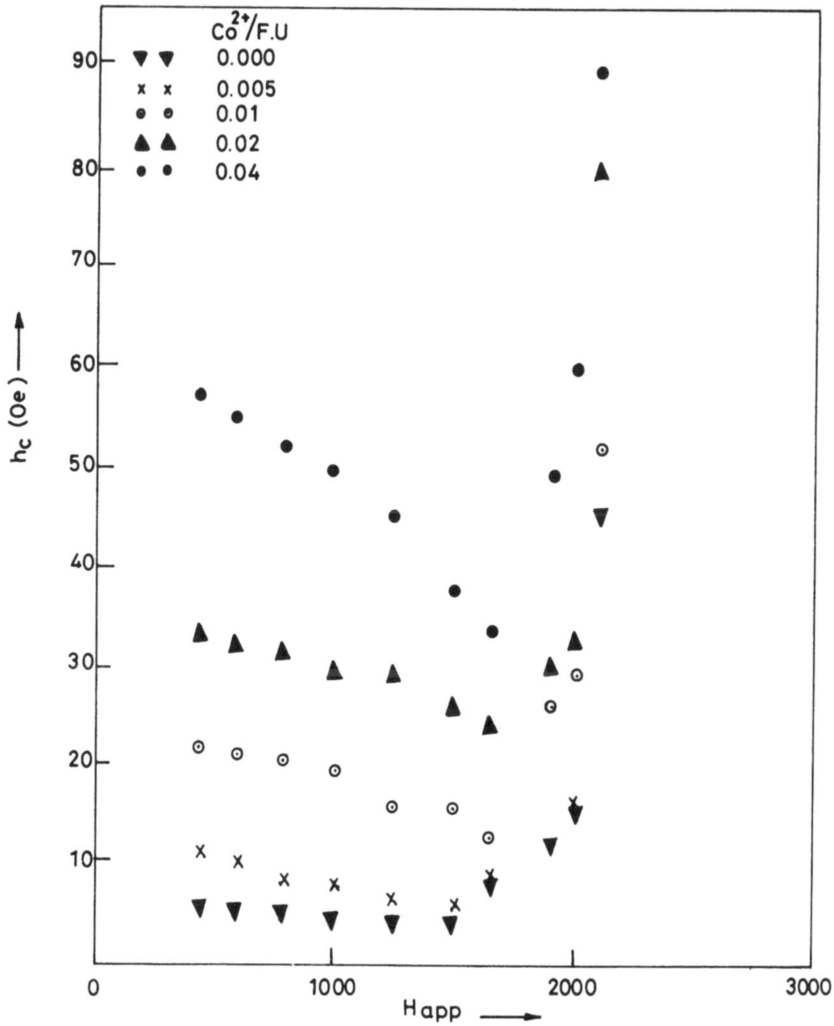


**Fig. 7** (*Contd.*)

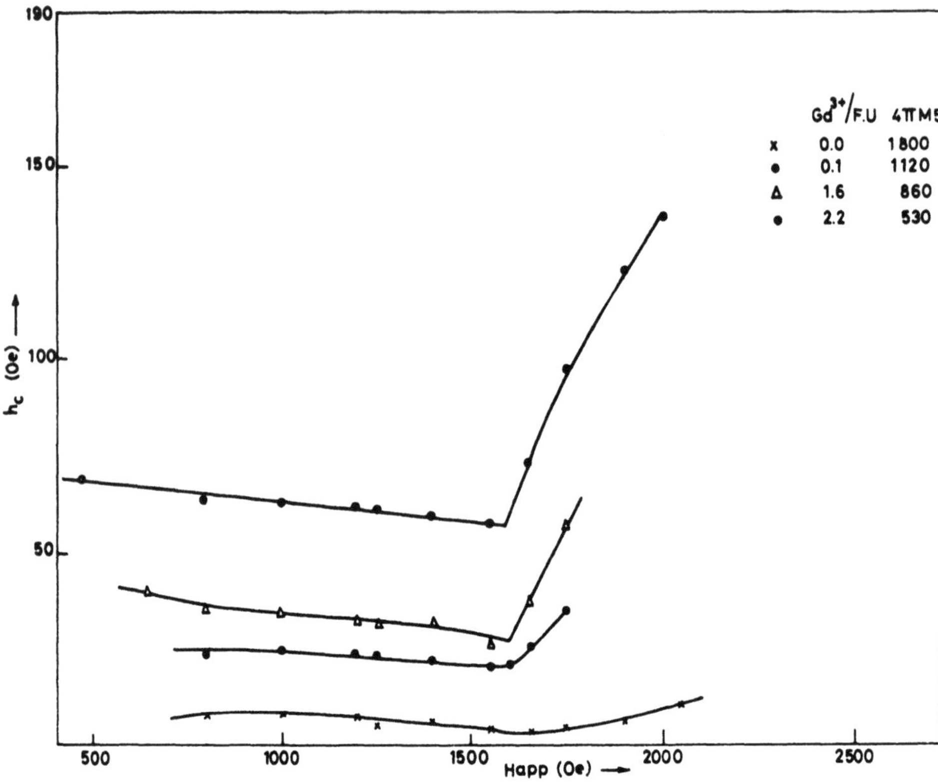


**Fig. 7.** Variation of critical fields as a function of static applied field for some garnets and ferrites [68].

$$H_{int} = -\left[\frac{4\pi M_s}{2}\right] + \left[\left(\frac{4\pi M_s}{2}\right)^2 + \left(\frac{\omega}{2\gamma}\right)^2\right]^{1/2}$$

where

$$H_{app} = H_{int} + 4\pi M_s \cdot N_z \tag{15}$$

$H_{app}$ is the applied static dc field, $N_z$ is the longitutional demagenetising factor and is equal to 1/3 for spheres. There is no change in the positions of the experimentally observed [68] butterfly curve minima i.e. the minimum positions of the butterfly curves for the samples of garnets and Co ferrites. But in case of Al ferrites the butterfly curves have been shifted towards higher static field than calculated from Eq. (14) by an amount approximately equal to the anisotropic field ($H_a$). The shifting of the butterfly minima due to anisotropy of the material with the addition of Al ions may be attributed in the following way. Under the influence of crystalline anisotropy the theoretical butterfly curve minimum should occur at an internal field of [67]

$$H_{int}^{anist} = -\frac{1}{2}(4\pi M_s + H_a) + \left[\frac{1}{4}(4\pi M_s + H_a)^2 + \left(\frac{\omega}{2\gamma}\right)^2\right]^{1/2} \quad (16)$$

and

$$H_{app} = (H_{int}^{anist} + H_a) - 4\pi M_s N_z$$

Using Eq. (16) the theoretical butterfly minimum position is calculated and observed are nearly the experimental value.

Above this minimum position threshold field increases rapidly with the increase of static field for all the three series of samples [68]. The low level substituted ferrites or garnets show smaller tendency for dipping at butterfly minimum before the steep rise. This is in agreement with other published data [31–33]. The experimental parallel pump data in these figures indicate that $\Delta H_k$ is indeed $k$ dependent. Moreover, increase of relaxing ions (like $Co^{2+}$ or $Gd^{3+}$) below butterfly curve minima, makes $\Delta H_k$ a strong $k$ dependent. Since $\omega$ the operating frequency is fixed, the threshold ($h_{crit}$) at each value of $H_{app}$ corresponds to a different $k$. These data clearly show that the threshold increases substantially as the $Co^{2+}$ or $Gd^{3+}$ content increases. The $h_{crit}$ for maximum $Co^{2+}$ content i.e. 0.04 $Co^{2+}$ ions/F.U. is over ten times greater than the threshold for the zero $Co^{2+}$ content sample. $Gd^{3+}$ also increase $h_{crit}$ substantially. But the increase of $h_{crit}$ by $Al^{3+}$ is the least. Borghese [39] observed the variation of $\Delta H_k$ for Mg-Mn ferrites with 2% cobalt with grain size. His results show that $\Delta H_k$ is virtually independent of grain size. In the present investigation the saturation magnetization is kept constant and the average grain size is also fixed for Co and Al ferrites. So the increase of $h_{crit}$ is only due to $Co^{2+}$ or $Al^{3+}$ in addition to the LiTi ferrites. These results suggest that the relaxation mechanism of low field loss contains no two magnon contribution, which is the major effect of the ferromagnetic resonance relaxation.

In this calculation $\theta_k$ is taken as $\pi/2$ which minimizes the critical field ($h_c$), and for a isotropic material for parallel pumping $h_c$ is given by

$$h_c = \left(\frac{\omega}{\omega_m}\right)\left[\frac{\Delta H_k}{\sin^2\theta_k}\right] \quad (17)$$

Under the influence of anisotropy the parallel pump threshold is given by [68]

$$h_c = \frac{\omega\,\Delta H_k}{\omega_m(\sin^2\theta_k)}\left[1 + \left(\frac{3Hk}{4\pi M_s(\sin^2\theta_k)}\right)^2 - \left(\frac{6Hk}{4\pi M_s(\sin^2\theta)_k}\right)(\cos^2\phi_k)\right]^{(-1/2)} \quad (18)$$

The square bracket term (SBT) represents the effect of anisotropy of the parallel pump coupling. Eq. (17) is applicable to the isotropic materials like Co ferrites and Gd YIG, but in case of Al ferrites where the crystalline anisotropy has a large effect Eq. (17) does not hold good. So for these samples $h_c$ is obtained by

Eq. (18). The experimental butterfly curves for Al ferrites do not show any double minima. Sethares et al [10] observed double minima butterfly curves for single YIG at 9.55 GHz. This double minima may be due to the spin-wave switching phenomena. Setheres et al ensure that only $\phi_k = 0$ or $\pi/2$ are to be selected.

In case of Al ferrites absence of any low field butterfly minimum ensures the instability of spinwave with $\phi_k = \pi/2$. The coupling of energy between fast relaxers like Co or Gd through long range dipolar interaction is very strong. Thus a large increase of instability is observed in case of Co ferrites and Gd YIG. But in case of Al ferrites the coupling due to anisotropic field is not very strong, thus a moderate increase of $h_c$ occur.

For high static field, i.e. above butterfly curve minima the range of available $\theta_k$ values becomes more restricted. At these external fields $\theta_k$ starts decreasing from $\pi/2$ to zero and $k$ remains nearly zero.

The theoretical verification of the experimental data [68] are made by the following models.

MODEL I

These parallel pump butterfly curves, for samples having large grain size ($a_0$) and small porosity ($p$), can be reproduced theoretically by considering the mean-free-path or the transist-time model given by Vrehen et al [65] and Patton [63]. According to this model a generated spinwave having frequency $\omega_k = \omega/2$ travels a distance 1 at a group velocity $v_g$ before being destroyed. $\iota$ is the mean-free-path and is related to the grain size $a_0$. It gives a life time for the spinwave,

$$\tau = \iota/v_g \tag{19}$$

and the mean-free-path related spinwave linewidth can be expressed as

$$\Delta H_k^{\iota} = \frac{1}{\gamma\tau} = \frac{|v_g|}{\gamma\iota} = \frac{2Dk}{\iota} \tag{20}$$

where

$$v_g = 2\gamma Dk\,\hat{k} + \gamma 4\pi M_s \sin\theta_k \cos\theta_k\,\hat{\theta}_k \tag{21}$$

$D$ is the exchange constant $5 \times 10^{-9}$ Oe-cm$^2$, $\gamma$ the gyromagnetic ratio is equal to $1.76 \times 10^7$ Oe$^{-1}$-sec$^{-1}$and $\theta_k = \pi/2$.

Polar component of $v_g$ i.e. the 2nd term in Eq. (21) is neglected, thus spin-wave linewidth can be expressed as proportional to the $k$ values.

Vrehen [65] also suggested that the mean-free-path was related to the grain size of the sample, by observing the inverse grain size dependence of the spinwave linewidth in fine grain samples. The application of this model with the present experimental data is not suitable, as the samples have a quite larger average grain size and porosity > 3%. The densification of ferrite is made through the addition of $Ba_2O_3$, as a sintesing aid at the time of initial mixing for sample preparation. This is due to the fact that the addition of bismuth oxide lowers the sintesing temperature to 1000–1050°C, where volatility of $Li_2O$ is minimal.

MODEL II

Scotter [57] has provided another model, which is a refinement of the transit-time model. According to this model the spinwave linewidth is affected by the size of the nonmagnetic pores, rather than the grain boundaries. Here the relaxation has the form

$$\tau = \frac{4R}{3\,p v_g^{(k)}} \tag{22}$$

where $R$ is radius of the pore and $p$ the porosity. The basic aim of Schotter's model is that the pores in ferrites act as the scattering center for the travelling spinwaves.

MODEL III

Sawado [62] has improved the pore scattering model by applying the screening effects to the pore dipole potential. This method is used explicitly to explain the experiment data here. The mean-free-path $\iota$ as obtained by Sawado [62] for a spinwave at $k$ and $\theta_k$ is given by

$$\iota = \beta / p\sigma(\theta_k) \tag{23}$$

where $\beta$ is the effective pore volume and is written as

$$\beta = \frac{3}{2\mu_s + 1}\beta_0 \tag{24}$$

The static permeability $\mu_s$ observed in this investigation at 10 KHz is 1000. This value of $\mu_s$ is used in calculating $\sigma(\theta_k)$. But for the calculation of mean-free-path $\iota$, the permeability used is the microwave permeability which is nearly equal to one.

$\beta_0$ is the volume of the pore and is measured from the SEM photographs, and is equal to $(4/3)\pi R^3$, where $R$ is the pore radius. This model has been used for calculating the theoretical $h_c$ values for all the ferrites and garnets. In case of Al ferrites effect of anisotropy is also added. The value of $R$ obtained for the sample is 0.17 to 0.20 microns which is of the same order obtained for YIG by Sawado. The scattering cross-section of a pore given by Sawado is

$$\sigma(\theta_k) = (4/9)\left(\frac{\omega_m \beta}{4\pi\gamma D}\right)^2 I(\theta_k) \tag{25}$$

where

$$I(\theta_k) = \int \sin\theta_k \, d\theta_k \, d\phi_k \int |P_2 (\cos\theta)|^2 \sin\theta_k \tag{26}$$

$I(\theta_k)$ is an integral which sums over all the angles into which the spinwave may be scattered, $\theta_k$ is the polar angle of the wave vector $k$. The spinwave linewidth is then given by

$$\Delta H_k(P) = \frac{P}{\gamma\beta} v_g(\theta_k)\, \sigma(\theta_k) \tag{27}$$

Sawado [62] obtained the spinwave linewidth by integrating overall the $\theta_k$ values and given the expression for $\Delta H_k$ as

$$\Delta H_k = \frac{P}{4\pi\gamma\beta} \int v_g(\theta_k) \sigma^T(\theta_k) \sin\theta_k \, d\theta_k \, d\phi_k \tag{28}$$

As the experimental data given here shows a strong $k$ dependence, the scattering cross section cannot be obtained by integrating overall allowed values of $\theta_k$. Because an integration overall spinwave angles provide a wave vector—independent linewidth. So the suggestions made by Silber and Patton [64] seems to be true for the samples investigated under present study. In this work $\Delta H_k$ is calculated by using Eq. (27) instead of Eq. (28).

Theoretical butterfly curves for parallel pumping conditions are obtained by including the wave-vector $k$ dependence of the linewidth, which can be expressed as

$$\Delta H_k = A + C \cdot k \tag{29}$$

where $A$ and $C$ are two constants. The value of these constants are calculated in the following way. The values of $A$ are calculated from the $h_c$ value at the butterfly minimum of the experimental data with $\theta_k = 90°$, $k = 0$ and at $H_{app}$ around the butterfly curve minimum position, by using Eq. (10). The values of constant $C$ are obtained from low field side experimental data of the butterfly curves, where $\theta_k$ corresponds to $\pi/2$ and for nonzero $k$ values.

The most striking feature of the present investigation is the rapid increase of $A$ values with the addition of $Co^{2+}$ and $Gd^{3+}$ contents except $Al^{3+}$ content samples. In addition, the parameter $C$ reflects the degree of $k$ dependence of $\Delta H_k$. From the calculated $C$ parameters it is clear that the increase of the relaxing ions ($Co^{+2}$ or $Gd^{+3}$) $\Delta H_k$ are more and more $k$-dependent.

The theoretical fitting of the low field experimental data are quite good for the threshold data below butterfly minimum.

But for higher static field the constant mean-free-path model does not fit the experimental data. So, the scattering of spinwave at grain boundaries cannot be the only mechanism which limits the spinwave life time in polycrystalline ferrites. In this investigation an attempt is made to fit the data with pore scattering approach given by Silber and Patton [64]. For the external static field beyond butterfly curve minima $\Delta H_k$ may be expressed in the form of

$$\Delta H_k = A + B \cdot \Delta H_k(p) \tag{30}$$

where $\Delta H_k(p)$ is given in Eq. (28). The parameter $B$ is calculated by setting $h_c$ equal to the experimental.

The agreement between the high field theoretical and experimental data, though not quantitatively perfect, are still good. The trend of both the curves are same. Only the difference is that the theoretical curves diverge more rapidly than the experimental curves. This may be due to the fact that the onset of absorption is not sharp for the static fields beyond 2000 Oe.

The results obtained for the samples under present investigations having large average grain size and porosity > 3% suggest that pore scattering contributes a large part to the spinwave threshold.

Thus the pore scattering theory originally given by Sawado [62] and latter on modification by Silber and Patton [64] presents a good theoretical fit to our experimental data. The present results show that the threshold minimum, $h_c$ (min) occurs only at $\theta_k = \pi/2$ and not at some value of $\theta_k$ less than 90°, as given by scotter [57]. The integral $I(\theta_k)$ appearing in the expression for scattering cross section $\sigma(\theta_k)$ shows the $\theta_k$ dependence of linewidth at higher static field i.e. above butterfly curve minimum. This integral is not only dependent on the porosity alone it depends on the size of the pore present in polycrystalline ferrite. Scanning Electron Microscope photrographs of these samples show that the inclusion of $Co^{2+}$ ion is in the composition increases the collective presence of smaller pores. The collective present of smaller pores increase the cross-sectional area $R/r$ times than a single larger pore, where $R$ and $r$ are the radius of the larger and small pores respectively. The increase of cross-section area has a marked affect on the spin-wave line width.

### 4.1.2 *Wave Vector (k) Dependence of* $\Delta H_k$

The wave vector $k$, dependence of $\Delta H_k$ are shown in Fig. 8. These curves are obtained from the parallel pump butterfly curves. The value of $k$ can be evaluated by using the dispersion relation for an isotropic material (like Co and Gd) and is expressed as

$$\omega_k^2 = \gamma[(H + Dk^2) \cdot (H + Dk^2 + 4\pi M_s \sin^2 \theta_k) \tag{31}$$

$H$ is the internal field = $H_{app} - N_z \cdot 4\pi M_s$ and $\theta_k = \pi/2$ for the parallel pump data.

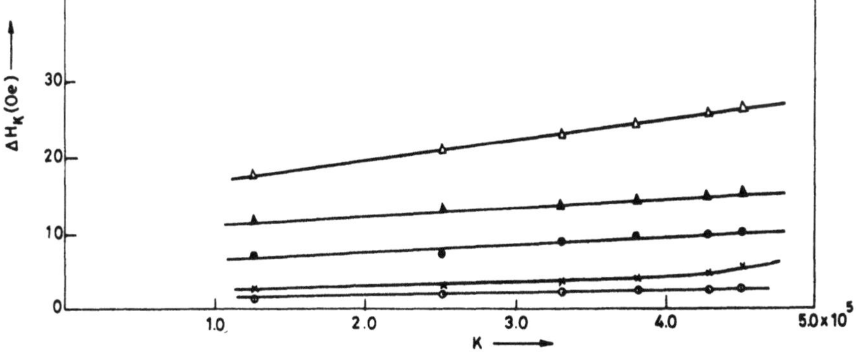


**Fig. 8.** Wave vector ($k$) dependance on $\Delta H_k$ [68].

For an anisotropic ferrite (Al-ferrite here) the dispersion relation is

$$\omega_k^2 = \gamma[(H + H_a + Dk + 4\pi M_s \sin \phi_k\, \theta_k)\cdot(H + Dk^2 + 4\pi M_s \cos \phi_k^2 \sin^2 \theta_k) - \{(4\pi M_s) \sin^4 \theta_k \sin^2 \phi_k \cos^2 \phi_k\}] \quad (32)$$

for $\phi_k = \pi/2$

$$\omega_k^2 = \gamma[(H + Dk^2) \cdot (H + H_a^2 + Dk^2 + 4\pi M_s \sin^2 \theta_k)] \quad (33)$$

The principal effect of anisotropic field on Al ferrite is that $\omega_k$ is dependent on the azimuthal angle, $\phi_k$, whereas in isotropic Co ferrite or Gd-YIG $\omega_k$ is independent of $\phi_k$. The gradient of $\Delta H_k$ vs $k$ curves which are approximately straight lines, are observed to increase rapidly with the increase relaxing ions only. In this investigation the inverse $k$ dependence was not observed. The results are in agreement with the data observed for large grain samples earlier [60, 61]. The linear dependence of $k$ with $\Delta H_k$ can be seen by visualizng the transit-time or mean-free-path model. The modulous of group velocity of the spin wave without the polar component is

$$v_g = 2\gamma Dk \quad (34)$$

i.e. $v_g \propto k$

which implies that,

$$\Delta H_k \propto k \quad (35)$$

Furthermore, the sharp increase of the stope of $\Delta H_k$ vs $k$ curves for Co-ferrite and Gd-YIG may be attributed due to the fact that, the addition of relaxing ions makes $\theta_k$ to increase from the unsubstituted value. From Eq. (8), the expression of $v_g$ the increase of $\theta_k$ increases the second term in this equation. Hence, according to the relation

$$\Delta H_k = |v_g|/\gamma l \quad (36)$$

there is a sharp increase of $\Delta H_k$ value which can be expected, because the values of $\gamma$ and $l$ are constant.

The observed spinwave linewidth data can be qualitatively explained with the idea of a constant mean-free-path model. When this model is applied to the relaxation due to fast relaxing impurities, the strength of the relaxation interaction for a particular relaxing ion should be taken into account. The spinwave linewidth may be expressed as

$$\Delta H_k = \Delta H_{k \to 0} + C \cdot k \quad (37)$$

The values of the constant $C$ are calculated from the $h_{c(min)}$ experimental values from parallel pumping configuration. From energy and momentum considerations the term on the right hand side of Eq. (37) is related to three magnon splitting processes and can only contribute to the parallel pump linewidth. $\Delta H_{k \to 0}$ attributed to $k = 0$ modes at $\theta_k = \pi/2$. The theoretical data have good agreement with the experimental data.

The results obtained here for the $k$ dependence of spinwave linewidth is due to a transist-time phenomena. Such dependence leads to the high power capability of ferrites, since relaxing ions are more effective at increasing the $\Delta H_{k \to 0}$ values.

### 4.1.3 *Spinwave Linewidth* $\Delta H_k$ *with the Variation of* $Co^{2+}$ *and* $Al^{3+}$ *Ferrite and* $Gd^{3+}$ YIG

Room temperature $X$-band spinwave linewidth, for $k = 0$ spinwaves calculated from parallel pump butterfly by curves data, are shown in Fig. 9. Using standard

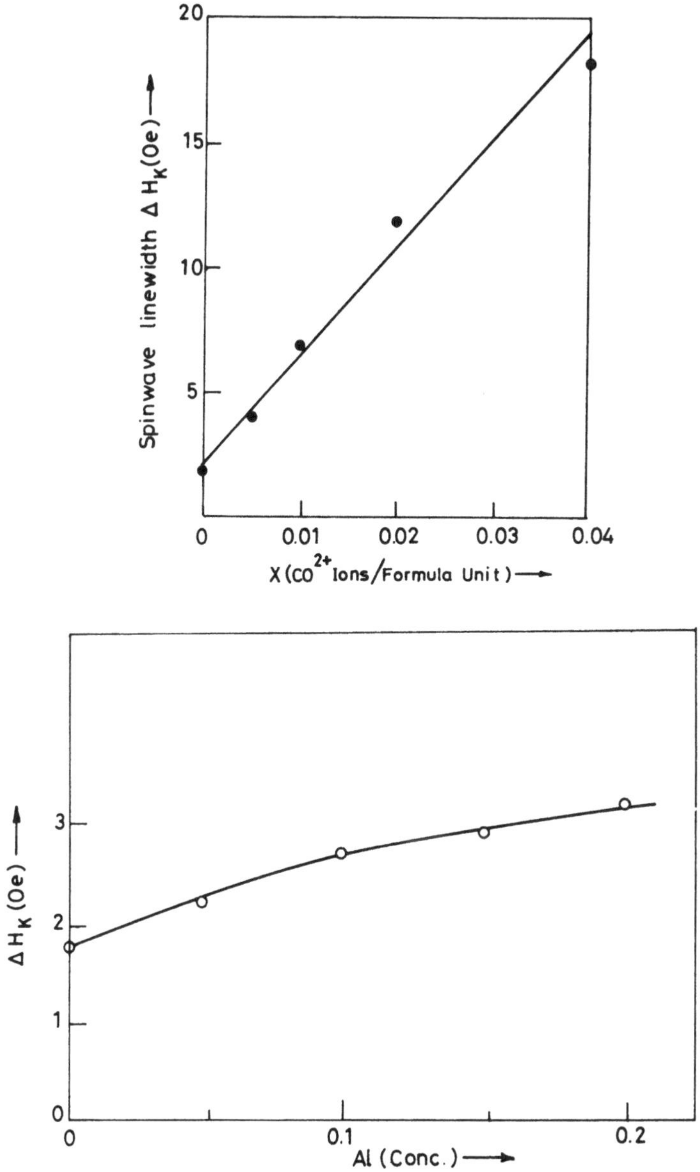


**Fig. 9.** Room temperature $X$-band spinwave linewidth, for $k = 0$ spinwaves calculated from parallel pump butterfly curves data [68].

theories of Suhl [40] and Schlomann [41] the $h_{crit}$ data obtained from parallel pumping experiment is used to calculate the value of $\Delta H_k$ for $k = 0$ ($\theta_k = \pi/2$) magnons. The standard relation between the critical microwave field amplitude for spinwave threshold $h_{crit}$ and spinwave linewidth $\Delta H_k$, for parallel pumping is given by

$$h_{crit} = \left(\frac{\omega}{\omega_m}\right)\left(\frac{\Delta H_k}{\sin \theta_k^2}\right)_{min} \tag{38}$$

where $\omega$ is the pump frequency, $\omega_m$ is $\gamma 4\pi M_s$ with $\gamma$ the gyromagnetic ratio. Fig. 9 dipicts that, for $Co^{2+}$ addition from zero to 0.04 ions/F.U. the X-band spinwave linewidth increases by more than an order of magnitude. $\Delta H_k$ increases approximately linearly with increase of $Gd^{3+}$ in YIG. But in case of $Al^{3+}$ the increase of $\Delta H_k$ though linear but increase is very slow and also at higher concentration of $Al^{3+}$, $\Delta H_k$ appears to attain a saturation value. The sharp increase in the value of $\Delta H_k$ arises due to the fast relaxation effect of $Co^{2+}$ ions in octahedral sites, which is based on their crystal field stabilization energy. The higher charge $Ti^{4+}$ ions on neighboring octahedral sites and low charge $Li^{1+}$ ions on adjacent tetrahedral sites help in the stabilization of $Co^{2+}$ in octahedral sites. The clustering effect of $Ti^{4+}$ ions around octahedral sites, if any, is expected even in the absence of $Co^{2+}$ ions and is not going to affect the material behaviour on addition of the fast relaxing $Co^{2+}$ ions. The observed data can be explained by considering the strong relaxation the effect of $Co^{2+}$ on the anisotropy of the material and the enhancement of spin-lattice relaxation rates. The anisotrophy cancellation effects attributed to $Co^{2+}$ ions do not appear where $Ti^{4+}$ ions are used as the dilutant in the basic composition of lithium ferrites. $Co^{2+}$ ions in octahedral sites provides a large positive contributions to the negative anisotrophy of the iron sublattices. Through magnetic superexchange interactions, the $Co^{2+}$ anisotrophy is communicated to the iron sublattices. This caused rotation of the $Fe^{3+}$ spins in the adjacent tetrahedral and octahedral sites. If the $Co^{2+}$ spins are decoupled from the $Fe^{3+}$ by the presence of $Li^{1+}$ in tetrahedral and $Ti^{4+}$ in octahedral sites, then the isolated $Co^{2+}$ anisotropy effects is undetected at these small concentrations. This give rise to the fact that $Co^{2+}$ now behaves as a paramagnetic ion, and this coupled to the $Fe^{3+}$ ions only through dipolar fields. The exchange isolation cannot prevent the occurrence of increased spin-lattice relaxation rates, due to the cross relaxation between paramagnetic ions, which can transfer energy to fast relaxers like $Co^{2+}$ by means of longer range dipolar interactions. This leads to a sharp increase of spinwave linewidth.

In case of Al ferrite the increase of $\Delta H_k$ is very small but still appreciable to use this material in medium power microwave devices. This slow increase of $\Delta H_k$ may be attributed to the coupling of energy to $Fe^{3+}$ ions via crystalline anisotropy. The strong point of Al addition over Co in LiTi ferrite is that $Al^{3+}$ does not raise the magnetic losses unlike $Co^{2+}$. So a fine control of aluminium can make this group of ferrite to best use for medium power microwave devices.

The large increase of $\Delta H_k$ in Gd-YIG over the concentration range studied may be attributed [68] in the following way. Substitution of $Gd^{3+}$ in YIG occupies the sites reserved for nonmagnetic yttrium. For garnets the principal relaxation mechanism is provided by the rare-earth ions. Gadolinium like other rare-earth ions relax the magnetisation in a two stage process. In the first stage energy is coupled from the ferric ion system into the spin system formed by $Gd^{3+}$ ions. This energy is then dissipated to the crystal lattice in a second stage. The exchange interaction provides the energy required for coupling between the ferric ions and $Gd^{3+}$ ions. This exchange interaction is an order of magnitude weaker than the interaction between ferric ions. $Gd^{3+}$ will behave as a paramagnetic ion. The cross relaxation between paramagnetic ions channel energy to fast relaxer through the fluctuating field it produces on the ferric lattice. Hence there is an increase of $\Delta H_k$.

## 4.2 Low Power Microwave Studies

The variation of $X$ band resonance linewidth ($\Delta H$) with Gd-YIG and Co and Al ferrites at room temperature are given in Fig. 10. It depicts that gadolinium substitution has a much smaller effect on $\Delta H$. In case of Co-ferrite the increase is slow, but in case of Al ferrite the increase in significant.

Upto 53% of gadolinium content the linewidth has increased to 190 Oe from 125 Oe for 33% $Gd^{3+}$ content sample. For the highest concentration studied here (2.2 $Gd^{3+}$/F.U.) $\Delta H$ has increased to 420 Oe. The small increase of $\Delta H$ for low contents may be due to the increase of porosity alone, but for higher concentration anisotropy broadening begins to contribute much. In higher concentration range the linewidth can be calculated by the coupling of the uniform precession with degenerate spinwaves of medium $k$-values. These couplings are due to the variation in magnetocrystalline anisotropy due to the granular structure of the material. The $Gd^{3+}$ ion contains seven $4f$ electrons which

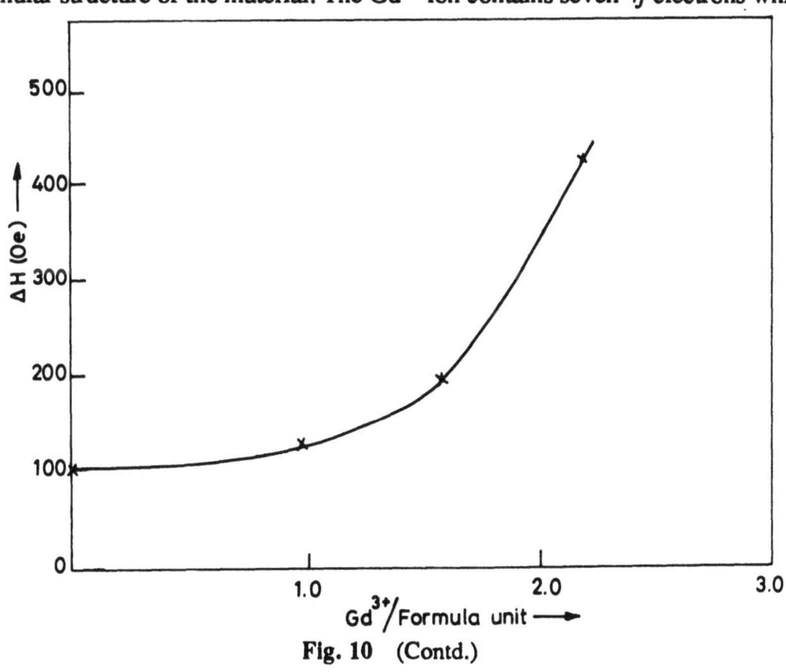


**Fig. 10** (Contd.)

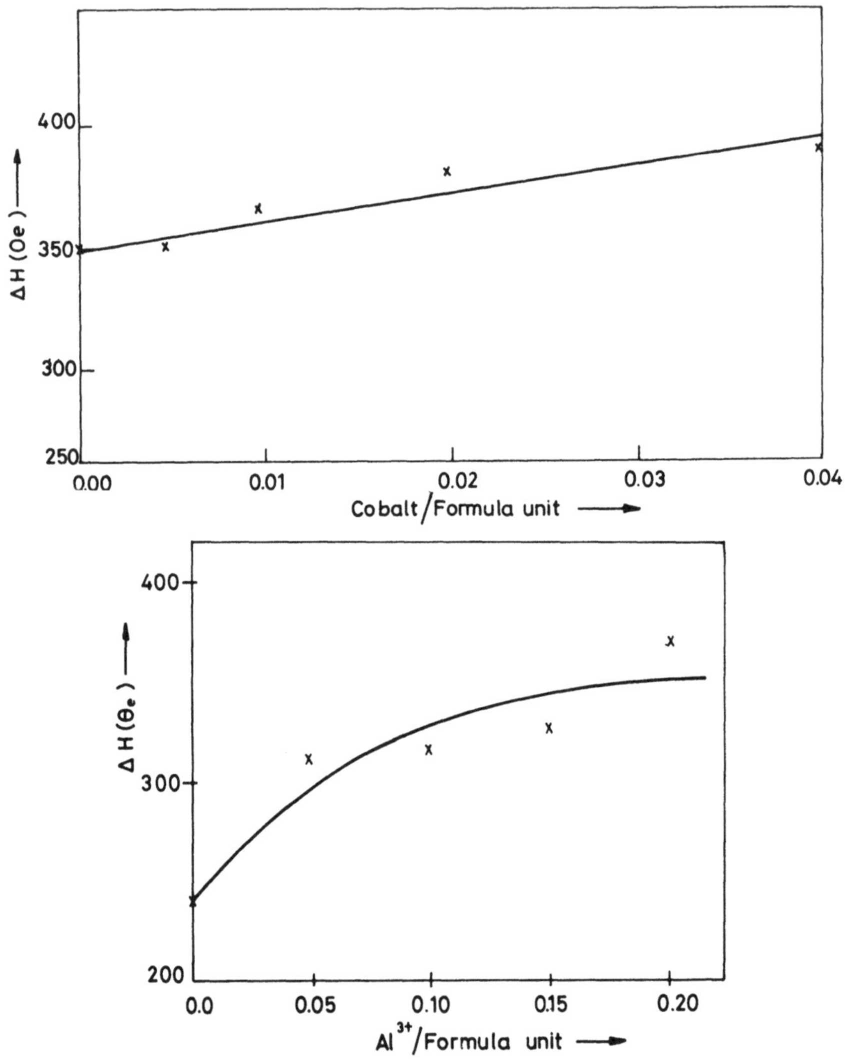


**Fig. 10.** The variation of X-band resonance linewidth ($\Delta H$) with Gd-YIG and Co and Al ferrites at room temperatures [68].

combine according to Hund's rule in such a way that the ion has no orbital angular momentum. The spin orbit interaction is therefore very small and cannot lead to a rapid relaxation as in the case of the other rare earth ions.

In case of $Co^{2+}$ ferrite the relaxation effect vary according to $(4\pi M_s)^{-1}$. The slope of the $\Delta H$ vs $Co^{2+}$ content reflect the lower magnetisation of the samples and the small increase of $\Delta H$ with cobalt content is caused by the fast relaxation mechanism of $Co^{2+}$ ions. The increase of $\Delta H$ with $Al^{3+}$ may be due to poor microstructure, increasing anisotropic field and also by increasing porosity. The large value of $\Delta H$ can be understood in terms of spinwave scattering. Due to crystalline anisotropy the distribution of resonance field for the randomly oriented crystallites induces the broadening of linewidth.

**REFERENCES**


1. J.L. Snoek, Phil. Tech. Rev., 8 (1946) p. 363.
2. L. Neel, Ann. Phys., 3 (1948) p. 137.
3. J.L. Snoek, "New Developments in Ferromagnetic Materials"; (Elsevier Press Inc., New York 1949).
4. E.W. Gorter, Philips Research Reports, 9 (1954) p. 295.
5. J.B. Goodenough, Magnetism and the Chemical Bond. (Interscience-Wiley, London, 1962).
6. G. Blasse, Philips Research Reports, Suppl. No. 3 (1964).
7. P.B. Braun, Nature, 170 (1952) p. 1123.
8. P.B. Baba, G.M. Argentine, W.E. Courney, G.F. Dionne and B.H. Temme, Trans. IEEE 'Mag-6 (1972) p. 83.
9. G.O. White, C.A. Edmendson, R.B. Goldforb and C.E. Patton, J. Appl. Phys. 50 (1978) p. 2381.
10. J. C. Sethares, M.H. Seavey, T.C. Purnhagen and M.R. Stiglitz, J. Appl. Phys., 35 (1964) p. 898.
11. J.H.E. Griffithis, Nature (London) 158 (1946) p. 670.
12. C. Kittel, Phys. Rev., 73 (1948) p. 155
13. C. Kittel, Phys. Rev., 76 (1949) p. 743.
14. D. Polder, Phys. Rev., 73 (1948) p. 1116.
15. J.H. Van Vleck, Physica, 15 (1949) p. 197.
16. S. Geller and M.A. Gilles, J. Phys. Chem. Solids 3 (1957) p. 30.
17. R.K. Wangshers, Phys. Rev., 93 (1954) p. 68.
18. M. Sparkes, Ferromagnetic Relaxation Theory (McGraw-Hill, New York) 1964.
19. J.H. Van Vleck, Phys. Rev., 78 (1950) p. 266.
20. F. Brown and D. Park, Phys. Rev., 93 (1954) p. 381.
21. H.G. Beljers and D. Polder, Nature 165 (1950) p. 800.
22. W.A. Yager, J.K. Galt, F.R. Merritt and E.A. Wood, Phys. Rev., 80 (1950) p. 744.
23. J.F. Dillon Jr., S. Geschwind and V. Jacoarino, Phys. Rev., 100 (1955) p. 75.
24. P.E. Tannenwald, Phys. Rev., 100 (1955) p. 1713.
25. R.W. Teals, J. Appl. Phys., Suppl. to 33 (1962) p. 129S.
26. E.A. Turov, "Ferromagnetic Resonance Ch-Vi S.V. Vonsovskii (ed.) (Pergamon Press Ltd, London, 1966) p.191.
27. W.A. Yager, J.K. Galt, R.F. Merritt, Phys. Rev., 99 (1955) p. 1203.
28. R.C. Lecraw and E.G. Spencer, J. Phys. Soc. Japan, 17 suppl B.I, (1962) p. 401.
29. F. Keffer, Phys. Rev., 88, (1952) p. 686.
30. A.M. Clogston, H. Suhl, L.R. Walker and P.W. Andersnon, Phys. Rev., 101 (1956) p. 903. J. Phys. Chem. Solids, 1 (1956) p. 129.
31. E. Schlomann, J. Phys. Chem. Solids, 6 (1958) p. 242.
32. L.R. Walker, Phys. Rev., 105 (1957) p. 390.
33. Geschwind and Clogston, Phys. Rev., 108 (1957) p. 49.
34. Schlomann, J. Phys, Chem. Solids, 6 (1958) p. 242.
35. R.W. Damon, Revs. Modern Phys., 25 (1953) p. 239.
36. N. Bloembergen and S. Wang, Phys. Rev., 93 (1954) 72.
37. H. Suhl, J. Phys. Chem. Solids, 1 (1957) p. 207.
38. H. Suhl, Proc., IRE, 44 (1956) p. 1270.
39. C. Borghese and R. Roveda, J. Phys. Colloque Cl, Tome 32 (1971) p. 150.
40. H. Suhl, J. Phys., 29 (3) (1958) p. 416.
41. E. Schlomann, J. Phys. Chem. Solids, 6 (1958) p. 242.

42. E. Schlomann, J.J. Green and U. Milano, J. Appl. Phys. 31 Suppl. (1960) No. 5 p. 386S.
43. R.C. Lecraw, E.G. Spencer and C.S. Porter, J. Appl. Phys., 29 (3) (1958) p. 326.
44. J.J. Green and E. Schlomann, J. Appl. Phys., 33 (S) No. 3 (1962) p. 1358.
45. J.H. Saunders and J.J. Green, J. Appl. Phys. Suppl., 33 (3), (1963) p. 1372.
46. E. Schlomann, J.J. Green, J.H. Saunders, IEEE Trans. on Magnetics, MAG-1 (1965) p. 168.
47. J.R. Samuel Dixon, J. Appl. Phys., 38 (1967) No. 3 p. 1417.
48. J.J. Green and C.E. Patton, J. Appl. Phys., 40 (i) (1969) p. 172.
49. C.E. Patton, J. Appl. Phys., 40 (7) (1969) p. 2837.
50. C.E. Patton and J.J. Green, IEEE Trans. Mag., MAG-5 (3) (1969) p. 626.
51. C.E. Patton, Proceedings of the ICF, (1970) p. 524.
52. C.E. Patton, J. Appl. Phys., 41 (1) (1970), p. 431.
53. C.E. Patton, J Appl Phys., 41 (4) (1970) p. 1637.
54. D.G. Scotter, J. Appl. Phys., 32 (10) 1971) p. 4088.
55. Yu. M. Yakovlev, Yu N. Burdin, Yu R. Shilnikov and T.N. Bushueva, Soviet Phys-Solid State, 12 (10) (1971) p. 2475).
56. C.E. Patton, IEEE Transactions on magnetics (1972) p. 433.
57. D.G. Scotter, J. Appl. Phys, 43 (11) (1972) p. 4813.
58. G.O. White, C.E. Patton and C.A. Edmendson, J. Appl, Phys. 50 (3) (1979) p. 2118.
59. C.E. Patton and Wolfgang Jantz, J. Appl., Phys., 50 (11) (1979) p. 7082.
60. C.E. Patton, Phys, Stat. Sol. (b) 92 (1979) p. 171 and 93 (1979) p. 7082.
61. Om Prakash and C.M. Srivastava, Bull. Mater. Sci., 2 (3) (1980) p. 181.
62. Eikichi Sawado, Proceedings of ICF (1980) p. 818.
63. C.E. Patton, Proceeding of ICF (1980) p. 807.
64. L.M. Silber and C.E. Patton, IEEE Trans. Mag., MAG-18 (6) (1982) p. 1630.
65. Q.H.F. Vrehen, H.G. Beljers and J.G.M. delau, IEEE Trans. Magn. MAG-5 (1969) p. 617.
66. J.J. Green and T. Kohane, SCP and Solid St. Tech. 7 (1964) p. 46.
67. Isidore Baby and Ernst Schlomann, J. Appl. Phy. Suppl. 33 (1962) p. 1377.
68. Bijoy Kumar Kuanr; Ph.D. Thesis, University of Delhi, 1992.

# 6

# Microwave Lithium Ferrites

Pran Kishan
Solid State Physics Laboratory, Delhi, India

## 1. INTRODUCTION

Among the various magnetic oxide materials used in devices operating at microwave frequency ranges lithium ferrite is presently occupying a prominent position. Although its potential as rectangular loop material was being utilised for long in memory devices the exploitation of its attractive feature as a microwave material was much delayed due to excessive losses observed in early samples. Many researchers working in the field put up a good deal of effort in the early 70s to obtain lithium ferrites with improved dielectric properties. Preparation of low loss lithium ferrite was first reported in 1971 by Collins and Brown [1]. Considerable interest has persisted in lithium ferrites since then. With the establishment of reproducible procedures for the preparation of lithium ferrites, there has been no looking back and at this juncture lithium ferrites are among the most wanted materials for the microwave frequency ranges. The family of lithium ferrites has expanded to such an extent that it now covers a very wide spectrum of properties for utilisation in microwave devices. The frequency range of its use extends even to millimeter wave bands.

Braun [2] was first to show that lithium ferrite with chemical formula $Li_{0.5}Fe_{2.5}O_4$ is an inverse spinel. The divalent ion in this case being a composite ion ($Li^{1+}_{0.5}$ $Fe^{3+}_{0.5}$). The compositional variations made possible by incorporation of a variety of other metal ions in its formula unit have given rise to lithium ferrite family as a class of material with many special features [3, 4]. These features are:

1. Unsubstituted lithium ferrite ($Li_{0.5}Fe_{2.5}O_4$) has the highest Curie temperature (Tc ≈ 640 °C) among the ferrimagnetic oxide materials. High Tc signifies temperature stability of its magnetic properties. The substituted compositions of lithium ferrites also inherit this property from the parent material. A case of materials belonging to different ferrite families and a comparison of Tc values for a typical value of magnetisation ($4\pi Ms$ = 1000 Gauss) is illustrated in Fig. 1.
2. The lithium ferrites possess excellent rectangular loop characteristics. From its earlier use in memory core industry the present application in latching microwave devices is a natural extension.

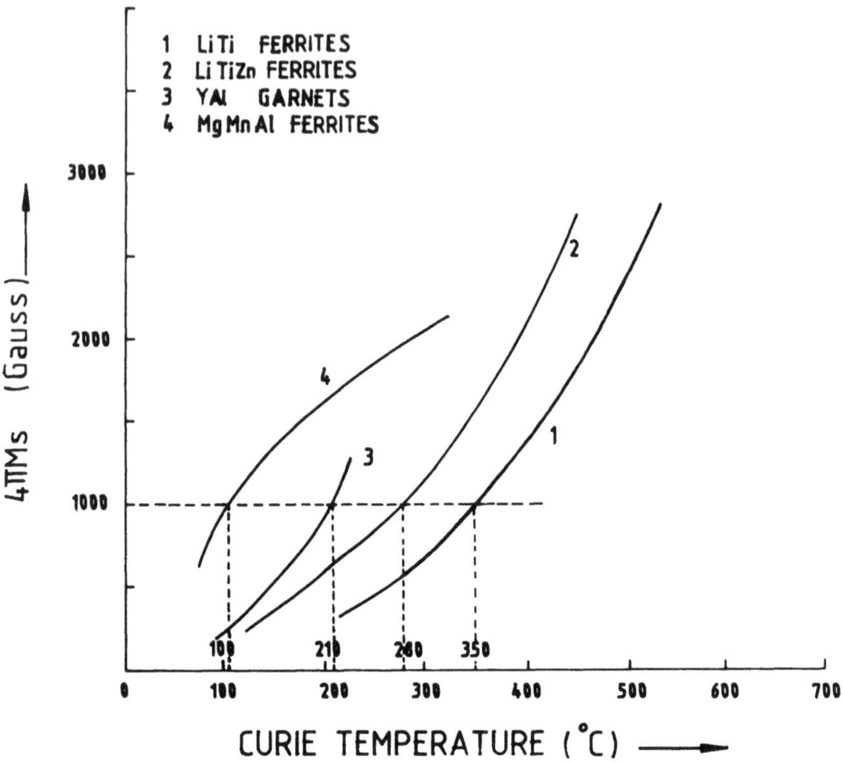


**Fig. 1.** Saturation magnetisation ($4\pi Ms$) vs Curie temperature (Tc) for different microwave ferrites.

3. The saturation magnetisation ($4\pi Ms$) of unsubstituted lithium ferrite at room temperature is ~ 3750 Gauss. It has become possible to vary this parameter to cover values upto 5000 Gauss by suitable compositional adjustments. Before this NiZn ferrite system was the only one available for such high values of $4\pi Ms$ for microwave applications. Other properties of technical importance such as remanence ratio, coercive force, resonance linewidth etc. are possible to tailor in lithium ferrites by incorporating appropriate substituents and controlling their concentration in the composition. A notable feature is a series of low linewidth lithium ferrites which have replaced high cost garnets in several microwave applications.
4. Lithium ferrites are known to have a low ratio of magnetostriction to magnetocrystalline anisotropy which means low stress sensitivity of remanence. This property is further improved by the compositional adjustment and is of great importance for microwave latching applications.
5. The dielectric constant ($\varepsilon'$) of the lithium ferrites with various substitutions fall in the range 14 to 20, the value being 15 for the parent member i.e unsubstituted lithium ferrite. These values are higher than that of the other micro-wave ferrite series (Garnets 15, MgMn ferrites 10 to 13 and

NiZn ferrites 9 to 12). High $\varepsilon'$ values are significant from the viewpoint of size and weight considerations of the device.

6. The dielectric loss tangents of lithium ferrites presently available are comparable with other microwave ferrite series ($\tan \delta_e < 0.0005$). It has become possible with the advancements made in the processing methods and from the compositional considerations.

The above features have contributed in the establishment of lithium ferrites to their present status in microwave magnetics.

## 2. MATERIAL PROPERTIES

The choice of the ferrite for a microwave device is decided by the working frequency, power level of operation and whether it works at resonance, far off resonance or at remanence. The properties of the ferrite optimised in various aspects determine the feasibility of the device and its level of performance. In this connection the following parameters of the material are of technical importance from the viewpoint of its utilisation in devices.

### 2.1 Dielectric Properties

It is the high resistivity of a ferrite which makes it a ceramic dielectric. The microwave frequency electromagnetic waves can penetrate and propagate with low dielectric loss in insulating materials. The permittivity of the ferrite is expressed as

$$\varepsilon = \varepsilon' - j\varepsilon''$$

where $\varepsilon'$ is the dielectric constant of the ferrite material. It lies in the range 9 to 20 for all ferrites. The dielectric loss tangent given by

$$\tan \delta_e = \varepsilon''/\varepsilon'$$

has values less than 0.001. For lithium ferrites $\varepsilon'$ has a high value in the range 14 to 20 and typical dielectric loss factors of 0.00025 are obtained.

The dielectric loss occurs primarily due to electrical conduction in the material. The conduction by hopping mechanism takes place between $Fe^{2+}$ and $Fe^{3+}$ ions present on equivalent crystallographic sites in the structure of the ferrite. The desired values of resistivity of ferrite material for low dielectric loss at microwave frequencies is more than $10^7$ ohm cm. In lithium ferrite, represented as

$$Li^{1+}_{0.5}\, Fe^{3+}_{2.5}\, O^{2-}_4$$

there are no $Fe^{2+}$ ions and a low loss factor is expected for it. However the preparation of the ferrite requires high temperature firing at 1150–1250 °C for obtaining dense material. At these temperatures oxygen dissociation and lithia volatility occurs resulting in formation of non-stoichiometric composition. The consequence is reduction of $Fe^{3+}$ ions into $Fe^{2+}$ ions and high dielectric losses. In addition the control of impurity ions of higher and lower valence in the structure and reduction of macroscopic and microscopic heterogeneities (e.g., pores, non-magnetic inclusions etc.) are also very essential. Suitable techniques have been developed to overcome it, namely sintering at lower temperatures and employment of appropriate substitutions in the composition to inhibit $Fe^{2+}$ formation. Annealing of sintered samples in oxygen atmosphere has also been found to be beneficial in some cases.

## 2.2 Magnetisation

The magnetisation of ferrites arises from the antiparallel alignments of the magnetisation of the two sublattices. The magnetic moments on each sublattice are arranged parallel to each other and the difference between the magnetisation of the two sublattices results in a net magnetic moment. The saturation magnetisation ($4\pi$Ms) is defined as the magnetic moment per unit volume when all its individual megnetic domains are aligned by the application of an external magnetic field. However the magnetic ordering is disturbed as the temperature of the material is raised. Magnetisation decreases and ultimately disappears at a characteristic temperature of the material called its Curie temperature (Tc).

It is the specific chemical formulation of the ferrite which primarily controls its saturation magnetisation. The process parameters involved in the preparation of the material also play an important role and for some cases the distribution of ions on the two sublattices is affected by the firing schedule of the material. The $4\pi$Ms of $Li_{0.5}Fe_{2.5}O_4$ which is ~ 3750 Gauss at room temperature can be varied by substituting appropriate other metal ions in the crystal lattice. It has become possible to prepare substituted lithium ferrite compositions with $4\pi$Ms value covering a wide range from low value of a few hundreds to 5000 Gauss. The Curie temperature, 640°C for the unsubstituted lithium ferrite, however decreases with the increasing substitution levels.

Several series of lithium ferrites for microwave applications are available with Tc varying from 80 to 620°C depending upon the $4\pi$Ms and other properties. The temperature variation of the magnetic properties of lithium ferrites are superior to other ferrites due to their inherent property of high Tc. Only some grades of garnets possess better temperature characteristics but their high cost and limited range of magnetisation restrict their utility.

## 2.3 Magnetic Losses

The basic phenomena underlying the propagation of microwave frequency electromagnetic waves in ferrites is the interaction between the magnetic field of the incident wave and the magnetisation of the ferrite. Energy extracted from the field in this process is called the magnetic loss and goes to the lattice as heat. The microwave losses of magnetic nature are characterised by the imaginary part, $\mu''$ of the complex permeability ($\mu = \mu' - j\mu''$) of the ferrite which exhibits a resonance behaviour at the angular frequency given by the relation

$$\omega = \gamma H$$

where $\gamma$ is the gyromagnetic ratio = 2.8 MHz/Oe and $H$ is the corresponding total internal field. The magnetic loss factor ($\tan \delta_m = \mu''/\mu'$) for microwave ferrites is desired to have low value of less than 0.001.

The total energy loss to the lattice due to magnetic interactions is also characterised by the ferromagnetic resonance linewidth ($\Delta H$), the half width of the resonance ($u''$) peak, and it includes all magnetic loss processes. The contributions to $\Delta H$ come from spin-lattice relaxation and magnetocrystalline anisotropy. Presence of non-uniformities in the ferrite medium such as pores, non-magnetic inclusions, cracks, surface roughness and variation in volume density of magnetic ions, all tend to increase the $\Delta H$ value. Suitable substitution of several ions like $Zn^{2+}$ in the chemical formulation of lithium ferrite is also quite effective in adjusting the $\Delta H$ of the material. The presence of relaxing ions like $Co^{2+}$ tend to increase it. Lithium ferrites of low linewidth values have been developed by the chemical formulation techniques. This has led to the development of lithium ferrites as low cost substitutes of garnets in several devices.

A number of microwave devices operate at fields away from the resonance e.g. a latched phase shifter at remanent magnetisation. For such devices off resonance linewidth is considered as the relevant loss parameter. The linewidth measured as a function of external magnetic field is expressed as effective linewidth ($\Delta H_{eff}$). The measurement of $\Delta H_{eff}$ with external magnetic field for lithium ferrite was first reported by Koelberger et al [6].

## 2.4 High Power Effect

The increase of incident rf power beyond a critical level causes the loss occurring in the ferrite to rise dramatically. The corresponding threshold is connected to the existence of a critical microwave field ($h_c$). This $h_c$ value puts a limit on the power handling capability of a particular microwave device. At high rf magnetic fields the spin waves are also excited at frequencies away from resonance resulting in the non-linear loss behaviour of the ferrite. The critical field is given as

$$h_c = C \frac{\omega}{\omega_m} \Delta H_k$$

where $\omega$ is the incident frequency and $\omega_m = \gamma 4\pi Ms$. The factor $C$ has value of the order of unity. The spinwave resonance linewidth $\Delta H_k$ is a material parameter and is considered to be a measure of the critical field. For normal microwave ferrites $\Delta H_k$ is 1 to 2 Oe. Higher values of $\Delta H_k$ for ferrites are obtained by incorporation of fast relaxing ions in the ferrite composition. However some increase in low-power magnetic loss accompanies it. Fine grained materials also exhibit improved spinwave linewidths and better peak power handling capability without any significant increase in low power loss.

## 2.5 Hysteresis Loop Parameters

A ferrite sample, is made-up of randomly oriented microscopic magnetic domains. Its magnetisation behaviour on the application of an external magnetic field is described by hysteresis loop and is characterised by parameters like coercive force ($H_c$) and remanence ratio ($R = 4\pi Mr/4\pi Ms$, defined at drive field of $5H_c$). A useful property of the hysteresis loop behaviour of the ferrite is that its magnetisation can be latched to a desired value without maintaining any continuous external field. The latched magnetization level can be easily adjusted at any value between $+4\pi Mr$ and $-4\pi Mr$ by the external field. In a toroidal sample this is done by driving a pulsed current through a threading wire wound on the toroid which can produce a field $H >> H_c$. Latched remanent positions at any in-between values can be reached by partially switching the toroid. Several types of phase shifters and switches are based on the latched magnetisation operation. The shape of the hysteresis loop plays an important role for such devices.

Ferrite materials with rectangular loop characteristics (high remanence ratio and low coercive force) are needed for this type of devices. The lithium ferrite materials employed in the fabrication of latching devices have $R > 0.85$ and $H_c \sim 1$ Oe. The switching coefficient and switching energy are two other important parameters for such a device and are governed by the hysteresis loop parameters of the material and the external circuit used in the device.

Stresses produced in the ferrite during shaping, machining processes and fabrication, affect the shape of the hysteresis loop and reduce the remanence ratio through magnetostriction effect as shown in Fig. 2 [5]. Ferrites with low stress sensitivity of remanence are essential for the microwave latching purposes.

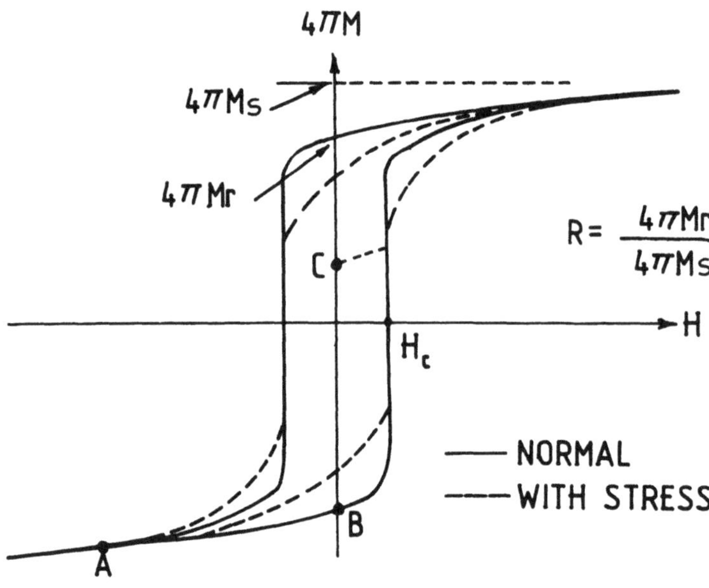


**Fig. 2.** Hysteresis loop.

This is achieved in lithium ferrites through appropriate chemical formulations. Sometimes thermal annealing of the ferrite samples after machining process is carried out to release the stresses.

## 3. CHEMISTRY OF LITHIUM FERRITES

Lithium ferrite ($Li_{0.5}Fe_{2.5}O_4$) is an inverse spinel with the cationic distribution $Fe_A\,[Li_{0.5}Fe_{1.5}]_BO_4$, where A and B represent tetrahedral and octahedral sites respectively. The lattice constant $a \approx 8.33$ Å. Its other properties of basic interest are: saturation magnetization, Curie temperature, loss factors and the rectangularity of the hysteresis loop.

The several series of lithium ferrites that have emerged out of the parent composition have been prepared by substituting a number of other metal ions in the ferrite composition. For microwave application the compositions that are prepared are represented by quite complicated chemical formulae. A typical composition includes several cationic constituents which can be divalent, trivalent and tetravalent ions, each included for one or more specific purposes. The beneficial and deleterious effects of some of these ions has been discussed by Baba et al [3]. Each ion has a different role to play in controlling the properties of the material. The resulting composition is derived keeping in view the desired properties of the product. This molecular engineering for lithium ferrites and its bearing on the properties is discussed briefly below.

### 3.1 Aluminium/Titanium Substitutions

The $4\pi$Ms of a ferrite is generally varied by incorporating non-magnetic ions for the magnetic ions in the formula unit. The substitution at octahedral site leads to reduction in $4\pi$Ms values. Aluminium seems to be an obvious choice for this purpose as it is often used in magnesium ferrites and garnets. The chemical formula of the ferrite becomes $Li_{0.5}Al_xFe_{2.5-x}O_4$. However several problems were encountered by the earlier workers in preparing the ferrite and the results were not satisfactory. Higher losses were observed and difficulties in controlling porosity and microstructure were experienced due to higher sintering temperatures required for synthesising the materials with the desired high densities. Dionne [7] and Yakovlev et al [8] found that in addition to reduction in magnetization, Al substitution results in increase of anisotropy. It has been observed that in small contents it has some beneficial effect on the power handling capability of the material. The effect of Al substitution in LiTi ferrites on spin wave resonance linewidth is discussed in Section 6.

Titanium substitutions were found to be much more attractive as it eliminated the problems experienced with Al. Improvements in sintering, control of porosity and microstructure and lowering of anisotropy with Ti content up to an appreciable level resulted in reduction of resonance linewidth and minimal effect on coercive force was obtained. However being a tetravalent ion its inclusion in the formula unit disturbs both the Li and Fe concentration

$$Li_{0.5+t/2}Ti_tFe_{2.5-3t/2}O_4$$

Ti ions occupy the octahedral sites and the cation distribution is described as [9]

$$Li_{t/2}Fe_{1-t/2}[Li_{0.5}Ti_tFe_{1.5-t}]O_4$$

The detailed analyses have indicated that a small amount of Ti on tetrahedral site cannot be ruled out. The titanium substituted lithium ferrites have a higher dielectric constant ($\varepsilon'$) and values upto 20 have been obtained for Ti rich compositions. It is apparent that the higher Ti concentration produces ferrites of lower $4\pi$Ms and Tc values.

### 3.2 Zinc Substitution

Zinc ions are often included in the composition for enhancing the $4\pi Ms$ values in ferrites as Zn enters predominantly the tetrahedral site. The formula of lithium ferrite with Zn substitution is represented by

$$Li_{0.5-z/2}Zn_zFe_{2.5-z/2}O_4$$

and the cation distribution is described by

$$Zn_zFe_{1-z}[Li_{0.5-z/2}Fe_{1.5+z/2}]O_4$$

For higher $z$ values (> 0.4) this substitution weakens the exchange interaction and does not result in any further increase in $4\pi Ms$, on the other hand a fall in magnetisation is observed. Similar behaviour of magnetisation is observed in all spinel ferrites on zinc substitution.

Zn substitution is also very effective in controlling several other properties. It promotes grain growth and densification during sintering and lowers the anisotropy. All this results in decrease in resonance linewidth and coercive force. Lower remanence ratio values are also obtained simultaneously [10, 11]. Resonance line widths of around 25 Oe have been achieved [12].

### 3.3 Simultaneous Zn and Ti Substitution

In lithium ferrites the practice being adopted is to include both Zn and Ti ions simultaneously in the composition

$$Li_{0.5+t/2-z/2}Ti_tZn_zFe_{2.5-3t/2-z/2}O_4$$

Studies on the effect of Zn and Ti substitutions on the static magnetic properties in lithium ferrites have been carried out by several authors [5, 13–19]. For a typical LiZnTi ferrite composition the distribution of cations over the two sublattices is represented as

$$Zn_zFe_{1-z}[Li_{0.5+(t-z)/2}Ti_tFe_{1.5+(z-3t)/2}]O_4 \qquad \text{for } z > t$$

and

$$Zn_zLi_{(t-z)/2}Fe_{1-(t+z)/2}[Li_{0.5}Ti_tFe_{1.5-t}]O_4 \qquad \text{for } z < t$$

The combined substitutional effect on material properties for Zn and Ti together is consistent with the individual characteristics of LiZn and LiTi ferrites. However the simultaneous incorporation of both non-magnetic Zn and Ti ions in the composition brings down the Curie temperature appreciably, while it becomes less effective in lowering the $4\pi Ms$ values compared to Ti substitution alone.

### 3.4 Manganese Substitution

A very essential requirement for microwave ferrites is their low conductivity and dielectric loss tangent ($\tan \delta_e$). Mn substitution is a well established practice being adopted for inhibiting the formation of any $Fe^{2+}$ ions. The role of Mn ions is described by the well known buffering reaction

$$Mn^{3+} + Fe^{2+} \rightleftarrows Fe^{3+} + Mn^{2+}$$

The reaction is favoured in the forward direction.

Another notable effect of Mn ions is in lowering the stress sensitivity of the material. The magnetostriction constant ($\lambda$) for $Li_{0.5}Fe_{2.5}O_4$ is $-8 \times 10^{-6}$ [20]. It was observed long back by Baltzer [21] that the system $(1-x)\ Li_{0.5}Fe_{2.5}O_4 \cdot xLi_{0.5}Mn_{2.5}O_4$ is said to have its saturation magnetostriction approach zero value at $x = 0.12$. Higher value of remanence ratio was also observed in the neighbourhood of this composition. Like other non-magnetic ions Mn also decreases the Curie temperature.

The Mn substitution is necessary in lithium ferrites particularly for latching applications where low magnetrostriction is needed. Material becomes less sensitive to stresses, which are introduced in the machining process of ferrite parts by mechanical pressure in device fabrication and by thermal effects, by Mn substitution in the ferrite composition.

### 3.5 Nickel Substitution

A small amount of Ni is also often included in the composition as it has been found useful in enhancing the remanence ratio. It is attributed to lowering of magnetostriction [22].

### 3.6 Cobalt Substitution

The loss of microwave power in ferrite devices is found to increase non-linearly beyond some critical incident power level. For high power devices the ferrites capable of handling desired power levels are designed. For this purpose the incorporation of divalent Co ions in very small content in the material has proved very effective. Inclusion of Zn in the composition in the presence of Co doped lithium ferrite is also found to increase $\Delta H_k$. Zn alone does not have any noticeable effect on this parameter [23]. An increase in $\Delta H_k$ value is, in general, always accompanied by an increase in magnetic losses.

As a consequence of the above discussion the substitution of the several cations in the composition leads to a chemical formula like

$$Li_{0.5+(t-z-b-c)/2}\,Zn_zTi_t\,Mn_aNi_bCo_cFe_{2.5-(z+3t+2a+b+c)/2}O_4$$

The level of substitution of various ions is chosen with the aim to obtain the material with desired properties. In practice the value of these cationic contents is varied in limited ranges only e.g. $z \le 0.35$, $t \le 1.0$, $a \le 0.1$ $b \le 0.1$ and $c \le 0.02$.

Some work has been reported on mixed lithium-magnesium ferrites. Individually both lithium ferrites and magnesium ferrites are considered to be good microwave materials having rectangular hysteresis loop properties. Mixed Li-Mg ferrite compositions containing Ti have been studied for their use in latching applications. Typical compositions are described by Wang et al [24]. It is also reported that these materials have found applications in a number of microwave devices.

The addition of $Bi_2O_3$ which essentially acts as a sintering aid is discussed separately in Section 4.

## 4. MATERIAL PREPARATION

Lithium ferrites were earlier finding application in memory cores due to its rectangular hysteresis loop properties. Their wider use particularly for microwave devices was restricted due to the difficulties experienced in sintering the material at the high temperatures employed to achieve high densities in stoichiometric form. The irreversible loss of lithia [25] and oxygen during sintering was the main cause that made lithium ferrites technologically difficult to prepare. Low resistivities of LiZn ferrites, sintered at 1175°C, were observed even in the case when firing was carried out in oxygen atmosphere [11]. This was attributed to the loss of oxygen and lithia occurring at the high sintering temperatures resulting in the reduction of $Fe^{3+}$ to $Fe^{2+}$.

The above said hindrances in making lithium ferrites useful for microwave frequencies were overcome by employing sintering aids like bismuth oxide, vanadium pentoxide niobium oxide etc. [26]. The first two are low melting materials and are effective by liquid phase sintering process. The addition of $Nb_2O_5$ enhances the sintering by increasing bulk diffusion occurring due to increased vacancy concentration caused by solubility of $Nb^{5+}$ in the ferrites [27]. $Bi_2O_3$ has been found to have an edge over others and has been adopted in manufacturing of lithium ferrites for microwave applications. The addition of $Bi_2O_3$ in small amounts has been reported by several authors [28, 29, 2] for the preparation of dense materials of both pure and substituted lithium ferrite composition. This aspect of the lithium ferrite preparation, which has played an important role in bringing up these materials to the present level, is discussed below.

A lithium ferrite composition is prepared by the conventional ceramic procedures. A small amount of $Bi_2O_3$ is mixed either at the starting stage or after the presintering. This makes the material sinterable at comparatively low temperatures (950–1050°C) where the loss of lithium and oxygen are minimal.

There have been speculations in the literature regarding exactly where and in what form $Bi_2O_3$ locates in the polycrystalline ferrite bodies. Baba et al [2] incorporated $Bi^{3+}$ in the formula unit of the ferrite replacing $Fe^{3+}$ ions. Similarly Peshev and Pacheva [30, 31] included $Bi^{3+}$ as a part of the formula unit although they considered it merely illustrative of the composition. Simonet and Hermosin [32] and Green et al [33] considered small quantities of $Bi_2O_3$ as a simple insoluble additions segregating mainly at the ferrite grain boundaries. Microstructural and thermal analyses carried out have confirmed this and have given a clear picture of the role played by $Bi_2O_3$ [29, 30, 31].

In the chemical formula unit containing several substituents, like the one given in Section 3, when a small amount of $Bi_2O_3$ is added, density upto 99% of the X-ray density is achieved. $Bi_2O_3$ content of 0.25 to 1.0 % by weight has been found to be most appropriate. Higer amount of $Bi_2O_3$ interferes with the crystallisation process and hampers grain growth. It leads to lowering of densities and the material properties are adversely affected. Smaller amounts are quite effective in lowering the sintering temperature and also improve several of the properties of interest from the point of view of its utility in microwave devices. Fig. 3 shows the properties of technical importance as a function of $Bi_2O_3$ content for a typical composition of microwave lithium ferrite. It is seen that density, average grain size, remanence ratio, coercive force and resonance linewidth are all affected by bismuth concentration. Addition of

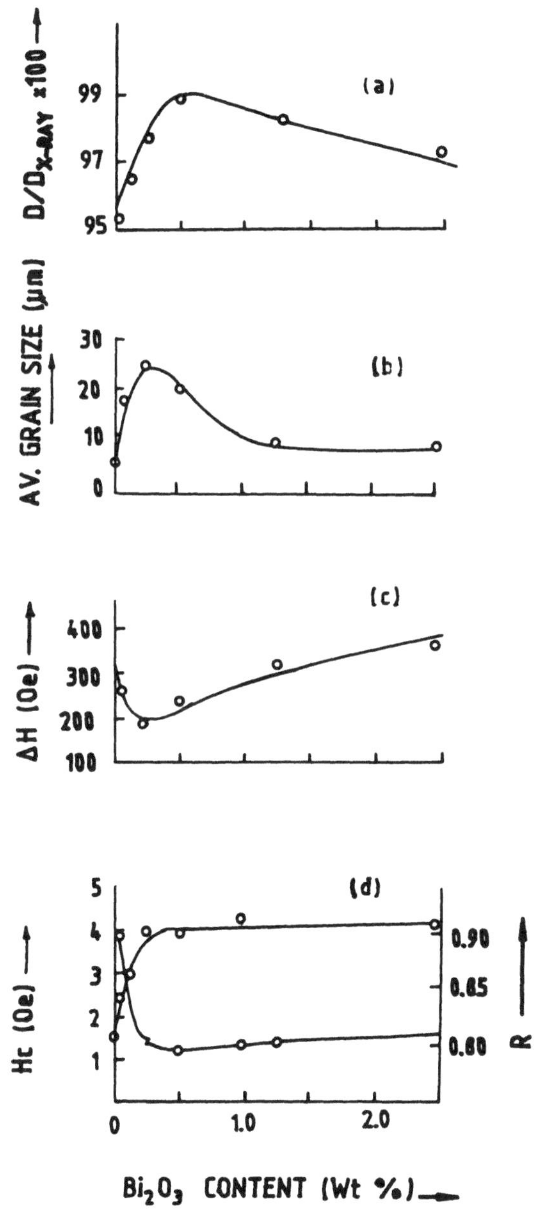


Fig. 3. Effect of bismuth oxide concentration on material properties.

$Bi_2O_3$ as a sintering aid also improves the mechanical properties of the sintered ferrite.

The thermal analysis of $Bi_2O_3$ rich mixtures with lithium ferrite shows that it melts eutectically at around 770°C while the melting point of $Bi_2O_3$ itself is 820°C. The microstructural studies by electron microprobe technique have depicted the grain boundary composition to be rich in bismuth while it does not appear in the grains. It is inferred that eutectically melting composition of $Bi_2O_3$ rich ferrite is responsible for lowering the sintering temperature and promoting grain growth by the liquid phase sintering process.

Presently addition of a small amount of $Bi_2O_3$ at the starting stage of the lithium ferrite preparation is being adopted on a routine basis. For certain cases such as low linewidth ferrites the concentration of $Bi_2O_3$ is kept at very low levels as it also acts as a unwanted non-magnetic impurity and proves deleterious for this property. All lithium ferrite compositions containing $Bi_2O_3$ are calcined at temperatures not exceeding 800°C followed by comminution and pressing. The material gets sintered to high densities in the temperature range 1000 to 1050°C with well developed grain structure.

The other aspect in the material preparation of the ferrites is the microstructure control. Several properties of the ferrites namely coercive force, remanent ratio and resonance linewidth depend to some extent on the grain structure of the material. Grain size effects on microwave ferrite magnetic properties have been described by Inui and Ogasawara [35]. Several new techniques are now being used in the ferrite preparation to achieve the desired control on the microstructure dependent parameters. For powder preparation besides conventional methods other processes like co-precipitation, cryochemical and freeze drying, solution spray draying, fluid bed etc. are used in practice. The primary aim in all these methods is to prepare thoroughly dispersed, uniform and homogeneous powders for carrying out most complete reaction to form the ferrite.

The second basic step in the preparation of polycrystalline ferrites is densification. The general approach is to compact the powder in the required shapes and fire at high temperatures in a suitable atmosphere. The firing schedule is accordingly adjusted for achieving optimal material properties. For lithium ferrites containing $Bi_2O_3$ as a sintering aid the firing is generally carried out in air. For some special purposes oxygen atmosphere is also used.

The techniques of isostatic pressing, hot pressing, isostatic hot pressing etc. are used for preparation of some special grades of materials like high power ferrites. These methods give a better control over density and size of the crystal grains.

## 5. TYPES OF LITHIUM FERRITES

As discussed in Section 3 a large variety of compositions of substituted lithium ferrites has become possible and its properties are varied over very wide ranges. For diverse microwave components operating at different frequency ranges there are several series of lithium ferrites available to the design engineer to chose from. The tailoring of material properties in lithium ferrites covers far more sets of combination of properties than is possible in any other ferrimagnetic oxide family. A number of manufacturers are now offering several different types of lithium ferrites. These types are based on the substituents employed and each one has a different set of properties with some special features.

The different grades of lithium ferrite material are broadly divided into the following categories.

### 5.1 Rectangular Loop Type

There are two series of lithium ferrite material which fall under this type. The major difference between them is the temperature stability of magnetic properties. On the basis of chemical compositions these are classified as (a) Lithium titanium ferrites and (b) Lithium titanium zinc ferrites

The LiTi series is characterised by high Curie temperatures and different grades with $4\pi Ms$ range upto 3500 Gauss are produced. All the members of the series exhibit rectangular hysteresis loop with moderately high resonance linewidths. These materials are employed in latching devices requiring high temperature stability of performance.

The LiTiZn ferrite family has a wider range of $4\pi Ms$. The higher range i.e. 4000 to 5000 Gauss materials are essentially LiZn ferrites while the lower part of series ($4\pi Ms < 3500$ Gauss) are LiTi with some low level Zn content. The addition of Zn to ferrites lowers its anisotropy and the members of this series have moderately low resonance linewidths. For obvious reasons the Tc of this series of ferrites is lower than of the corresponding LiTi ferrite series. The Tc value is highest (500°C) for the grade of material with $4\pi Ms$ around 3500 Gauss and decreases for both higher and lower $4\pi Ms$ grades. These materials are also used in latching devices and perform with a better loss characteristics.

Both LiTi and LiTiZn ferrites have low value of spin wave resonance linewidth and are suitable for low power devices.

Figure 4 gives the temperature variation of magnetisation of some of the members of the two series and bring out the difference in temperature stability in the range of interest for the device engineer.

### 5.2 High Power Lithium Ferrites

This series is derived from the LiTi and LiTiZn ferrite series modified by inclusion of some fast relaxing metal ions in the chemical composition. The spin wave linewidth is increased in this way but the accompanying increase in the magnetic loss factor and to some extent of coercive force cannot be prevented. The high remanence ratio of the parent LiTi series is retained to a large extent. High power lithium ferrites with $\Delta H_k$ upto 7 Oe with compatible loss characteristics are available.

These materials are specially suited for latching devices operating at high microwave peak power levels.

In Section 6, the high power lithium ferrites are described in a little more detail.

### 5.3 Low Linewidth Lithium Ferrites

The low resonance linewidth of lithium ferrite is obtained by lowering the

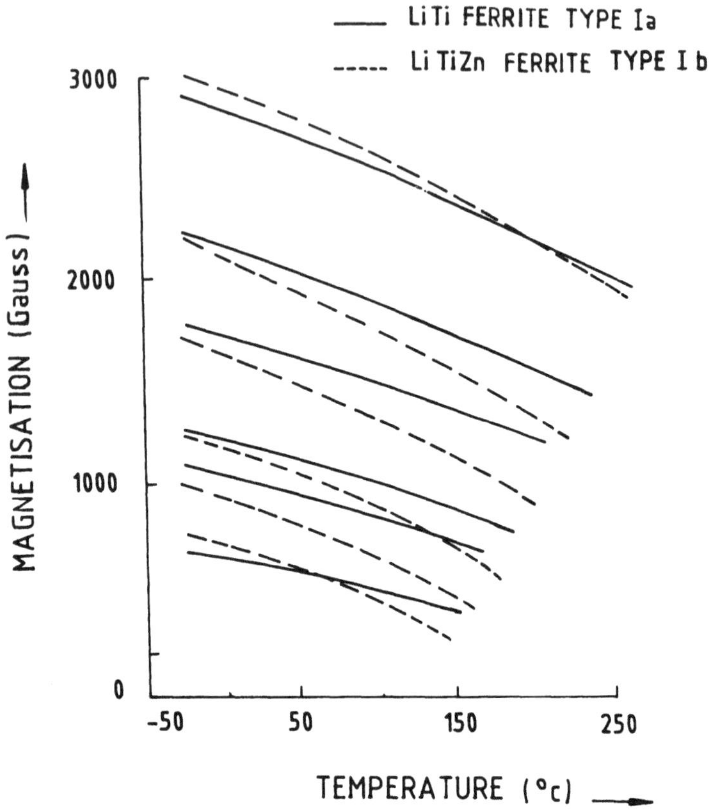


**Fig. 4.** Temperature variation of saturation magnetisation ($4\pi Ms$) for various grades of type Ia and Ib.

anisotropy of LiTiZn ferrites by appropriately enhancing the Zn concentration. For $4\pi Ms$ range upto 3000 Gauss, materials with $\Delta H < 100$ Oe have been prepared and are commercially available. Very low $\Delta H$ (25 Oe) has been achieved in low $4\pi Ms$ materials [12]. The materials of this series are no longer of rectangular loop type and the curie temperatures are also quite low compared to the other types discussed above.

This set of lithium ferrites are suited for devices like circulators and isolators and have been established as equivalent to garnet materials for use in the fabrication of a number of microwave devices. Unlike garnets, lithium ferrites are easy to prepare and do not contain any costly ingredients.

A brief summary of the properties of the various types of lithium ferrites is given in Table 1.

## 6. FOR HIGH POWER DEVICES

For enhancing the high power handling capability of ferrites generally the fast relaxing divalent cobalt ions are substituted in the ferrite composition. The $Co^{2+}$ ions occupy the octahedral sites and cause strong spin lattice interactions arising from its unquenched orbital angular momentum. These ions also contribute

**Table 1. Types of microwave lithium ferrites**

| Type | $4\pi Ms$ (Gauss) | Tc (°C) | $\varepsilon'$ | $\Delta H$ (Oe) | $\Delta H_k$ (Oe) | $\tan\delta_e \times 10^{-4}$ | $\tan\delta_m \times 10^{-4}$ | $R$ | $H_c$ (Oe) |
|---|---|---|---|---|---|---|---|---|---|
| LiTi | 800–3500 | 300–620 | 15–19 | 350–500 | 2 | < 5 | < 5 | > 0.9 | 1-2 |
| LiTiZn | 250–3000 | 120–500 | 16–20 | 100–300 | 2 | < 5 | < 5 | > 0.85 | 0.6-1.5 |
| | 4000–5000 | 520–400 | 15 | 200–400 | 2 | < 10 | — | > 0.9 | 2-4 |
| High Power | 600–2000 | 200–450 | 17–19 | 300–700 | ≤ 7 | < 5 | < 10 | > 0.85 | 1–2 |
| Low Linewidth | 800–3000 | 150–350 | 15–19 | < 100 | 2 | < 5 | ≤ 10 | — | — |

strong positive anisotropy. The spin wave resonance linewidth ($\Delta H_k$) of the ferrite is found to increase by almost an order with extent of the substitution of $Co^{2+}$ as small as 0.05 ions/formula unit (FU) in the chemical composition (Fig. 5).

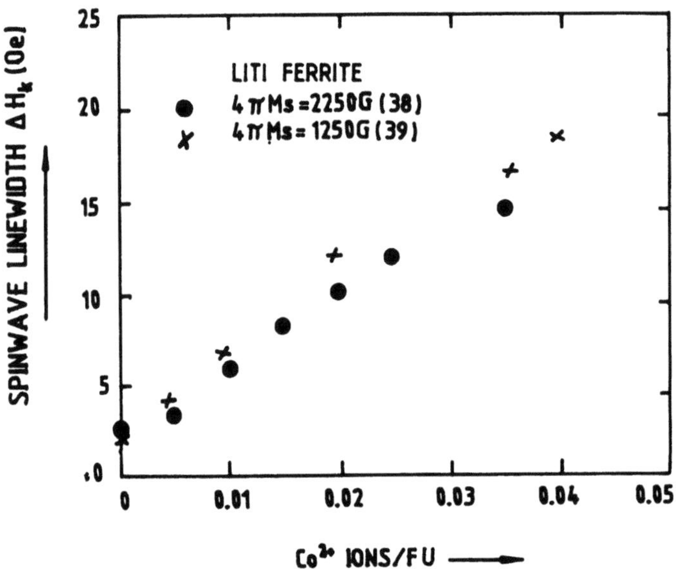


**Fig. 5.** Variation of spinwave resonance linewidth ($\Delta H_k$) with cobalt content.

Bannerjee et al [36], Brower and Patton [37] and later Dionne [38] have observed an apparent paradox in the properties of $Co^2$+ substituted lithium ferrites. The presence of $Co^{2+}$ in the ferrite has simultaneously two types of effects namely through anisotropy and relaxation. An anisotropy cancellation is expected at a particular cobalt concentration, and a minimum in resonance linewidth should be observed. The hysteresis loop properties are also supposed to reflect this anisotropy effect. These authors have reported that such a situation is not observed in titanium substituted lithium ferrites. A new model of cation clustering by $Co^{2+}$ has been put forward for understanding the observed behaviour of anisotropy effects [38].

Figure 5 shows that $Co^{2+}$ concentration of 0.01 ions per formula unit increases $\Delta H_k$ by more than two fold [38, 39]. A simultaneous increase of magnetic losses also occurs. For devices working at moderate power levels, it would be desirable to introduce less than 0.01 ions/FU ($Co^{2+}$) for synthesising the optimum material. The control of cobalt content in these small concentrations on reproducible basis in the preparation of ferrites is found to be rather difficult, $\Delta H_k$ values being extremely sensitive to $Co^{2+}$ concentration. In the recent work [40, 41] an alternate course of aluminium substitution in LiTi ferrites has been suggested. $Al^{3+}$ content upto 0.2 ions per formula unit have been used and results obtained on several properties like spin wave resonance linewidth, resonance linewidth, remanence ratio and coercive force are given. Fig. 6 depicts

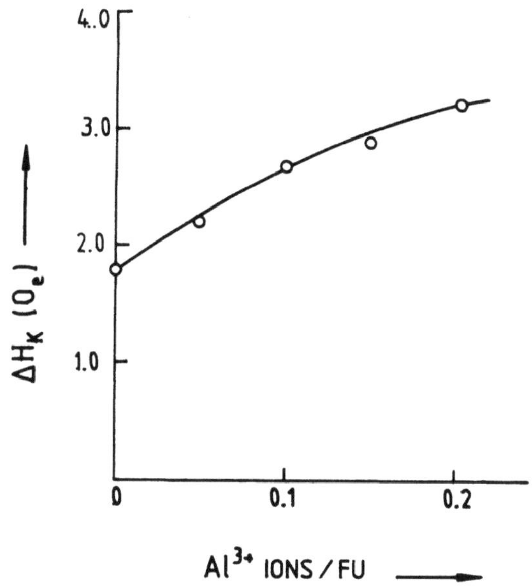


**Fig. 6.** Variation of spinwave resonance linewidth ($\Delta H_k$) with aluminium content.

the variation of $\Delta H_k$ with the Al concentration, the value increasing in a controlled manner from 1.8 to 3.2 Oe for Al concentration varying from 0 to 0.2 ions/FU. It is also noticed that there is a tendency of $\Delta H_k$ to attain a saturated value for higher Al concentrations. In comparison the ferrite containing Co shows monotonic rise in $\Delta H_k$. The effect of aluminium concentrations on $4\pi$Ms and Tc is rather large and has to be taken into consideration while arriving at the appropriate composition by suitably adjusting the other cationic contents like titanium.

The aluminium and cobalt substituted lithium ferrites have been evaluated in reciprocal phase shifter configuration at moderately high peak power levels. Fig. 7 shows the experimentally observed results [41]. For a composition free of both $Al^{3+}$ and $Co^{2+}$ the onset of nonlinear effect occurs at around 100 Watt peak power while it is higher than 300 W for aluminium containing ferrite (0.1 ions/FU). The $Co^{2+}$ (0.004 ions/FU) is also effective in raising the threshold power level but it is accompanied by some increase in insertion loss.

## 7 SINGLE CRYSTALS AND FILMS

Single crystal garnets have found extensive use in the microwave components such as tunable oscillators, filters, limiters etc. YIG and its substituted compositions grown mostly by solution growth methods are generally used in the form of small spheres. Other single crystal materials like lithium ferrites, having higher magnetisation values and technologically important for high microwave frequencies, have not matched the quality of the single crystal YIGs. Schloemann and Blight [42] have made use of lithium ferrite single crystals for enhancing the usable bandwidths of stripline circulators.

The epitaxially grown films of ferrites are already finding applications in

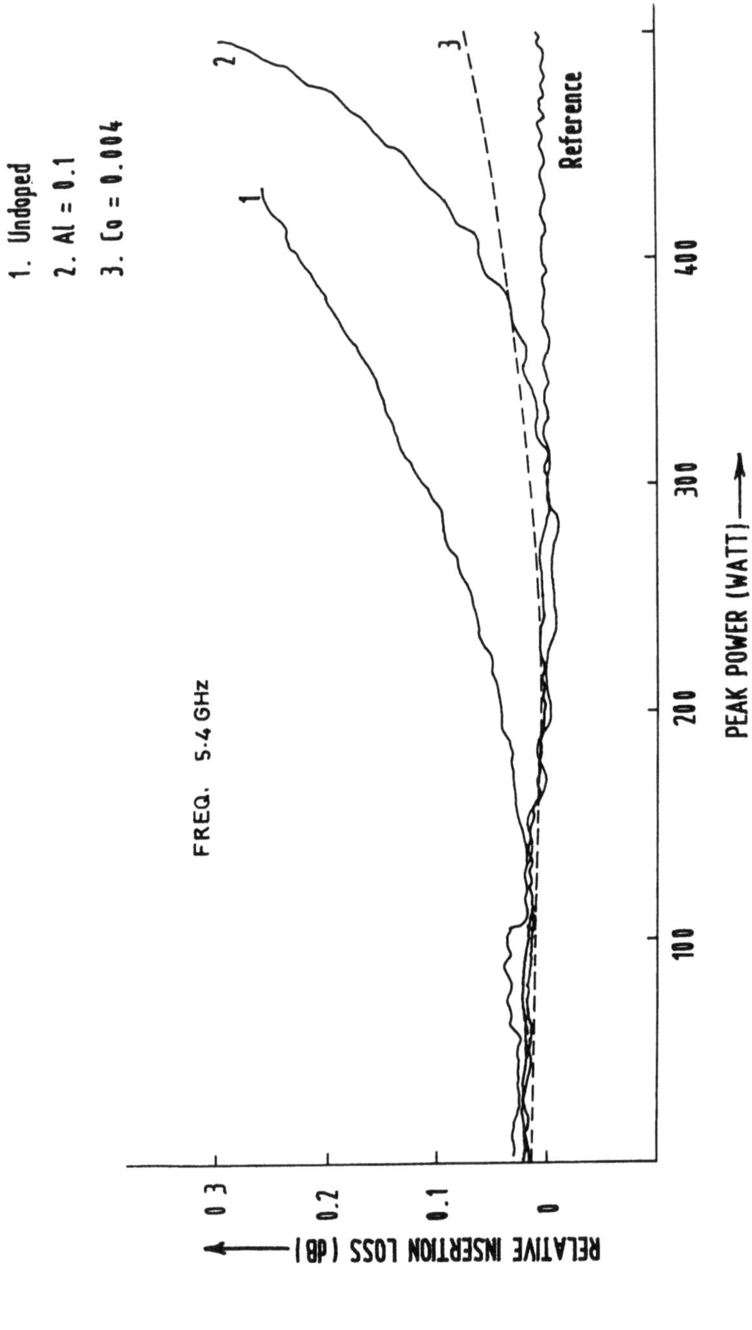


**Fig. 7.** Peak power threshold results.

magnetostatic wave and bubble memory devices. But here also it is the YIG material films, grown by liquid phase epitaxy (LPE) methods, which have dominated the scene. It has been established that thin film approaches are highly attractive for microwave and millimeter wave devices but the problems connected with producing films of required thickness and quality have restricted the desired advancement. The motivation for thin film ferrite devices is also provided by cost, size and weight considerations. There are several thin ferrite film microwave devices reported in literature having potential for competing with single crystal devices [42, 43]. The different aspects of ferrite thin films are described by Glass [44].

The growth of lithium ferrite and lithium ferrite-aluminate films by LPE has been investigated by Glass et al [45] and Van der Straten et al [46] respectively. Interest in ferrite films has also extended to polycrystalline materials. Thick films deposited from pastes containing lithium ferrite have been evaluated in microstrip edge guided isolators and circulators [47]. Other techniques attempted include ferrite plating [48] and arc plasma spray. Millimeter wave ferrite phase shifters have been demonstrated by employing the latter method [49]. Experiments have been carried out on deposition of ferrite films on GaAs substrates [50].

The rapidly growing field of monolithic microwave integrated circuits (MMICs) has created renewed interests in development of ferrite film devices. From the viewpoint of compatibility it is desired that both ferrite and semiconductro devices are fabricated on the same chip for realizing monolithic integration. In this connection the three possible approaches for deposition are-ferrite film on semiconductor substrate, semiconductor film on ferrite substrate or both semiconductor and ferrite films on a mutually compatible substrate. As mentioned above attempts have been initiated for deposition of ferrite films on GaAs substrates, but the results are not very satisfactory. Other approaches are also in a stage of infancy.

For deposition of single crystal films lattice match between the substrate and film material is a basic requirement. The lattice constant of ferrite spinels is ~ 8.4 Å and of III-V compounds is ~ 5.9 Å. The mismatch of lattice is quite obvious. Schloemann [50] has pointed out that the lattice can match along a [100] interface with the two lattices rotated by 45° with respect to each other. In such a case the ratio of the two lattices should be $\sqrt{2}$ for a perfect match. This requirement is coincidentally met by lithium ferrite (8.33 Å) and GaAs (5.64 Å)/InP (5.87 Å), the ratio being quite close to $\sqrt{2}$. It has been further suggested that any residual mismatch of a few percent between the two lattices can be absorbed by the deposition of suitable buffer layers. This approach is expected to provide some interesting results for MMIC compatible ferrite components.

## 8. FOR PHASE SHIFTER APPLICATIONS

Microwave ferrite phase shifters are extensively used for switching and control of microwave signals in radar and communication systems. Their major field of application is phased array antennas for electronic scanning at frequencies above 3 GHz. These devices provide greater precision of operation than semiconductor devices and simultaneously provide lower losses and higher power ratings. However greater size, higher weight and lower switching time have to be tolerated in ferrite based devices. A single axis scanning antenna is made up of tens to hundreads of phase-shifters while a two axes antenna comprises of a few thousand elements.

The three types of ferrite phase shifters which are considered suitable for use in phased array antennas are:

1. Non-reciprocal toroidal latching type in rectangular wave-guide [52–54]. It finds use in wideband, fast switching applications.
2. Reciprocal dual mode type in circular or rectangular wave-guide [55–56]. It is considered ideal for high frequency (above 5 GHz), short range two axes scanning radar.
3. Rotary field type in circular waveguide [57]. This phase shifter is suited for low side lobe, single axis scanning radar and is characterised by high accuracy and high rf power handling capability.

The first two types i.e. toroidal and dual mode are currently being employed in phased arrays. Both operate with remanent flux in the phasor section. The different remanent states are obtained by the application of a special driver circuit. An interface is used between the system and the device for converting the digital command into current/voltage pulse which drives it to different states. The phase changes are obtained by resetting to the reference limit point B on the hysteresis loop (Fig. 2) through the saturation point A by applying a large negative pulse. It is further driven up the major loop to a set point C for the desired phase state.

The performance of a phase shifter is generally characterised by parameters such as phase-shift range, insertion loss, frequency range, switching time, switching energy, phase accuracy, peak power rating and temperature stability besides cost, size and weight. The properties of the ferrite material used has a bearing on most of these parameters. Different types of ferrite materials employed in fabrication of these phase shifters are, in general, required to have rectangular hysteresis loop with low coercive force and low loss factors. Magnesium-manganese ferrites are often used but suffer from lack of temperature stability and low rf peak power handling capability. The polycrystalline garnets generally do not have high remanence ratio but with suitably substituted compositions good temperature stability and high power performance can be obtained. Lithium ferrites with the versatility of tailoring the properties over wide ranges by suitable compositional variations have emerged as the most suitable materials for these phase-shifter applications. The important features of the lithium ferrites responsible for this are (i) upper limit of $4\pi Ms$ value upto 5000 Gauss which makes them particularly useful for higher frequency ranges above the X-band region (ii) the high dielectric constant ($\varepsilon'$) values, (iii) better temperature stability due to high Curie temperatures, (iv) possible tailoring of properties for enhancement of peak power handling capability and (v) low cost. Some of these features have been described earlier in Section 1.

The LiTi ferrites of type Ia (Section 5) are more often used due to their better temperature stability. For phase shifter with higher peak power specifications lithium ferrites of type II are preferred. Based on the design consideration in a C-band phase-shifter lithium ferrite of $4\pi$Ms 1200 Gauss is used while an X-band phase shifter employs material with $4\pi$Ms around 2000 Gauss. As a thumb rule the magnetisation of a material selected for the phasor is about one-fifth of the centre frequency of the device consistant with the relation $4\pi Ms/\omega \ll 1$. In high frequency phase-shifters operating at 35 GHz and above ferrites of 5000 Gauss are used, limited by the highest value of magnetisation available among the ferrites.

The shaping and machining of the ferrite in the various forms like toroids, rods, yokes etc. for phase shifters is an important step in the use of ferrites for this application. The production methods have been established for uniform reproducibility alongwith good mechanical properties in the various configurations of lithium ferrite materials.

The applications of both reciprocal and non-reciprocal phase shifters have been extended as control elements in variable power dividers [57, 58]. Further advancement of these devices for antennas having polarization agility development of a simultaneous dual polarization phase shifter has been reported [59].

The toroidal non-reciprocal and dual mode reciprocal, largely based on lithium ferrites, presently represent the major ferrite device market. A number of phased array antenna programmes (e.g. PATRIOT and AEGIS) have boosted the production of ferrite phasors [60].

## REFERENCES


1. T. Collins and A.E. Brown, J. App. Phys., **42**, 3451, 1971.
2. P.D. Baba, G.M. Argentina, W.E. Courtney, G.F. Dionne and D.H. Temme, Trans. IEEE, MAG-**8**, 83, 1972.
3. G.F. Dionne Proc. IEEE, **63**, 777, 1985.
4. P.B. Braun, Nature, **170**, 1123, 1952.
5. G.O. White, C.A. Edmondson, R.B. Goldfarb and C.E. Patton, J. Appl. Phys., **50**, 238, 1978.
6. R.E. Koelberger, G.O. White and C.E. Patton, J. Phys. (Paris), **38** (Suppl. 4), Cl–255, 1977.
7. G.F.Dionne, J. App. Phys., **40**, 4486, 1969.
8. Y.M. Yakovlev, E.V. Rubalskaya and B.L. Lapovok, Sov. Phys.—Sol. State, **10**, 2301, 1969.
9. G. Blasse, Phil. Res. Rep. (Suppl.), **8**, 98, 1964.
10. R.G. West and A.C. Blankenship, J. Am. Cer. Soc., **50**, 347, 1967.
11. J.V. Young and J. Smit, J. Appl. Phys., **42**, 2344, 1971.
12. M. Sun and Y. Xiu, Trans. IEEE, MAG-**16**, 672, 1980.
13. G.F. Dionne, J. Appl. Phys., **45**, 3621, 1974.
14. S.M. Zhilyakov, V.V. Ivolga, V.I. Mal 'tsev and E.P. Naiden, Sov. Phys.—Sol. State, **19**, 1817, 1977.
15. J.L. Dorman, T. Merceron and M. Nogues, Proc. Int. Conf. On Mossbauer Effect (Jaipur, India) 193, 1981.
16. C.E. Patton, C.A. Edmondson and Y.H. Liu, J. App. Phys., **53**, 2431, 1982.

17. M. Rosenberg, P. Deppey, S. Dey, U. Jansen, C.E. Patton and C.A. Edmondson, Trans. IEEE, **MAG-18**, 1616, 1982.
18. Pran Kishan and K.K. Laroia, Adv. in Cer., **15**, 299, 1985.
19. S. Phanjoubam, D. Kothari, J.S. Baijal and Pran Kishan, Sol. State. Commn., **68**, 549, 1988.
20. J. Smit and H.P. Wijn, Ferrites, Wiley NY, 1959.
21. P.K. Baltzer, US Patent 3034987, 1962.
22. R.G. West, J. App. Phys., **34**, 1113, 1963.
23. M.P. Bunina, E.V. Lebodera, A.I. Pilshchikov and I.I. Silvestrobich, Sov. Phys.—Sol. State **12**, 341, 1970.
24. H. Wang et al Acta Electronica Simica, No. 1, 64, 1979.
25. G. Bandopadhyaya and R.M. Fulrath, J. Am. Cer. Soc., **57**, 483, 1974.
26. H.B. Im, Proc. ICF-3 (Kyoto, Japan), 281, 1980.
27. H.T. Kim and H.B. Im, Trans, IEEE, **MAG-18**, 1541, 1983.
28. V.G. Glotov, Porosh Met. **10**, 27, 1970.
29. K.D. Dugar Zabon, Porosh Met., **11**, 60, 1971.
30. P. Peshev and H. Pacheva, Mat. Res. Bull., **15**, 1055, 1980.
31. P. Peshev and H. Pacheva, Mat. Res. Bull., **15**, 1199 1980.
32. J.J.Green and H.J. Van Hook, Trans. IEEE, **MTT-25**, 155, 1977.
33. W. Simonet and A. Hermosin, Trans. IEEE, **MAG-14**, 903, 1978.
34. Pran Kishan, D.R. Sagar, S.N. Chatterjee, L.K. Nagpaul, Nitendar Kumar and K.K. Laroia, Adv. In Cer. **15**, 207, 1985.
35. T. Inui and N. Ogasawara, Trans, IEEE, **MAG-13**, 1729, 1977.
36. S.K. Bannerjee, P.D. Baba, B.J. Evans and S.S. Hafner, J. Phys. (Paris) Colloq. C. 145, 1971.
37. C.J. Brower and C.E. Patton, J. Appl. Phys., **53**, 2104, 1982.
38. G.F. Dionne, J. Appl. Phys., **57**, 3727, 1985.
39. B.K. Kuanr, P.K. Singh, Pran Kishan And G.P. Srivastava, Proc. ICF-5 (Bombay, India), 1005, 1989.
40. Pran Kishan, N. Kumar, B.S. Matheru, Chandra Prakash, K.K. Jain and G.P. Sharma, Proc. ICF-5 (Bombay, India), 949, 1989.
41. Pran Kishan, G.P. Sharma, Nitendar Kumar, K.K. Jain and B.K. Kuanr, Proc. Tenth Int. Conf. on Microwave Ferrites (Poland), Sept. 1990.
42. E. Schloemann and R.E. Blight, Trans. IEEE, **MTT-34**, 1394, 1986.
43. D.C. Webb, Trans. IEEE, **MAG-34**, 2799, 1988.
44. H.L. Glass, Proc. IEEE, **86**, 157, 1988.
45. H.L. Glass, F.S. Stearns and L.R. Adkins, Proc. ICF-3 (Kyoto, Japan), 39, 1980.
46. P.J.M. van der Straten and R. Metselaar, J. Cryst. Growth, **48**, 114, 1980.
47. C.K. Maiti, D. Bhattacharya and N.B. Chakravarti, Trans. IEEE, **CHMT-8**, 221, 1985.
48. M. Abe, Y. Tamaura, Y. Goto, N. Kiramura and M. Gomi, J. App. Phy., **61**, 3211, 1987.
49. E.W. Babbit and R.A. Stern, Trans. IEEE, **MAG-15**, 1744, 1979.
50. M. Abe, T. Itoh, Y. Tamaura, Y. Gotoh and M. Gomi, Tran. IEEE, **MAG-23**, 3736, 1987.
51. E.F. Schloemann, Proc. IEEE, **76**, 188, 1988.
52. M.A. Truehaft and L.M. Silber, Proc. I R E, **48**, 1538, 1958.
53. W.J. Ince and E. Stern, Trans. IEEE, **MTT-15**, 87, 1967.
54. G. Cattarin, C.D. Gregorio, U. Pattrucci, F. Pucci and R. Roveda, Rev. Tec. Selenia, **8**, 29, 1982.
55. C.R. Boyd, Trans. IEEE, **MTT-18**, 1119, 1970.

56. C.R. Boyd, Microwave J., **20**, 41, 1977.
57. D.S. Beese and D.E. Sharon, IEEE, MTT-S, Int. Microwave Symp. Digest, 333, 1983.
58. C.R. Boyd, IEEE, MTT-S, Int. Microwave Sym. Digest, 735, 1986.
59. W.E. Hord and C.R. Boyd, IEEE, MTT-S, Int. Microwave Sym. Digest, 727, 1986.
60. G.P. Rodrigue, Proc. IEEE, **76**, 121, 1988.

# 7

# Single Crystal YIG and Allied Materials

T.R.N. Kutty
Materials Research Centre, Indian Institute of Science, Bangalore, India

S. Sundaram
Defence Electronics Research Laboratory, Hyderabad, India

## 1. INTRODUCTION

During the course of a wide search for the preparation of a magnetic material of better performance than the conventional ferrites, Bertaut and Forrat [1] found in 1956, that a mixed oxide of yttrium and ferric iron had peculiar ferrimagnetic properties. It was found to have extremely low ferrimagnetic line width and high saturation magnetization. In addition, it was found to be a good insulator with comparatively very low electrical conductivity. This is an ideal combination of properties indeed, which enables its application in magnetically tuned devices useful throughout the microwave region.

It may be of interest to point out that such extraordinary and useful properties were observed with the mixed oxides ($Y_2O_3$ and $Fe_2O_3$) by Pauthenet (1956) [2] while he was working for his doctorate dissertation. He found very low magnetic line width ($\Delta H$) of 5.7 Oe at 3.93 GHz for sintered mixed oxides of iron and yttrium in the polycrystalline form. This unusual behaviour triggered intense activity in this area during the last three decades [3–5].

On examination, the mixed oxide polycrystalline material was found to have a structure identical with a garnet, in contrast to that of the spinel structure of conventional ferrites. Garnets are a class of compounds with the general formula $X_3Y_2(ZO_4)_3$. In natural garnets, X will be a divalent element such as magnesium $Mg^{++}$), iron ($Fe^{++}$), calcium ($Ca^{++}$), manganese ($Mn^{++}$), etc. Y will be invariably aluminium ($Al^{+++}$), iron ($Fe^{+++}$) and chromium ($Cr^{+++}$), Z is found to be silicon. To represent the oxidation state of the component elements in such natural garnets, they are called 2 : 3 : 4 garnets. In addition to such a combination as is found in natural garnets, it is possible to have cations with different oxidation states in synthetic garnets. One of the most important class of garnets containing rare earth elements correspond to 3 : 3 : 3. All the three X, Y, Z components will have the oxidation state of + 3 only.

Among the popular garnets, the following may be mentioned in connection with the application to the fields of optics, radar and communication.

| X | Y | Z | Formula | Acronym |
|---|---|---|---|---|
| Y | $Fe^{+++}$ | $Fe^{+++}$ | $Y_3Fe_2\ (FeO_4)_3$ | YIG |
| Y | $Al^{+++}$ | $Al^{+++}$ | $Y_3Al_2\ (AlO_4)_3$ | YAG |
| Gd | $Fe^{+++}$ | $Fe^{+++}$ | $Gd_3Fe_2\ (FeO_4)_3$ | GdIG |
| Gd | $Al^{+++}$ | $Al^{+++}$ | $Gd_3Al_2\ (AlO_4)_3$ | GdAlG |
| Gd | $Ga^{+++}$ | $Ga^{+++}$ | $Gd_3Ga_2\ (GaO_4)_3$ | GGG |

Varieties of combination of rare earth oxides and oxides of other metals of appropriate valence have been tried to prepare mixed oxide crystals. Immense amount of data have accumulated in literature concerning the preparation, properties and technical applications of garnets.

A range of single crystal substituted hexagonal ferrites [6, 7, 8] having a large number of phases has also been introduced in this decade for millimeter wave tunable applications.

This chapter deals with magnetic properties of single crystal and YIG allied materials, methods of preparation, their characterisation and typical applications.

## 2. APPLICATIONS OF YIG AND OTHER ALLIED MATERIALS

The usefulness of rare earth iron garnets (ReIG) as magnetic material started off on account of the low ferrimagnetic line width ($\Delta H$) which enabled realization of microwave components [9, 10] such as bandpass and band-stop filters. YIG tuned oscillators have found wide applications in modern electronic instrumentation and systems due to their broad tuning range, excellent tuning linearity and "clean" spectrum. ReIG have also found applications in harmonic generators, limiters and also as programmable frequency sources/counters for high resolution radar systems.

YIG spheres are commonly used as resonator elements in the above devices. The tiny YIG sphere is the solid state equivalent of the microwave cavity resonator. For realization of tunable filters above 30 GHz, conventional single crystal YIG materials have got limitations and new resonator materials based on hexagonal barium ferrites are used upto 75 GHz.

In an endeavor to harness its excellent acoustic properties (very low acoustic loss surpassing that of quartz), YIG is also shaped into rods and plates for delay devices.

Exceptional magnetostatic properties of YIG have been exploited for the wave propagation at microwave frequencies. Consequently many signal processing functions have been realised which include (i) Pulse compression, (ii) Variable delay, (iii) Storage, (iv) Parametric amplification, (v) Pulse recall and (vi) Delay lines with built-in isolation.The magnetostatic mode in YIG anteceded indeed practical magneto acoustic behaviour. Optically, rare earth iron garnets have interesting Faraday rotation properties. Apart from their application in the form of spheres, rods and plates, rare earth iron garnets have found extensive applications in the form of thin layers in several devices. A layer of iron garnet of a few micron thickness is grown epitaxially on a non-magnetic single crystal substrate such as gadolinium gallium garnet (GGG). This is used in the fabricaion of magnetic memory domain devices. Such thin layer garnets have been extended for the fabrication of a variety of surface devices.

In addition to the above practical applications, rare earth iron garnet crystals provide basic parameters for theoretical predictions. It is possible to substitute various metal ions (both magnetic and non-magnetic) at unambiguous sites in the crystal and measure magnetic properties. These results could be verified experimentally with the theoretically predicted values such as g-values, ferri-magnetic resonance line widths and giant anisotropy. Thus, it is possible to prepare tailor-made magnetic crystals of desired saturation magnetisation ($4\pi Ms$), compensation temperature and magnetic anisotropy.

## 3. CRYSTAL STRUCTURE OF GARNET

Garnet crystals belong to body centered cubic system [11–13], corresponding to the space group Ia3d. There are 8 formula units per unit cell. It means that there are 160 atoms in the unit cell. The crystal structure affords three types of non-equivalent sites in the oxygen lattices which are occupied by metal ions. The formula of a garnet may accordingly be represented by $C_3A_2$ $(DO_4)_3$. Figs. 1 and 2 give the general arrangement of coordination polyhedra in the garnet structure. The C ions are arranged on a bcc sublattice; they are 8-coordinated ions, occupying the center of a distorted dodecahedral polyhedra of oxygen atoms.

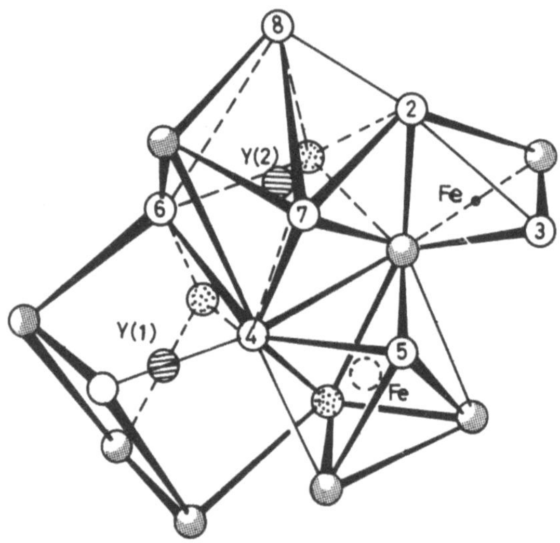


**Fig. 1.** Portion of the garnet structure with cation and oxygen atoms.

There are 24 C sites in each unit cell of garnet and have the point group 222 ($D_2$). A metal ion in a-site is surrounded by 6 oxygen ions which are at the corners of an octahedron. There are sixteen a-sites in a unit cell which have the local symmetry of point group 3($S_6$). The 4-coordinated D ions have a distorted tetrahedral symmetry and these (d) sites correspond to the point group 4 ($S_4$).

There are 24 d-sites in one unit cell. These, together with 96 h-sites occupied by oxygen ions with triangular coordination, constitute 160 atoms of the garnet

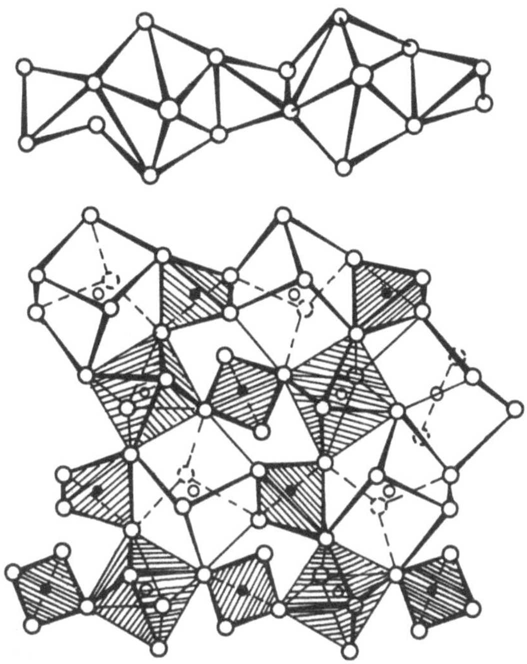

**Fig. 2.** (a) Corrdination polyhedra in garnet structure; (b) Alternating edge sharing tetrahedra and dodecahedra chains in garnet.

cell. We can take yttrium iron garnet as a typical example and we find individual ($FeO_4$) tetrahedra share corners with ($FeO_6$) octahedra. Both these polyhedra share edges with ($YO_8$) dodecahedra.

## 4. MAGNETIC PROPERTIES OF GARNETS

### 4.1 Saturation Magnetisation ($4\pi M_s$)

The basic requirement for a garnet to be magnetic is that it should have magnetic catiohs in sufficient proportions at d and a-sites. The d-site cations are grouped into a sublattice and they are coupled ferromagnetically. The a-site and c-site cation sublattices are similarly coupled ferromagnetically. On the other hand, the d-site cation sublattice is antiferromagnetically coupled to a-site sublattice. The net magnetic moment is given by the algebraic sum of all the contributions made by the ions in three different sublattices.

$$M = M_d - M_a - M_c \qquad (1)$$

where $M_d$, $M_a$ and $M_c$ represent the respective sublattice magnetic moments.

In the case of yttrium iron garnet the yttrium ion is diamagnetic, so that the magnetism is due to the trivalent iron ions only. The iron occupies both 16 (a) sites and 24 (d) sites while yttrium occupies the 24(c) sites. The net magnetic moment of YIG per molecule will therefore be:

$$M = 3Fe^{3+} - 2Fe^{3+} = 5\mu_B \qquad (2)$$

This net magnetic moment of YIG shows typical Brillouin functional behaviour. Substitution of paramagnetic or diamagnetic ions at different sites modifies the magnetic properties of garnets. Any of the rare earth ions with an atomic number above 61 (the rare-earth ions below samarium apparently have ionic radii too large to fit into the garnet structure) can be substituted for yittrium. For instance, if $Gd^{3+}$ is substituted at c-sites, the term $M_c$ enters in Eq. (1). The magnetic moment will be aligned antiparallel to the moment of the ions on the (d) sites. It will diminish the net magnetic moment $M$. It is, therefore, reasonable to expect that the saturation magnetisation $4\pi M_s$ for gadolinium doped YIG will be lower than that of pure YIG and it has been found to be so.

In addition, such doped YIG crystals are observed to show variation in the saturation magnetisation ($4\pi M_s$) with temperature. There is a temperature at which $M = 0$ and this is called compensation temperature ($T_{comp}$). This $T_{comp}$ shifts to higher values with increased substitution of $Gd^{3+}$ for yttrium. Above $T_{comp}$, the contributions from c-sublattice diminishes.

Substitution of non-magnetic ions such as $Al^{3+}$ at d-sites of YIG, provides a means to regulate the shape of $4\pi M_s$ vs $T$ curves (Fig. 3).Thus, garnets with

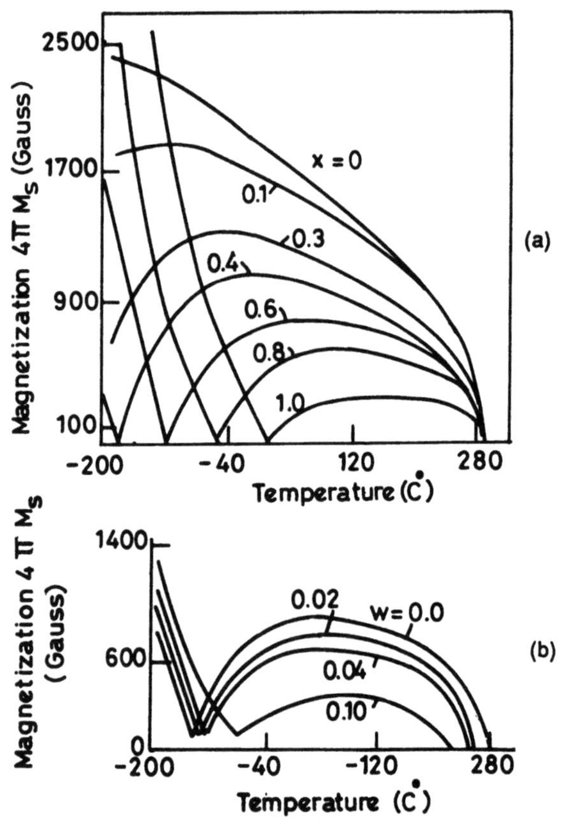


Fig. 3. Saturation magnetization of substituted garnets: (a) Saturation magnetization of yttrium gadolinium iron garnet $\{Y_{3-3x}Gd_{3x}\}$ $[Fe_2]$ $(Fe_3)$ $O_{12}$; (b) Saturation magnetization of yttrium gadolinium aluminium iron garnet $\{Y_{1.5}Gd_{1.5}\}$ $[Fe_2]$ $(Fe_{3-5}wAl_5w)$ $O_{12}$.

minimum temperature coefficient for $4\pi M_s$ values can be obtained by the combined substitution at c and d-sites. It is also possible to substitute a-sites with metals like gallium, indium and other trivalent elements. Such substitutions also will modify the magnetic properties. With appropriate substitutions it is possible to prepare a variety of garnet magnetic materials suitable for specific purposes. A list of such materials reported in literature is given in Tables 1 and 2.

**Table 1. Reported major substitutions in garnets**

| Ion | Probable ionic radii (Å) | Example |
|---|---|---|
| (a) Substitution at d-sites (Coordination No. 4) | | |
| $Al^{3+}$ | 0.39 | Al-garnet |
| $Fe^{3+}$ | 0.49 | Fe-garnet |
| $Ge^{4+}$ | 0.40 | Ge-garnet |
| $Si^{4+}$ | 0.26 | Si-garnet |
| $Ga^{3+}$ | 0.47 | Ga-garnet |
| $V^{5+}$ | 0.355 | $Ca_3Fe_2\ (Fe_{1.5}V_{1.5})\ O_{12}$ |
| $P^{5+}$ | 0.17 | $(NaCa_2)\ Mg_2P_3O_{12}$ |
| $As^{5+}$ | 0.335 | $(NaCa_2)\ Mn_2\ (As_3)O_{12}$ |
| $Li^{+}$ | 0.59 | $Na_3Al_2Li_3F_{12}$ |
| (b) Substitution at a-sites (Coordination No. 6) | | |
| $Ce^{3+}$ | 1.03 | YFe-garnet |
| $Dy^{3+}$ | 0.91 | Ge-garnet |
| $Ho^{3+}$ | 0.90 | " |
| $Y^{3+}$ | 0.89 | " |
| $Er^{3+}$ | 0.89 | " |
| $Tm^{3+}$ | 0.87 | " |
| $Yb^{3+}$ | 0.86 | " |
| $Ln^{3+}$ | 0.85 | " |
| $Mn^{2+}$ | 0.82 | $(CaY_2)\ (Mn_2)\ (Ge_3)\ O_{12}$ |
| $In^{3+}$ | 0.80 | YFe-garnet |
| $Fe^{2+}$ | 0.78 | YFe-garnet containing Si |
| $Zn^{2+}$ | 0.75 | $Gd_3Zn_2\ (GaGe_2)\ O_{12}$ |
| $Sc^{3+}$ | 0.73 | Ge-garnet |
| $Mg^{2+}$ | 0.720 | $(MgGd_2)\ (Mg_2)\ Ge_3O_{12}$ |
| $Zr^{4+}$ | 0.71 | $Ca_3Zr_2\ (V_{0.5}Ga_{2.5})\ O_{12}$ |
| $Sn^{4+}$ | 0.69 | $Ca_3Sn_2\ (Ga_2Sn)\ O_{12}$ |
| $Mn^{3+}$ | 0.65 | $(MnGd_2)\ (CoMn)\ Ge_3O_{12}$ |
| $Fe^{3+}$ | 0.64 | Fe-garnet |
| $No^{5+}$ | O.64 | $Ca_3\ (NbHf)\ Ga_3O_{12}$ |
| $Ga^{3+}$ | 0.62 | Fe-garnet, Ga-garnet |
| $Cr^{3+}$ | 0.615 | YFe-garnet |
| $Al^{3+}$ | 0.53 | Al-garnet |
| $Co^{3+}$ | 0.53 | $Ca_3\ (Sc_{1.8}Co_{0.2})\ Ge_3O_{12}$ |
| (c) Substitution at c-sites (Coordination No. 8) | | |
| $Sr^{2+}$ | 1.25 | $Sr_3Sc_2Ge_3O_{12}$ |
| $Ca^{2+}$ | 1.12 | Si-, Ge-garnet |
| $Bi^{3+}$ | 1.11 | YFe-garnet |

*(Contd.)*

| | | |
|---|---|---|
| $Sm^{3+}$ | 1.09 | Fe-, Ga-garnets |
| $Eu^{3+}$ | 1.07 | Ga, Fe-garnet |
| $Cd^{2+}$ | 1.07 | Si, Ge, Fe-garnets |
| $Gd^{3+}$ | 1.06 | Al-, Ga-, Fe-garnets |
| $Tb^{3+}$ | 1.04 | Al-, Fe-, Ga-garnets |
| $Dy^{3+}$ | 1.03 | Al-, Fe-, Ga-garnets |
| $Ho^{3+}$ | 1.02 | " |
| $Y^{3+}$ | 1.02 | " |
| $Er^{3+}$ | 1.00 | " |
| $Tm^{3+}$ | 0.99 | " |
| $Yb^{3+}$ | 0.98 | " |
| $Ln^{3+}$ | 0.97 | " |
| $Mn^{2+}$ | 0.93 | Si-garnet, $(MnY_2)$ $(Mn_2)$ $Ge_3O_{12}$ |
| $Mg^{2+}$ | 0.89 | $Mg_3Al_2Si_3O_{12}$ |
| $Co^{2+}$ | Unknown | $(CoGd_2)$ $Co_2Ge_3O_{12}$ |
| $Cu^{2+}$ | " | $(CuGd_2)$ $Mn_2Ge_3O_{12}$ |
| $Fe^{2+}$ | " | $Fe_3Al_2Si_3O_{12}$ |

## 4.2 Ferrimagnetic Resonance Line Width ($\Delta H$)

It has already been mentioned that in a ferrimagnetic material (which is also a good electrical insulator) like YIG, a net magnetic moment exists as a result of the coupling between various sublattices. The resultant magnetic moment can be aligned with an externally applied quasi static magnetic field ($H_0$). If an alternating magnetic field (such as microwave – RF frequency) is simultaneously applied perpendicular to the static field, the magnetic moment of the crystal will precess around the static field. The rate at which the magnetic moment of the material precess depends on the fundamental constants of the material and magnitude of the static field. If the frequency of the RF field coincides with the internal precession frequency, there is a strong interaction called 'Ferromagnetic Resonance' (see Figs. 4 and 5). The precession angle is controlled by the strength of the RF field, coupled to the material, but is limited to a maximum. The resonance (absorption)peak will have a line width characteristic of the chemical composition as well as the physical condition of the specimen.

This parameter is the important guiding property for its applications in devices. Ferromagnetic resonance absorption line width is expected to be the lowest for the best performance of the material. Such materials have important engineering advantages for certain microwave ferrite devices.

Narrowness of ferrimagnetic resonance line width $\Delta H$, is the major factor which makes rare earth iron garnets more useful in microwave devices. This is in general due to the perfect ordering of the constituent atoms in the lattice. There are three important factors which contribute to the resonance line width and they are:

(1) Scattering of nonuniform precession into degenerate modes of magnons arising from the porosity and from the surface finish. For example, polycrystalline YIG specimen of 79% density has a value of $\Delta H \approx 420$ Oe whereas that of 99% density has a value of 15 Oe. For a single crystal specimen, polished with 5µm grit, the $\Delta H$ value is only 10 Oe, while the best polished YIG specimen has a line width of less than 1, viz. 0.1 Oe.

**Table 2. Summary of YIG and hexagonal ferrites parameters**

| Material | | $4M_e$ (Gauss) | $H_a$ (Oe) | $O_u$ | Frequency of $O_u$ measurement (GHz) | Minimum useful frequency (GHz) | Temperatures dependance of resonant frequency (MHz/°C) | Curie temp. (°C) |
|---|---|---|---|---|---|---|---|---|
| YIG | $Y_3Fe_5O_{12}$ | 1,800 | 70 | 9,500 | 9 | 3.4 | Can be made 0 to first order by sphere alignment | 277 |
| GaYIG | $Y_3Fe_4Ga_1O_{12}$ | 400 | 50 | 1,500 | 9 | 0.75 | | 167 |
| Hexagonal Ferrites | | | | | | | | |
| M-phase | $BaFe_{12}O_{10}$ | 4,800 | 17,000 | 950 | 63 | 52 | + 15 | 450 |
| M-phase | $BaFe_{11.1}Sc_8O_{12}$ | 3,800 | 8,000 | 430 | 32 | 26 | – 0 | 340 |
| M-phase | $SrFe_{12}O_{12}$ | 4,700 | 18,700 | 950 | 62 | 57 | + 16 | 460 |
| M-phase | $SrFe_{10.6}Al_{1.5}O_{10}$ | 3,300 | 25,500 | 670 | 31 | 75 | + 25 | 410 |
| Y-phase | $Ba_2Zn_2Fe_{12}O_{22}$ | 2,850 | 9,900 | 1,250 | 17 | 9 | — | 101 |
| U-phase | $Ba_2Zn_2Fe_{32}O_{12}$ | 3,890 | 9,300 | 500 | 26 | 30 | — | 400 |
| Z-phase | $Ba_2Zn_2Fe_{24}O_{41}$ | 3,900 | 4,500 | 310 | 35 | 17 | — | 380 |

Notes: 1. Sc and Al doplings can be varied continuously.
2. Y phase has planar anisotropy, $t_r = (2.8\ \mathrm{MHz/Oe} \times He(He + H_a))^{1/2}$ when the anisotropy plane is aligned with both dc and RF fields.

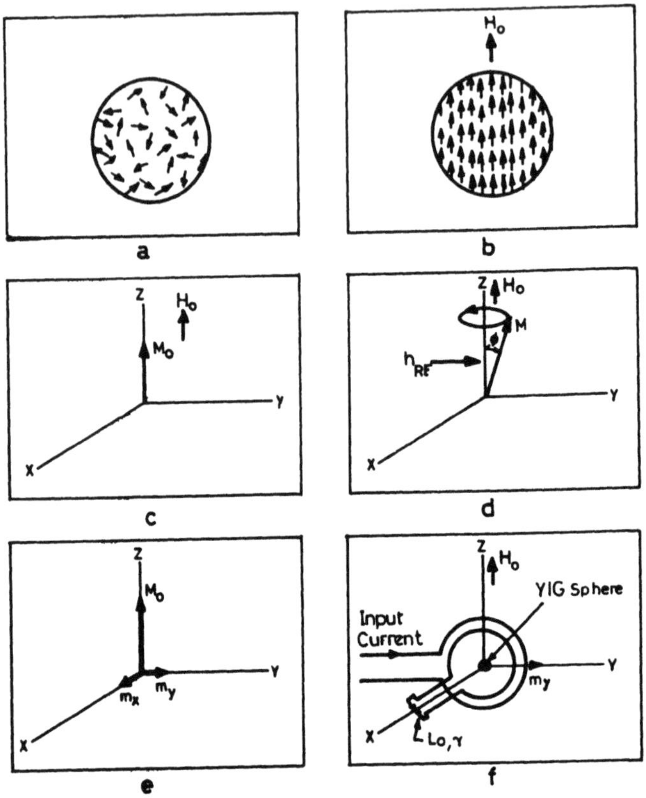


**Fig. 4.** Ferrimagnetic resonance (a) Randomly oriented magnetic dipoles in the unmagnetized ferrite, (b) Magnetic dipoles aligned under the influence of a magnetic field, (c) An equivalent representation of *b* showing the combined effect of the aligned dipoles, (d) Precession of the net magnetization vector due to RF magnetic excitation, (e) Equivalent representation of precessing magnetization vector, and (f) Tuned bandpass filter consisting of YIG sphere at center of two mutually orthogonal loops (extracted from 'Microwave Theory and Application' Hewlett Packard Company, p. 163).

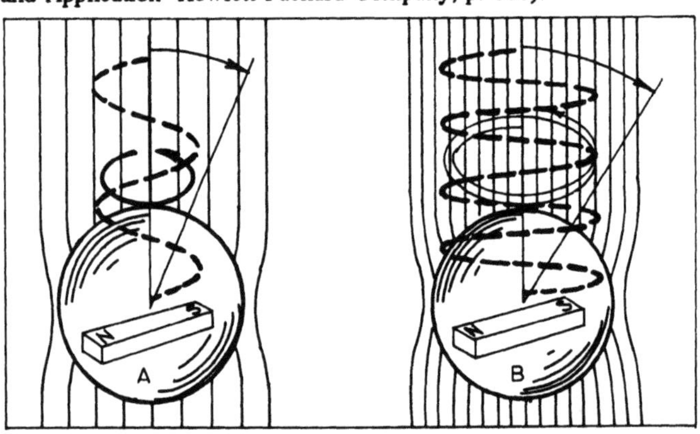


**Fig. 5.** Electron precession rate versus flux density (extracted from 'APPLICATION Notes on YIG RF Components', YIG-TEK Corporation, p. 17).

(2) Impurities, substituted in the lattice greatly influence the linewidth, particularly those rare earth ions that occupy the yttrium positions. For instance, a sample of YIG, containing 0.6 atom percent of terbium has as high a $\Delta H$ value as 200 Oe. Samarium doping in YIG increases $\Delta H$ to the extent of 22 Oe per atom percent of Sm.

(3) $\Delta H$ can also be influenced by the slow relaxation in the lattice through the redistribution between the ions. For example, when YIG is prepared in a nonoxidising condition, a fraction of the $Fe^{3+}$ ions can have a lower oxidation state ($Fe^{2+}$). It is also possible to generate $Fe^{2+}$ ions deliberately in presence of a bivalent impurities. For instance, if silicon is present in YIG as an impurity, it occupies the position of $Fe^{3+}$ in the d-sites, forming $SiO_4$ tetrahedra. This creates charge imbalance in the crystal which is compensated by the formation of an equivalent amount of $Fe^{2+}$ ions in the a-site. This exchange of an electron between $Fe^{2+}$ and $Fe^{3+}$ greatly increases the $\Delta H$ values.

It is needless to emphasis that YIG devices used for microwave engineering should have minimum $\Delta H$ values and this could be achieved only with single crystals of highest purity, except when impurities are introduced with a definite purpose.

## 5. CRYSTAL GROWTH OF YIG

Single crystal YIG materials are generally grown by either flux or hydrothermal or epitaxial techniques. Among all the methods described in literature for the crystal growth of YIG, the flux method stands unique as the surest and quickest method for preparing single crystals. The different techniques are described below.

### 5.1 Flux Method

Molten inorganic salts (or 'Flux' as it is generally known) have high solvating power for refractory crystals. The common molten solvents include KF, PbO, $PbF_2$, $B_2O_3$ and mixtures of these solvents and owing to the corrosive nature of the flux, preparation is carried out in platinum vessels only. In order to fix the composition of the charge (flux material and the mixed oxides, yttrium oxide and iron oxide) and also temperature limits for crystal growth, it is necessary to have some idea on the $Y_2O_3$-$Fe_2O_3$ phase system. The phase diagram was first investigated by Nielsen and Dearborn [5] and later, in detail, by Van Hook [14]. The system can be described only by the use of additional components $Fe_2O_3$–$Fe_3O_4$–$YFeO_3$, because the partial reduction of $Fe_2O_3$ to $Fe_3O_4$ takes place. In fact, for the same reason, $Y_3Fe_5O_{12}$(YIG) melts only incongruently around 1550°C at 1 atm. of $O_2$ pressure. The phase diagram of Van Hook [14] is given in Fig. 6. Above 1385°C, garnet phase coexists with magnetite. The composition of the melt is different from the coexisting solids and is oxygen deficient. Therefore, $Fe^{2+}$ concentration in the garnet phase will be higher and is a function of oxygen pressure as well as the temperature. Thus, from the phase diagram, it follows that pure $Y_3Fe_5O_{12}$ cannot be produced by the melting technique. Besides, flux method, when adopted, should be carried out at temperatures corresponding to the stability field of hematite + garnet.

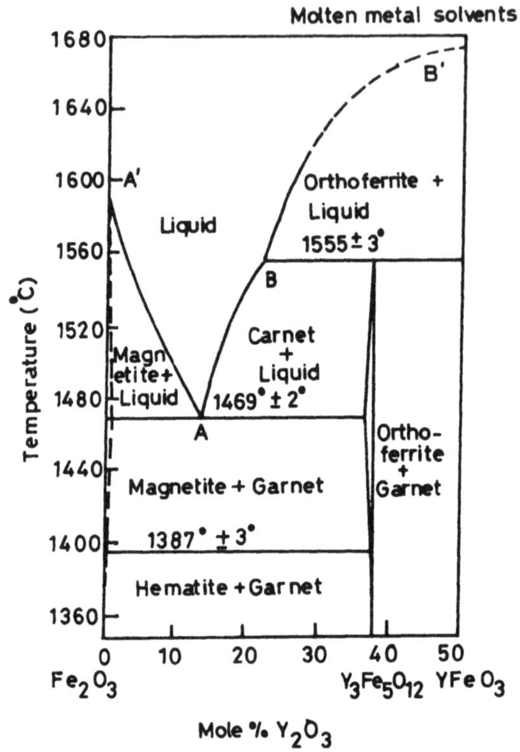


Fig. 6. Details of YIG phase diagram.

*5.1.1 Composition of Charge*

The composition of charge depends upon type of flux used viz. $PbO$-$PbF_2$-$B_2O_3$ or $BaO$-$B_2O_3$.

5.1.1.1 $PbO$-$PbF_2$-$B_2O_3$

The first flux material ever tried for growing single crystals of YIG happens to be lead oxide, PbO. The phase diagram worked out by Nielsen and Dearborn [5] showed that YIG is incongruently melting. The liquids region for crystal growth tends towards the region rich in ferric oxide. It can be made out from this that crystallization of YIG is possible only in the narrow region as shown in Fig. 7.

The solvent PbO has certain disadvantages because of its (a) high viscosity, (b) high density and (c) high melting point. Comparatively high temperatures are required to dissolve the oxide, and at this temperature, lead oxide has a tendency to decompose and get reduced to metal and attacks platinum.

Attempts have been made to improve the performance of the above described system by adding other compounds to the flux material, PbO. Addition of lead fluoride ($PbF_2$) brings down the temperature of fusion. The lowest eutectic is found to be around 500°C. Lead fluoride will resist the reduction to give rise to lead, and hence the flux becomes less corrosive to platinum. The density and viscosity are also reduced considerably which improves the flow property of

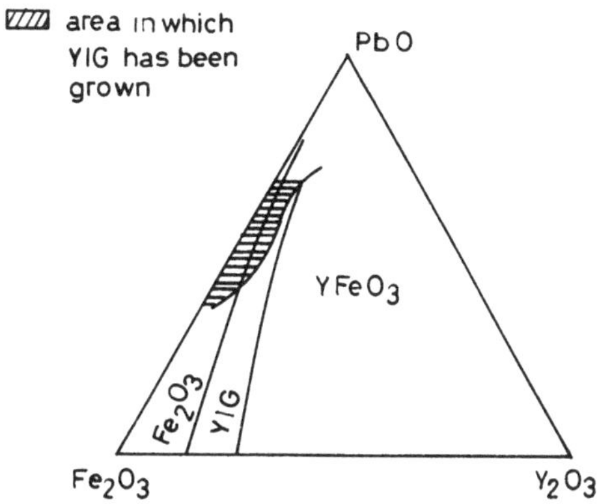


**Fig. 7.** A partial phase diagram suggested for the system $PbO$-$Y_2O_3$-$Fe_2O_3$.

the molten mass. Crystals will sink to the bottom and it is possible to separate flux more easily.

It may be pointed that lead fluoride is highly volatile under experimental conditions. The loss due to evaporation has to be prevented to maintain the composition and also protect the furnace from the fluoride attack. Addition of boric oxide (or boric acid) will bring down the vapour pressure. The presence of calcium oxide has been found to improve the size and quality of YIG crystals. The solubility curve [15] for YIG in $PbO$-$PbF_2$-$B_2O_3$ is given in Fig. 8.

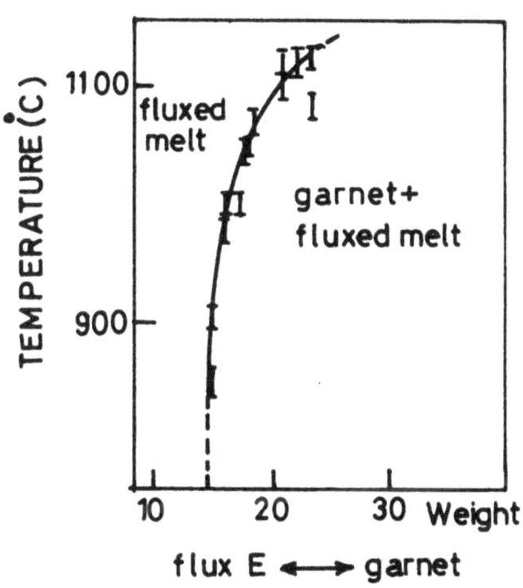


**Fig. 8.** Saturation concn. of $Y_3Fe_5O_{12}$ in $PbO$-$PbF_2$-$B_2O_3$ flux.

5.1.1.2 $BaO$-$B_2O_3$ FLUX

In spite of the addition of boric oxide (or boric acid), the above flux material is found to have appreciable amount of vapour pressure. In addition to the corrosive character of lead compounds towards platinum and its alloys, lead compounds are highly poisonous and injurious to health. It is, therefore, reasonable to search for a noncorrosive, nonpoisonous and less volatile flux material for growing the crystals of YIG. Such a system has been found to be barium oxide-boric oxide mixture. Another advantage of such a system is its comparatively lower density. The stability region for such a system has been worked out by previous investigators [16–18] and is shown in Fig. 9. The figure shows that YIG has a narrow region of stability corresponding to the ratio of barium oxide to boric oxide 3.6 to 1 (BaO : $B_2O_3$ :: 3.6 : 1) and (4.55 : 1).

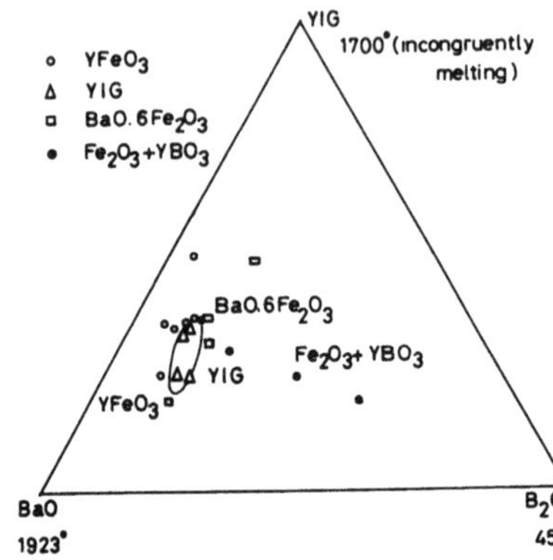


**Fig. 9.** Approximate phase diagram for the system BaO-$B_2O_3$-yttrium iron garnet in weight percent.

From the study of solubilities, it has been found that the stability region for YIG at high temperature is also narrow. YIG is stable between 1005 and 1190°C. Above 1190°C the stable composition corresponds to $YFeO_3$ and below 1005°C, the composition corresponds to $BaFe_{12}O_{19}$. From these considerations, it becomes obvious, that the composition of the charge should be such that YIG is stable over a convenient range of temperatures and the formation of other phases is avoided.

### *5.1.2 Methods of Preparation of Pure Chemicals and Reagents*

YIG and other crystals are to be grown with materials of highest purity. The level of impurities present in such materials should be very much lower than in the conventional reagents of analytically pure grade (below a few parts per million). Yittrium oxide ($Y_2O_3$), free from any paramagnetic impurities and other detectable rare earth impurities, should be of 'five nines' purity (99.999). Pure ferric oxide ($Fe_2O_3$) can be prepared by either precipitation of ferric hydroxide and subsequent dehydration or by the decomposition of ferric ammonium sulphate at elevated temperature. In the first method, ferric hydroxide when, freshly precipitated, has a great tendency to adsorb considerable amounts of a variety of impurities owing to high surface activity and it is difficult to remove them either by washing or dialysis. The second method yields good quality ferric oxide but recrystallisation is necessary to purify further.

It is needless to emphasize that the flux materials for YIG crystal growth should also be of highest purity (i) Lead monoxide (PbO) is prepared in polypropylene vessels (to avoid contamination with silica from glassware), starting from analytical grade hydrated lead acetate Pb $(CH_3COO)_2$ $3H_2O$.

(ii) Starting material for the preparation of lead fluoride ($PbF_2$) is either lead acetate or lead nitrate ($Pb(NO_3)_2$) of analytically pure grade, via lead carbonate. (iii) Boric oxide ($B_2O_3$), one of the ingredients added to improve the performance of the flux, can be replaced by boric acid which is added to the charge and heated slowly to the required temperature. (iv) Barium borate ($BaOB_2O_3$) corresponding to the composition $BaO : B_2O_3 :: 1 : 0.6$, employed as an alternate flux material, is prepared by heating barium carbonate ($BaCO_3$) and boric acid over a selected temperature range for several hours and cooling the molten mass. It is also necessary to prepare barium carbonate starting from very pure barium chloride and ammonium carbonate. (v) Calcium oxide, a minor constituent, is obtained by heating calcium carbonate in a platinum dish to 900°C. (vi) Vanadium pentoxide, used as an oxidative additive, is prepared by slow decomposition of ammonium vanadate. (vii) Lead peroxide is prepared by the oxidation of Pb (II) salt, such as lead nitrate, with sodium hypochlorite solution.

### *5.1.3 Mixing Up of the Charge Material*

Each component is finely powdered and ground in an agate mortar. The ground mass is transferred to a polythene bottle. The mixture in the bottle is agitated for 24 hours and mixed intimately in a mechanically operated tumbling device. The operation is repeated two to three times for every charge. The charge is then transferred to the platinum crucible and kept at different temperatures successively at 120, 350 and 500°C. Such a procedure will bring out uniform composition through out the bulk and consolidate the charge.

### *5.1.4 Methods for Flux Growth*

#### 5.1.4.1 Soaking and Cooling

Premelting of the charge, is carried out be slowly heating the Pt-crucible to the melting point of the flux (500°C in the case of $PbO\text{-}PbF_2$) and then to 850°C, in a period of six hours. This allows water and carbon dioxide, if any, to escape freely and to compact the mass in the platinum crucible. Premelting reduces the bulk volume of the charge to about 25% and this enables more of the charge to be added, whenever necessary. After premelting, the platinum vessel with its contents is weighed. This is to check the extent of flux evaporation at the end of the processing.

Prior to crystallization by slow cooling, it is necessary that the constituents of YIG, namely $Y_2O_3$ and $Fe_2O_3$, completely dissolve in the molten flux. Any individual solute particle acts as nucleating centre, when crystallization occurs, thereby causing the formation of a large number of small crystals. Therefore, the melt is maintained at a temperature well above the crystallization point for long periods. The crucible is heated to 1275°C. The temperature is maintained for 15–20 hr. This is called the soaking period.

After the soaking period, the melt is cooled to lower temperatures. The cooling is not uniform in the early stages. Instead, a cooling rate of 5 deg per hour is maintained, till the temperature is lowered by 50°C. This will initiate nucleation and growth at a faster rate. However, not all nuclei will grow at the same rate. Therefore, initial rapid cooling will result in a large number of crystals of varying dimensions of which a few will be bigger ones. After lowering the temperature by 50°C, the furnace (Fig. 10) is reheated to near soaking temperature (1270°C). This cycling in temperature enables the dissolution

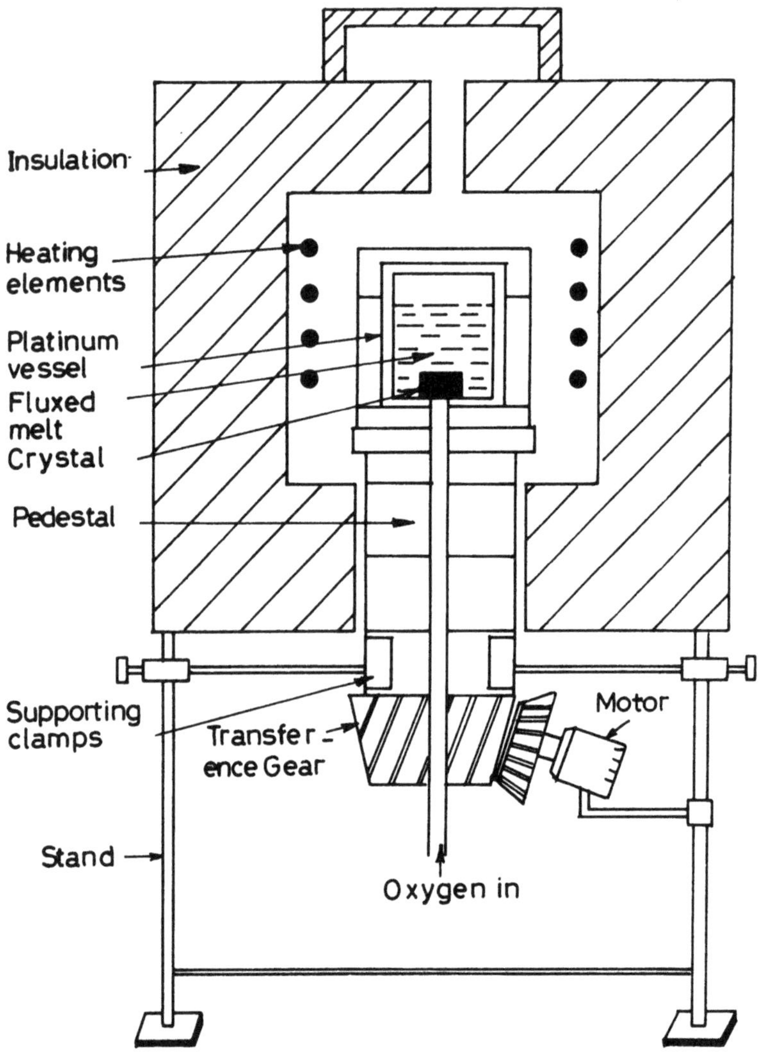


**Fig. 10.** Furnace used for flux growth of YIG crystals by acceleratory crucible rotation technique.

of smaller crystals leaving behind a few larger ones, which must have dissolved only partially. The furnace is now cooled at a much slower rate (0.5–1 deg/hr). The partially dissolved bigger crystals will now act as growth centres, thereby restricting the development of newer nuclei. As the growth continues, the surface area of the crystals increases. The larger amounts of constituents from the solution can get deposited on the growing crystal. This brings about a continuous change in the supersaturation, which can be balanced by an increased cooling rate towards the end of the processing. However, it has to be mentioned that the optimum growth rate does not increase to high values so as to maintain larger cooling rates. This means that the cooling rate, after the initial cycling may be 0.5–1 deg/hr which can be increased to 3 deg/hr when the temperature of the furnace goes down below 1100°C. However, cooling rate of 5 deg/hr or above can still be deterrent to the quality of the crystals.

Faster cooling rates are always found to result in crystals with lot of flux inclusion. This is due to the fact that the number of the growing nuclei will be more at the beginning of the processing. The number keeps on increasing as the temperature is lowered at a faster rate. These fast growing nuclei interfere by coalescing and most probably will contain in themselves a part of the growing medium. When the growth is complete, the crystals so produced will have small pockets of solidified flux materials in them. Also, the fast growth of nuclei encourages dendritic growth. When the dendrites join arms, small amounts of flux materials get entrapped and the crystals acquire a defective core. Therefore, slow cooling, especially at the early stages is essential.

#### 5.1.4.2 Annealing and Recovery of Crystals

Studies on YIG-PbO-$PbF_2$ system have shown that the YIG phase is not stable below around 950°C. The garnet dissolves in the flux and yttrium orthoferrite, $YFeO_3$, is formed. Below 900°C magnetoplumbite, $PbFe_{12}O_{19}$ is formed as an additional phase. The redissolution of garnet in the molten flux can be prevented and crystals recovered by more than one method.

(i) The container can be taken out of the furnace at 950°C and the melt can be poured off. The crystals stick to the bottom or sides of the platinum vessel which is immediately placed back in the furnace. The crystals are then annealed to room temperature.

(ii) In another method, crucible can be inverted in the furnace itself using supporting rods. In the inverted position, the crystals remain at the top of the platinum vessel stuck up to the sides and the melt flows down to the bottom. Such inversion is possible with crucible assembly shown in Fig. 11. The vessel has to be cut with the help of a hydrogen flame to collect the crystals.

(iii) A third method is to cool the melt below 950°C, comparatively rapidly (50 deg/hr) so that little time is given for the dissolution of YIG and development of $YFeO_3$ or $PbFe_{12}O_{19}$ phases. Once the melt is below the freezing point, the solidified mass is annealed to room temperature. The crystals in this case will be embedded in the matrix of the flux and they can be recovered by the preferential leaching.

The first method of recovering the crystals results in serious thermal shock, however quickly the pouring off step can be completed. Besides, this procedure exposes the operator directly to the vapours of the molten lead compounds and to the radiating heat from the furnace. Therefore, only the other two methods

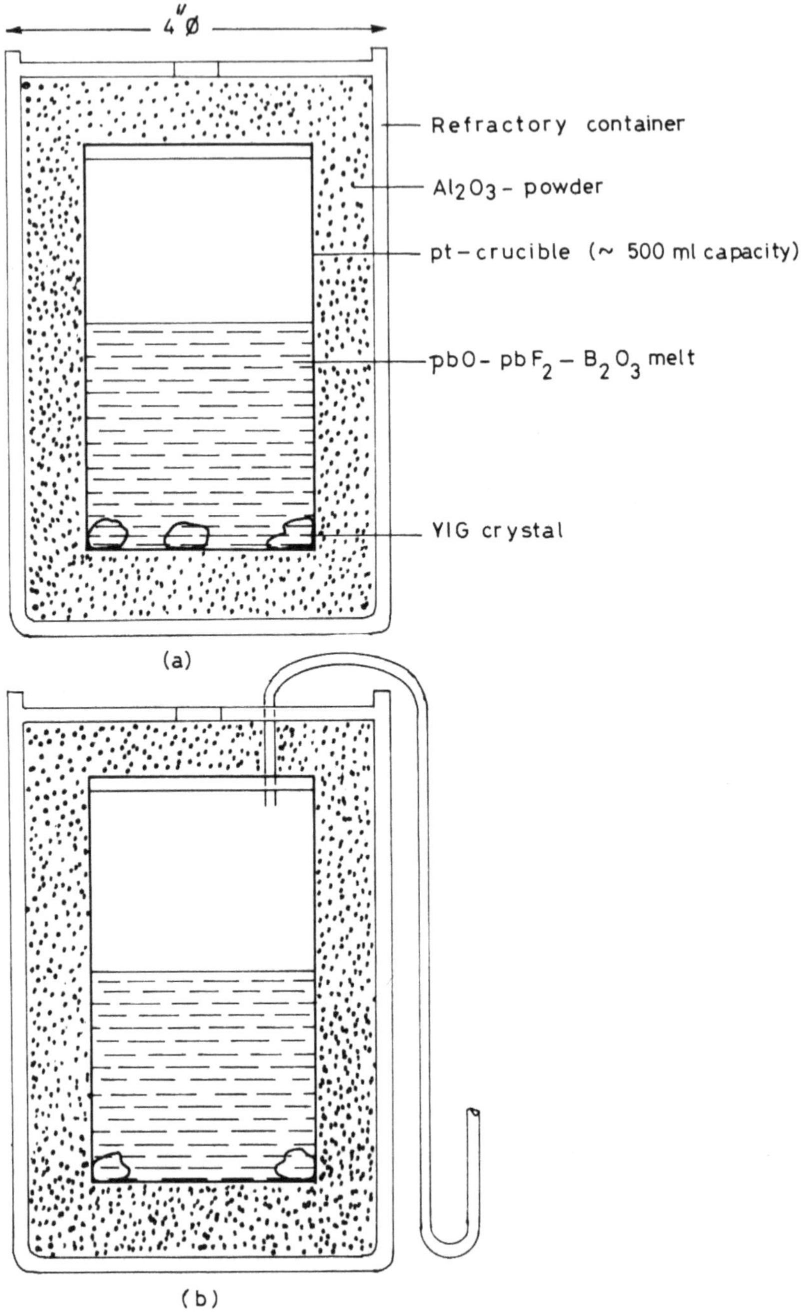


**Fig. 11.** Crucible assembly.

are generally adopted. It has been mentioned in literature that the third method results in development of strain in the crystal while being embedded in the solidified matrix of the flux. Investigations have shown that there is no difference in the properties of crystals when method (ii) or (iii) is adopted, other parameters being kept constant. Since the third method is simpler for larger batch productions, this procedure is generally followed.

The crystals have to be separated from the solidified flux. This is done by leaching with dilute acids. Nitric acid is an ideal choice, since it dissolves the solidified flux relatively faster and at the same time garnet crystals are not attacked. The crucible is kept in boiling dilute nitric acid (3N). A charge of about 500 g can be dissolved out by continuous boiling for two to three days. In the case of procedure (i) and (ii), 2–3 hr of treatment with boiling nitric acid (3N) removes the coating of the flux.

#### 5.1.4.3 Conditions for flux growth

It is necessary to maintain oxidizing conditions in the furnace during the flux growth of YIG. First, it will minimise the formation of $Fe^{2+}$ ions in garnet which, if present will increase conductance of YIG crystals through $Fe^{2+} \rightarrow Fe^{3+}$ hopping mechanism. Due to $Fe^{2+} \rightarrow Fe^{3+}$ exchange coupling, there will be considerable increase in the ferrimagnetic resonance linewidth. Secondly, the oxidizing conditions prevent the formation of lead through the reduction of Pb ions. Lead will attack platinum crucibles as mentioned earlier.

Many of the earlier workers maintained oxidizing conditions by carrying out crystal growth in oxygen atmospheres under pressure (5–100 atm). However, this method is quite elaborate and tedious. On the other hand, proper additives can be incorporated in the flux which will maintain oxidizing conditions.

(i) Addition of $V_2O_5$ (0.1–0.5 wt % of the total charge) sets up the $V^{5+} \rightleftarrows V^{4+}$ equilibrium which prevents the formation of $Fe^{2+}$ ions. Vanadium even if incorporated in YIG will result only in materials of low loss. Actually it has been reported that the simultaneous substitution of Ca and V in YIG gives material of extremely low loss.

(ii) Addition of $PbO_2$ can be the other alternative. Since PbO is a major component in the flux, a part of it can be replaced by $PbO_2$ (1–2% by wt) which will set in the $Pb^{4+} \rightleftarrows Pb^{2+}$ equilibrium. Sometimes, $PbO_2$ is added in the charge initially. Or otherwise the charge is kept in oxygen atmosphere at 400°C for over 12 hrs. This will generate $Pb_3O_4$ through the oxidation of PbO. At higher temperatures, disproportionation of $Pb_3O_4$ takes place as

$$Pb_3O_4 \rightleftarrows 2PbO + PbO_2 \tag{3}$$

$PbO_2$ thus formed will be effective as mentioned earlier.

(iii) A stream of oxygen is directed to the bottom of the crucible during the early stages of nucleation. This will act as the 'cold finger' for nucleation to take place preferentially at the bottom. Besides, oxygen prevents the reducing conditions to set in.

#### 5.1.4.4 Parameters that Influence Size and Quality of Crystals

Insufficient mixing and homogenization of the charges, prior to melting is found to affect the quality of crystals. Crystals will be comparatively smaller

(1–3 mm size) even when given reasonable soaking periods (12–15 hr). Co-precipitation of yttrium, ferric iron and lead from solutions as carbonate is found to be an alternative method to attain easy homogeneity. Cooling rate greatly influences crystal size as well as the number of inclusions they contain. A continuously changing cooling rate, as shown in Fig. 12, is preferable to a linear rate. Better crystals result by inducing nucleation at the bottom of the crucible. This preferential nucleation can be effected by means of a stream of oxygen projected to the bottom of the crucible. The crystals resulting from the growth runs in which considerable $PbF_2$ has evaporated are found to be smaller in size with more inclusions. The size and yield of crystals have increased. The inclusions also come down with the addition of $PbF_2$ to PbO.

### 5.2 Hydrothermal Method

Solubility of many of the compounds which are insoluble in water under ambient conditions can be increased considerably at higher temperatures under pressure. This can be further enhanced by the addition of suitable reagents, generally named as 'mineralisers'. By controlling the supersaturation at different parts of the pressure vessel, which is achieved by adjusting the temperature gradient, crystal growth can be carried out from aqueous solutions very much above the boiling point of water. This technique called "hydrothermal method", can be employed to grow single crystals at much lower temperatures than required for the flux method. Hydrothermal method has been successfully employed in the industrial preparation of quartz crystals. Quartz is grown from dilute sodium carbonate or sodium hydroxide solutions around 400°C and about 1000 atm. Crystal growth of YIG is an extension of this process. However, the solubility of the constituent oxides, viz., $Y_2O_3$ and $Fe_2O_3$ is very low in dilute alkali. Therefore, the concentration of the mineraliser should be increased to much higher value to obtain profitable growth rate. [19] Since the growth can be carried out around 400°C, the problem of ferrous iron formation can be considerably reduced.

The autoclave used in this method is schematically shown in Fig. 13. The lower part is filled with polycrystalline YIG which is kept at a higher temperature. The seed crystals are suspended from the closing piece of the liner, by means of silver wire. In hydrothermal method, proper temperature conditions have to be maintained so that the fluid phase fills the entire space available inside the autoclave. It is necessary to keep a temperature gradient between the two ends of the pressure vessel, such that the nutrient zone (lower part) is hotter than the growth zone (upper part). A chromel-alumel thermocouple connected to a potentiometric controller, regulates the furnace temperature. Thermocouples positioned at the top and bottom of the autoclave are used for measurements. The temperature gradient can be altered to any desired value by positioning the base of the autoclave and adjusting the opening of the furnace. Temperature gradient between 0 and 75°C can be obtained when the average working temperature is around 400°C.

Crystal growth is carried out in 20% (by weight) sodium hydroxide solution. The nutrient is mostly polycrystalline YIG. When the constituent oxides are

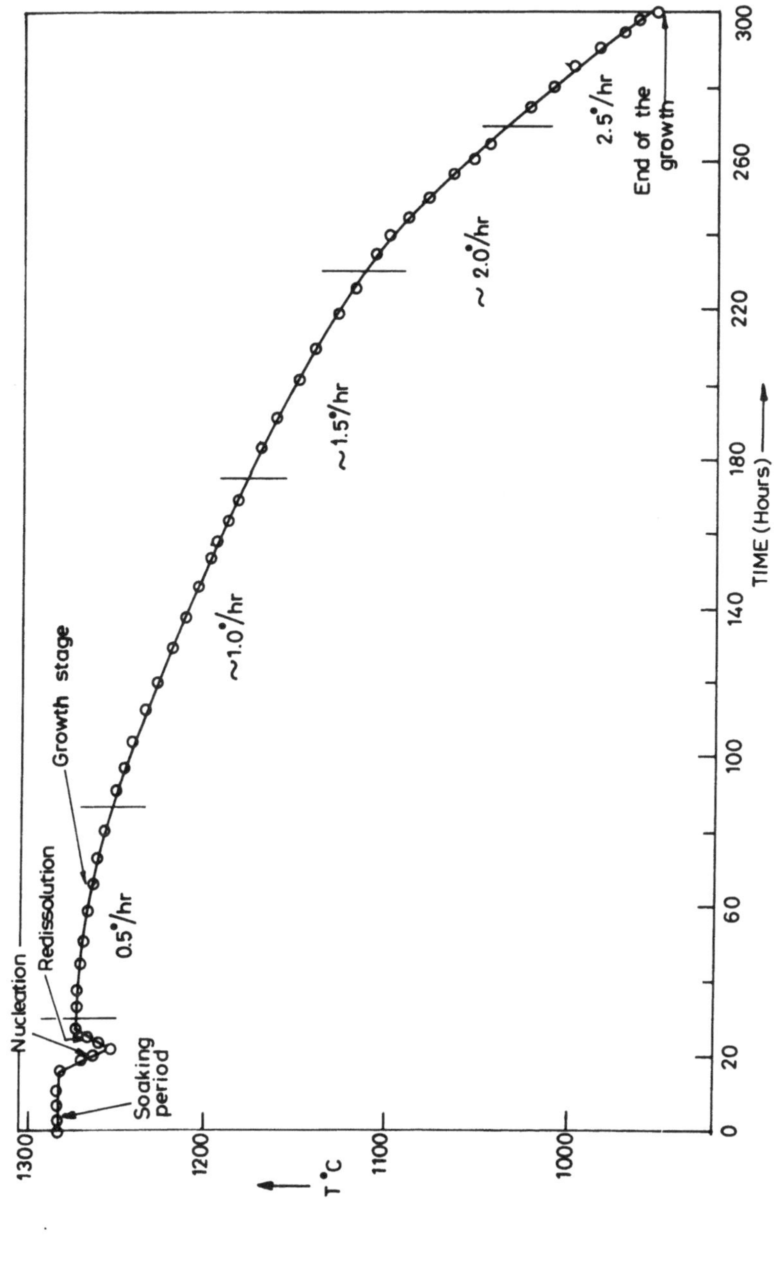


**Fig. 12.** A typical cooling curve.

directly taken, the mole ratio of $Y_2O_3$ and $Fe_2O_3$ is kept close to 3:5. 75% of the available volume inside the liner is filled with sodium hydroxide solution. The seed crystals used are those obtained from the flux experiments. The baffle opening is maintained around five percent. The space between the liner and the autoclave is filled with distilled water (65–70% fill). Since sodium hydroxide solution has lower pressure than that of pure water, lower percentage fill is adequate to balance the pressure developed inside the liner.

The autoclave is assembled as shown in Fig. 13. It is closed with the cap, followed by torquing the tightening screws. It is positioned in the furnace and

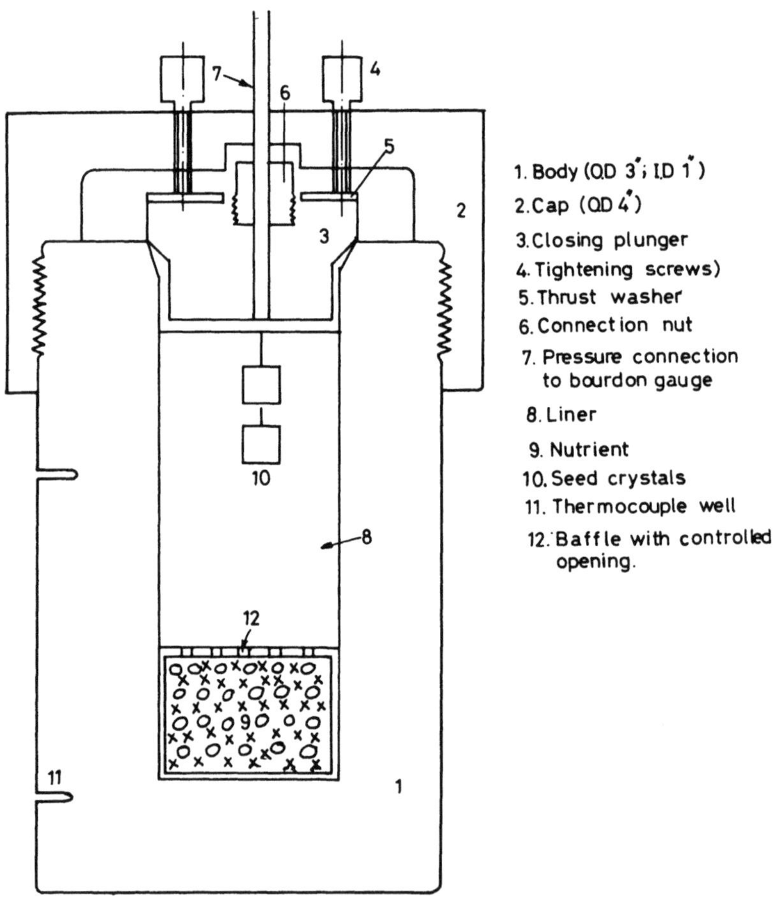


**Fig. 13.** High pressure vessel for hydrothermal growth.

heated to 400°C. The temperature gradient is adjusted such that the bottom temperature is 400°C, while the top thermocouple reads 380°C (temperature gradient, $\Delta T \sim 20°C$. The pressure within the liner is around 600 atm, while outside the liner it is more then this value. The pressure vessel is maintained for over 15 days, at the end of which the autoclave is cooled to room temperature and the crystals recovered. Growth of crystals is rather slow in the case of YIG from 20 wt % sodium hydroxide. The average growth rate is less than 5 mils per day. The growth rate increases with larger temperature gradient, but coherent and uniform growth is obtained with $\Delta T \sim 20°C$.

YIG crystals grown by the hydrothermal method have minimum $Fe^{2+}$ iron content (invariably less than 0.02 atoms per formula unit). However the greatest drawback is the presence of (OH) groups incorporated in the crystal. Infrared spectra of hydrothermally grown YIG crystals show a broad band in the region of 2800–3650 $cm^{-1}$. This broad absorption is invariably absent for the flux-grown specimens. Presence of (OH) group broadens the ferrimagnetic resonance linewidth of the material. Therefore, its reduction is of considerable importance. It has been suggested that potassium hydroxide, though more corrosive, is a better mineraliser [20]. The addition of calcium is also said to reduce the (OH) content.

### 5.3 Epitaxial Technique

The adoption of 'Liquid phase Epitaxy (LPE)' techniques (generally used for the fabrication of 'magnetic bubble memory' devices) for the growth of single crystal YIG materials in the form of 'thinfilms' [21] has opened a new vista in 'MICROWAVE MAGNETICS' which offers wide potential applications [22] in the VHF and Microwave Frequency bands extending to millimetric waves. This is the thermal growth of single crystal thin layers of magnetic garnets on non-magnetic substrates by LPE.

YIG films can be grown on highly polished gadolinium gallium garnet (GGG) utilizing vapour phase epitaxy (V.P.E.) in a T-shaped reactor made of fused silica. Source materials, yttrium and iron, are in the form of volatile chlorides contained within platinum crucibles in the vertical leg of the *T*. Their vapour pressures are controlled independently. These materials are carried into the reaction zone, the downstream of the horizontal segment by dry HCl and helium for the yttrium chloride and by helium for the ferrous chloride. Oxygen, in a helium carrier, is introduced into the upstream side of the horizontal segment. GGG is chosen in the substrate since it is chemically inert and provides a good lattice match (within 0.007 Å) and thermal expansion match from the deposition temperature (approx. 1200°C) to room temperature.

V.P.E. [21] gives rise to four oxides including YIG at different locations within the reactor. It is possible to adjust the position and size of the YIG zone to coincide with the location of the GGG substrate. This shift of the reaction zone is carried out by observing the colour of the oxide deposits on a fused silica test plate introduced for a short time at the normal deposition location. The length of the YIG zone is 5 to 7.5 cm and the film growth rate is 6 $\mu m$ per hour. The V.P.E. process suffers from the disadvantage of restricting the thickness of YIG films to 20 $\mu m$, due to depletion of source materials in the reactor. This thickness is inadequate for useful microwave filters. However, ferrimagnetic resonance linewidths ($\Delta H$) of less than 1 Oe have been measured at X-band.

V.P.E. has been surpassed by liquid phase epitaxy (L.P.E.) which is an extension of the bulk crystal flux growth technique. Garnet films grown by L.P.E. have defect densities determined principally by the substrate defect density which for good quality GGG is less than 5 per $cm^{-1}$. Film thickness can be held within 10% across 90% of the substrate area, with no detectable composition variations or second phases. Films can be grown at growth rates of 0.1 $\mu m$/min to more than 3 $\mu m$/min and to thickness of greater than 100 $\mu m$.

Epitaxial films of YIG have been grown by Adam et al [22] on GGG by L.P.E. from a $Y_2O_3$ (1.4 g) – $Fe_2O_3$ (11.82 g) – $B_2O_3$ (4.04 g)–PbO (182.74 g) flux. The following details have been reported by them. The growth technique used horizontal dipping with axial rotation. The 200 g melt was contained in a platinum crucible, and has a saturation temperature of 920°C. The 1.5 cm diameter substrate was held by a platinum vacuum chuck, and was rotated at 100 rev/min during growth and at 1000 rev/min, immediately after removal from the melt. Using this technique, films of up to 100 $\mu m$ thickness were grown at a rate of approximately 2 $\mu m$/min at 900°C. The versatility of the L.P.E. system allows the addition of rare earth ions or diamagnetic ions such as $Ga^{3+}$.

## 6. CHARACTERISATION OF YIG CRYSTALS

Following physicochemical methods are employed to characterise the YIG crystals: (1) Crystal morphology, (2) Density, (3) X-ray diffraction pattern, (4) Infrared absorption spectrum, (5) Estimation of lead and ferrous iron contents, (6) Determination of Curie temperature, (7) Saturation magnetisation and (8) Ferrimagnetic resonance line width of YIG specimens of spherical geometry. This involves the fabrication of highly polished spheres. Such a characterisation will be helpful as guide to determine whether the YIG crystals could be used in the devices. In turn, from the results of these characterizations, it is possible to optimize the method of crystal growth and the fabrication of spheres to meet the specific requirements. The quality of the crystals and the surface finish of the spheres could thus be improved.

### 6.1 Crystal Morphology

Almost all well-formed crystals are rhombic dodecahedra, modified to various degrees by icositetrahedral (211) faces [23]. The sharpest corners of the rhombic dodecahedra result from the intersection of four (110) plane and point in six equivalent (100) directions. The 8 remaining corners result from the intersection of three ⟨110⟩ planes and point in ⟨111⟩ directions.

Chemical etching techniques provide a simple estimate of crystal perfection with regard to dislocation. Suitable etchants for YIG include cold 20% (HCl), 85% ($H_3PO_4$), $HNO_3$ (8 N)–acetic acid –$H_2O$) mixtures and dilute $H_2SO_4$ (10 N).

Polished spheres of YIG can be etched in HCl for 1 hr. Etch figures showed ⟨100⟩, ⟨110⟩ and ⟨111⟩ holes [24]. Dislocation densities range from $10^2$ to $10^4/cm^2$.

### 6.2 Density

The gross density of YIG crystals is used as a rough measure of the extent of inclusions present. The measured density can be compared with X-ray density. Gross difference between these values indicates the presence of other phases e.g. orthoferrite ($YFeO_3$). Density of the crystals is measured by the displacement method using specific gravity bottle. This involves the determination of the volume of a liquid displaced by a known mass of the crystal. Bromoform is used as the displacing liquid (density = 2.80 gm/cm$^3$, at 25 C), so as to increase the accuracy of measurement. The specific gravity bottle along with the liquid is maintained at 25 C using a thermostat.

## 6.3 X-ray Diffraction Pattern

X-ray patterns of crushed single crystals are obtained by Debye–Scherrer method using Cr-$K_\alpha$ radiation (V-filtered) from a Rich-Seifert Unit. Philips Camera of 114.6mm diameter is used. In order to obtain the accurate cell edge value ($a_0$), those calculated from the high angle reflections are plotted against the Nelson-Riley function (Fig. 14). It gives a straight line and is extrapolated to $\theta = 90$ deg to read the actual $a_0$ value. For the specimen used for this plot (Fig. 14),

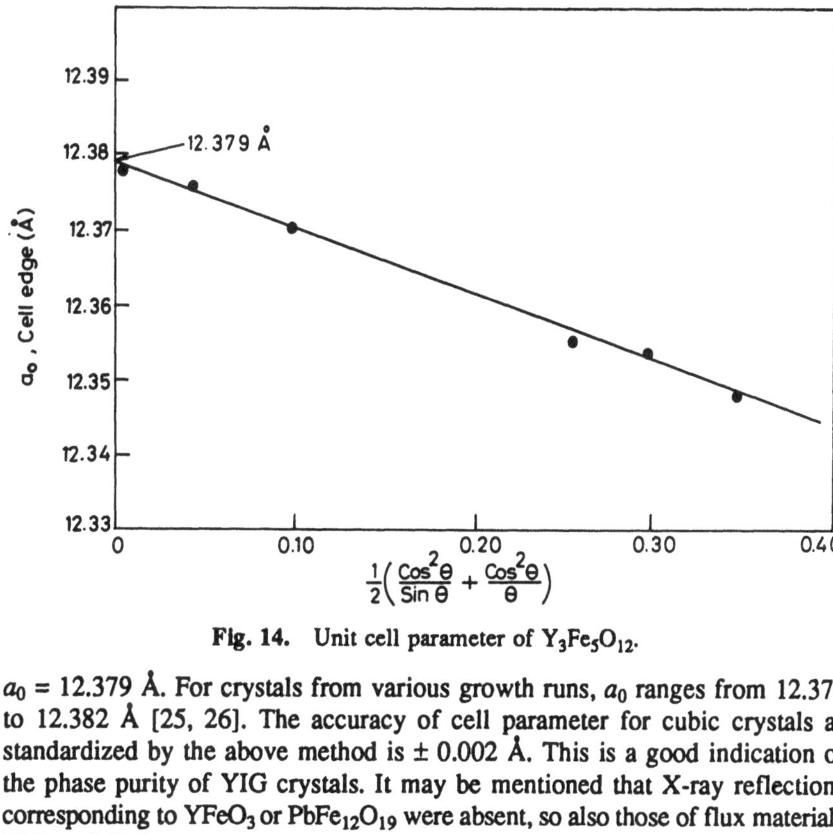


**Fig. 14.** Unit cell parameter of $Y_3Fe_5O_{12}$.

$a_0$ = 12.379 Å. For crystals from various growth runs, $a_0$ ranges from 12.374 to 12.382 Å [25, 26]. The accuracy of cell parameter for cubic crystals as standardized by the above method is ± 0.002 Å. This is a good indication of the phase purity of YIG crystals. It may be mentioned that X-ray reflections corresponding to $YFeO_3$ or $PbFe_{12}O_{19}$ were absent, so also those of flux materials PbO and $PbF_2$.

The quality of single crystals are tested from the X-ray rotation photographs. Crystals shaped in the form of spheres (0.8 mm dia) are used for this purpose. The sphere is oriented in the [100] axis. The reflections were found to be undistorted. This indicates the strain-free nature of the crystal.

## 6.4 Infrared Absorption Spectrum

Infrared absorption spectra of rare earth iron garnets are fairly diagnostic of the purity of them. The important absorption bands for YIG is around 600 and 380 $cm^{-1}$. The band around 600 $cm^{-1}$ can be tentatively assigned to asymmetric stretching modes of ($FeO_4$) tetrahedra and has the first overtones around 1150 $cm^{-1}$ [27]. However, it is more complex in the sense that lattice modes also influences these absorption bands. The important point to be mentioned is that infrared spectrum serves to detect the silicon impurities. As mentioned earlier, silicon impurity cannot be completely eliminated. Since rare earth iron garnets and natural silicate garnets are isomorphous, the silicon will be incorporated in iron garnets as $SiO_4$ tetrahedra. In Fig. 15, the characteristic absorption spectrum of a silicate garnet is compared with those of two YIG samples: (1) containing

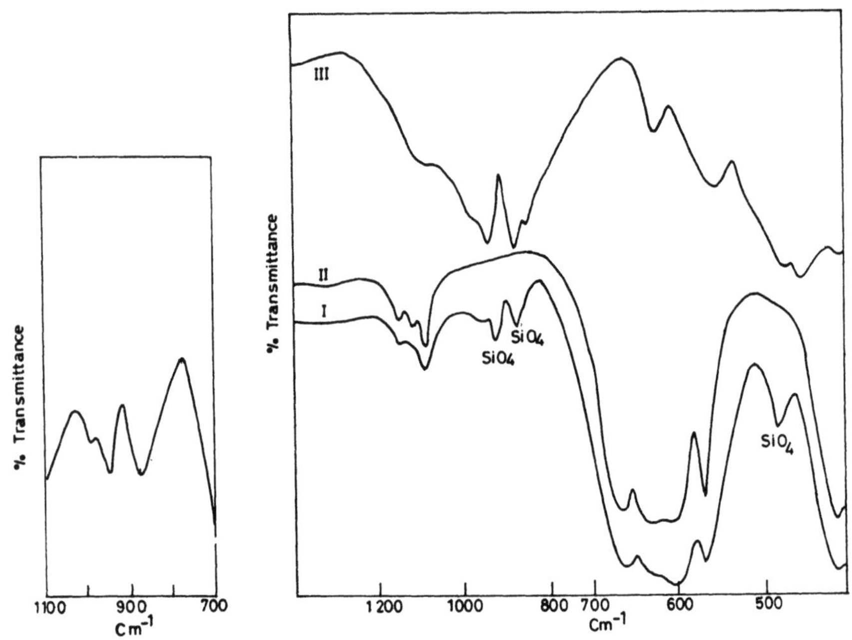


**Fig. 15.** Infrared spectra of garnets.

considerable amount of Si impurity, (2) another crystal of minimum Si impurity. The extra bands of sample (1) around 875 and 915 $cm^{-1}$ [28] coincides with those of silicate garnets. The infrared spectrum of sample (2) also indicates that it has minimum divalent impurities which could be mostly $Fe^{2+}$ and to a less extent $Pb^{2+}$. Incorporation of $Pb^{2+}$ in the garnet lattice is more difficult due to ionic size difference. Fig. 15 also gives the spectrum in the region 700–1100 $cm^{-1}$ for silicate ions in YIG taken at higher concentration of the material in the matrix (25 mg of YIG in 500 mg KCl).

### 6.5 Estimation of Lead and Ferrous Iron Contents

The crushed single crystals are analysed for lead in the following way. A known weight of the well powdered sample (– 320 mesh) is fused with potassium hydrogen sulphate to fuming temperature. The solidified mass is digested with

2N nitric acid and then evaporated to near dryness. The residue is boiled with water when all the salts excepting $PbSO_4$ dissolve. $PbSO_4$ formed is estimated gravimetrically. The amount of lead present is directly proportional to flux inclusion. The lead content has been found to be minimum (< 0.15%) in well-grown crystals. However, the percentage of lead is higher in those crystals grown with a faster cooling rate, reaching upto 1.3%.

From the filtrate of the above experiment, yttrium and iron are determined. High quality crystals gave the analysis $Y_2O_3$ = 43.92 and $Fe_2O_3$ = 55.89% (calculated values for YIG are $Y_2O_3$ = 44.08 and $Fe_2O_3$ = 55.92%). The analytical data, again, confirm the phase identity.

Ferrous iron in YIG is estimated by the vanadate method. For this purpose, the single crystals are crushed and powdered to – 400 mesh in an agate mortar under acetone. This method is normally adopted to prevent oxidation during grinding. A known weight of the dry powder is treated with 10 ml standard ammonium vanadate solution and 5 ml concentrated sulphuric acid. The solution is kept at near boiling point on a sand-bath with minimum evaporation. After about 2 hr, the solution is cooled, diluted with 4N $H_2SO_4$ and titrated against standard ferrous ammonium sulphate solution. The difference between direct and the test titration values is proportional to the reducing capacity—which is due to $Fe^{2+}$ content in the sample. The duration of the experiment can be reduced using a pressure vessel at 150°C. The well grown crystals with low Si content have FeO content < 0.28%. This value increases for products from crystal growth runs where oxidizing additive has not been added. FeO content increases to 1.5% in some cases. The lowest FeO content (0.025 atom per formula unit of YIG) observed corresponds to experiments in which $V_2O_5$ is added to the charge. It has been found that $Fe^{2+}$ content increases when Si is intentionally added. It may be mentioned that Si-doped YIG has giant anisotropic properties.

## 6.6 Determination of Curie Temperature

It is the temperature at which the ferrimagnetic crystal becomes a normal paramagnetic compound. The Curie temperature of YIG crystals is determined by the standard technique. An iron rod of about 12" long is held by a magnet, such that the rod remained suspended in a tubular electric furnace. A few crystals of YIG are attached to the bottom end of the rod and the position of the rod is so adjusted that the crystals remain in the middle of the furnace. A chromel-alumel thermocouple is positioned at the same level. The temperature of the furnace is gradually increased at a rate of ~ 0.5°C/min. At a certain temperature, the YIG crystals have fallen down due to the loss of magnetisation. This temperature corresponds to the Curie point. The experiment can be repeated with the same crystals. The Curie temperature is found to vary between 276 and 278°C. The reported value for YIG is 279–280°C [28].

## 6.7 Saturation Magnetisation

The most commonly used methods for evaluation of $4\pi M_s$ [29] are the (i) ballistic method (ii) vibration coil magnetometer (iii) the vibrating sample magnetometer

(iv) various force methods and (v) microwave methods. The Curie temperature can be determined from the plot of $4\pi M_s$ vs temperature. There are three practicable ways of estimating $4\pi M_s$ by microwave technique. Measurement of the shift in the ferrimagnetic resonance field for a non-spherical specimen [30], observation of magnetostatic modes [31] which is applicable to narrow line width single crystals and study of magnetostatic cavity modes in the bulk materials.

Saturation magnetisation ($4\pi M_s$) can also be determined by measuring the mass susceptibility of powdered YIG crystal at a field strength of 10,000 Oe (around $290 \times 10^{-6}$ cgs units/g). At this field, magnetisation saturation is achieved and the $M$ values correspond to $M_{\text{saturation}}$. The $4\pi M_s$ is related to the mass susceptibility in the following way.

$$B = H + 4\pi M_s \tag{4}$$

$$B = H + 4\pi\psi d \cdot H \tag{5}$$

Therefore,
$$4\pi M_s = 4\pi\psi d \cdot H \tag{6}$$

where $\psi$ = mass susceptibility, $d$ = density, $H$ = field strength and $B$ = magnetic permeability

The $4\pi M_s$ calculated in this manner is around 1750 Oe [32].

## 6.8 Fabrication of Spheres of YIG for the Evaluation of Ferrimagnetic Resonance Linewidth

Spheres are to be fabricated from YIG crystals for the evaluation of magnetic properties as well as for use in various devices. The spheres are to have good sphericity and maximum possible surface finish. The crystals are first cut using diamond blades, the thickness of which should be as small as possible to minimise loss of material. The width of the cut is 0.1–0.2 mm, small cubes of 2.0–2.5 mm edge length are thus made by this method.

The cubes are ground into spheres by the air cyclone cavity method. Before they are ground in the cavity, the cubes are tumbled in a metal cylinder containing a slurry of 200 mesh carborundum powder in water for 24 hr. The edges and corners of the cubes are rounded off by this procedure. The cubes are then placed in the cavity, one at a time, the wall of which is lined with 400 grade emery paper. The air is admitted into the cavity at a pressure of 50 psi. The nozzle through which the air enters, is so adjusted that the air stream takes a circular turn in the cavity (Fig. 16) which gives the crystal a swirling random motion forcing it to graze the abrasive paper frequently. The air escapes through the central holes. The continuous motion and random abrasion help the crystal to attain spherical shape [33, 34].

After the crystal specimen has attained the spherical shape, the size regulation and polishing are carried out by changing to finer emery papers in the order: grade 1, 1/0, 2/0, 3/0 and 4/0. As the finer emery is used, proper cleaning of the cavity as well as the sphere is necessary, when the emery paper is changed. Proper adjustment of air pressure and time of grinding result in very good surface finish. The maximum reduction in size of the specimen takes place

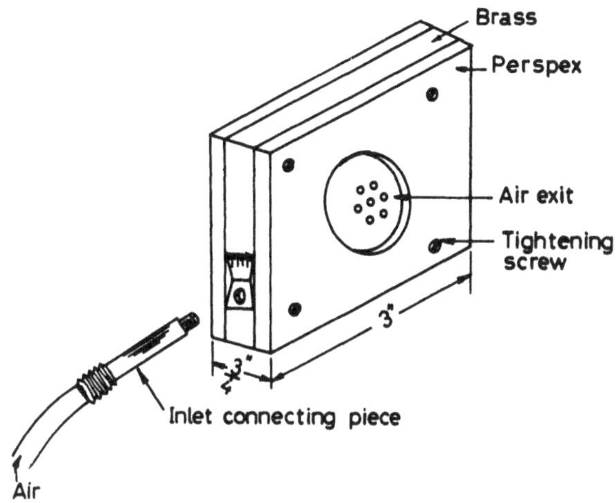


**Fig. 16.** Cavity for grinding and polishing YIG spheres.

when emery paper of 1/0 and 2/0 grit size are used. Strict control of the grinding time is necessary at this stage, to control the size. The size of the sphere is checked at 1/2 hour intervals at 1/0 and 2/0 stages. With 3/0 and 4/0 emery, the surface acquires higher polish and air pressure is reduced to 5–10 psi. After the use of 4/0 emery, the spheres will be shining in reflected light.

The spheres are further polished with diamond paste of finer grit size in steps of 2-4, 0-2 and 0-1/2 μm grit. These abrasive compounds are first applied to a soft carrier medium like wood, hard variety of paper or canvas cloth. The diamond particles get embedded in the medium. The wall of the air cyclone cavity is lined with this abrasive carrier.

Final polishing is carried out by a mechanochemical method using a suspension of colloidal silica (< 0.05 μm grit) in 2 N KOH. This polishing mixture is available in the trade name "Syton". The slurry is placed on a pellon polishing pad and spheres are polished in it by simultaneous pressing and rolling for 6-10 hr. The finish of the surface is checked by viewing under reflecting microscope (magnification, 50), whereby no scratches or pits should be seen.

## 6.9 Ferrimagnetic Resonance Linewidth ($\Delta H$) of YIG Materials

Linewidth of YIG materials can be determined by the following methods [9, 35, 36]:

(i) Resonant cavity method.
(ii) Shorted line method.
(iii) Transmission method.

(i) Resonant cavity method is the most widely used technique for evaluation of $\Delta H$ of polycrystalline ferrites. The basic measurements to be made are of resonant frequency and $Q$ of the cavity containing the sample as the function of the internal steady field. When a sample of the ferrite is placed in the cavity at a position of maximum magnetic field and minimum electric field, the real part of the permeability gives rise to a shift in the resonant frequency and imaginary component to a fall in $Q$. The line width $\Delta H$ is measured by measuring the change in bias field corresponding to the 3 dB bandwidth. The exact solution to the problem of ferrite loaded cavity is only possible in certain cases provided the volume of the sample is very small compared to that of the cavity and perturbation theory may be applied. The assumption is made that the fields within the cavity are substantially unchanged by introduction of the specimen.

(a) However, very narrow linewidths as of single crystal YIG, cannot be easily measured by the resonant cavity technique because, in the neighbourhood of ferrimagnetic resonance the coupling, which is proportional to $(\Delta H)^{-1}$ becomes so strong that perturbation theory no longer holds good.
(b) Another disadvantage of the cavity technique is that the linewidth can be measured only at a single frequency.

(ii) In shorted transmission [37] line method, the sphere is kept in a position of maximum rf magnetic field, for instance a half wave length away from the short-circuited end. The frequency is kept constant and the applied magnetic field is varied to obtain ferrimagnetic resonance. Linewidth is determined by measuring the change in bias field corresponding to the 3 db bandwidth.

The disadvantages of this method are:

(a) The linewidth can be measured only at a fixed frequency for a given shorted line.
(b) Sophisticated instrumentation is required to observe precisely the 3 dB bandwidth and a high order of signal frequency and amplitude stability is essential.

(iii) In the transmission method [38], the 'band stop filter' principle is made use of in measuring the linewidth. The sphere is placed in a RF transmission line and transverse dc magnetic field is applied. At resonance, depending upon the linewidth of the sphere, power will be absorbed by the sphere. This power loss at resonance is made use of in measuring the linewidth of the sphere. By varying the dc field, the resonant frequency can be varied so that linewidth can be measured at any frequency of interest.

In evaluating the linewidth of a sphere, the required data are (i) the insertion loss (Fig. 17) introduced by the sphere at resonance and (ii) the frequency of resonance in MHz which can be measured accurately.

## 7. BASIC PARAMETERS OF YIG RESONATORS

The design of YIG tuned devices is centered around high quality resonators fabricated out of single crystal ferrimagnetic materials. Highly polished spheres of single crystal yttrium iron garnet (YIG) and gallium doped YIG (Ga YIG) exhibit ferrimagnetic resonance.

The two significant characteristics of YIG described earlier, which dominate the design of any magnetic, tunable device are:

(i) Saturation magnetisation ($4\pi M_s$—gauss)
(ii) Resonance linewidth ($\Delta H$—oersted).

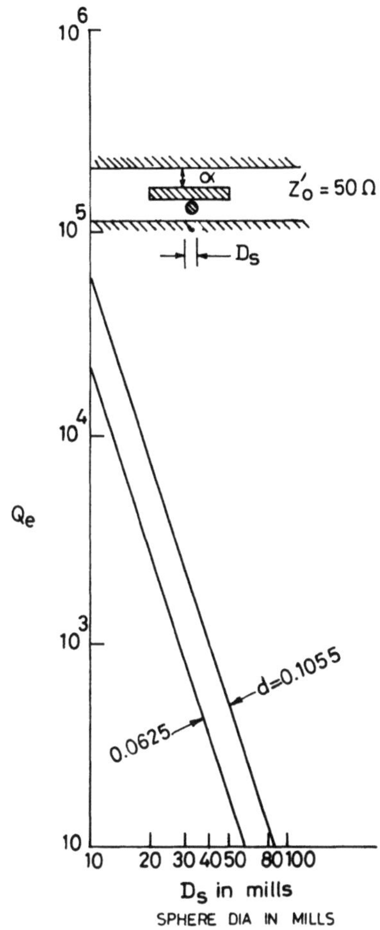


**Fig. 17.** Carter's chart of $Q_c$ vs sphere dia for a spherical YIG resonator.

The saturation magnetisation ($4\pi M_s$) is a function of the number of electron spins in the material per unit volume. Larger the $4\pi M_s$, easier it is to couple the RF field to the resonator.

Linewidth is the width of resonance in oestered as the RF frequency is held constant and applied dc field is varied. It can also be defined as the difference between the two values of biasing field (at each side of resonance) for which the imaginary part of the intrinsic susceptibility of the material equals the real part, while the frequency is held constant. This factor depends upon the material composition and the shape and surface finish of the resonator.

*$Q_u$ Factor*: The unloaded $Q_u$ of the spherical resonator is given by

$$Q_u = \frac{f_0 \times 10^{-6}}{2.8 \Delta H} \tag{17}$$

The unloaded $Q_u$ of the YIG resonator increases with frequency ($f_0$) for a given value of $H$. It is desirable to have a low linewidth, to derive a high $Q_u$.

## 7.1 Resonant Frequency

The resonant frequency for any ellipsoidal YIG resonator (neglecting anisotropy effects) is given by

$$f_0 \,\mathrm{MHz} = 2.8\sqrt{[H_0 - (N_z - N_x)(4\pi M_s)][H_0 - (N_z - N_y)(4\pi M_s)]} \tag{8}$$

where $N_x$, $N_y$ and $N_z$ are demagnetizing factors in the $x$, $y$ and $z$ directions, $H_0$ is the biasing field in oersted and $4\pi M_s$ is in gauss.

For a sphere, $N_x = N_y = N_z = 1/3$. Hence,

$$(f_0) = 2.8\, H_0 \tag{9}$$

Thus, for a *spherical resonator* which is easier to prepare with precision, the resonant frequency is independent of saturation magnetization and *linearly dependent only on biasing field* $H_0$. In any other shape for the resonator, for example, thin disc in the XY plane where $N_x = N_y = 0$ and $N_z = 1$, the resonant frequency

$$f_0 = 2.8\,(H_0 - 4\pi M_s) \tag{10}$$

In this case, resonant frequency is influenced by saturation.

### 7.1.1 *Resonance Limitations*

An important factor in the phenomenon of Ferrimagnetic resonance, is critical frequency $f_c$ which is determined by the value of $4\pi M_s$ according to the equation

$$f_c = 2\gamma N_t\, 4\pi M_s \tag{11}$$

where $N_t$ is transverse demagnetization factor. For a sphere, $N_t = N_x = N_y = 1/3$ and hence

$$f_c = \frac{2}{3}\, 4\pi M_s \tag{12}$$

Non-uniform coupling between the rf field and the precessing magnetic dipoles occur below this frequency, giving rise to non-linear behaviour and increases in losses. At frequencies lower than $(1/2)\, f_c$, resonance cannot occur as it is not possible to achieve magnetic saturation in the YIG resonator.

Also, resonator operation at temperatures close to or above the Curie temperature $T_c$, of the material is not possible as the $4\pi M_s$ drops to zero at $T_c$.

## 7.2 RF Power Limitations

There are two major mechanisms which limit the amount of RF energy which may be coupled to the YIG resonator. 'High level limiting' (RF level of the order of 20 $\mathrm{dB_m}$) is caused by the fact that the angle of electron precession has a finite limit. The other RF power limiting mode is called 'coincident mode' or 'low level limiting'. This type of resonator is magnetically saturated by the biasing external magnetic field. An unlimited sub-harmonic mode absorbs all the energy above the amount which the main mode can transfer in this condition. The normal limiting levels for this mode are $-25$ to $-15$ $\mathrm{db_m}$. Because of these power level limitations, YIG tuned devices are restricted to low power applications.

### 7.3 Orientation and Temperature Effects on YIG Resonators

The internal field which adds to or subtracts from the external magnetic fields is a critical consideration in the design of YIG resonators. The internal magnetic field is very much influenced by two factors viz., orientation and temperature.

#### *7.3.1 Orientation Effects*

YIG and GaYIG materials, which have cubic crystal structure, have three principal crystal axes; the (100), the (110) and (111) axes. For a given external biasing field, $H_0$, YIG resonant frequency is influenced by given external biasing field, $H_0$, the YIG resonant frequency is influenced by the direction of these axes relative to the direction of applied field. For a given YIG sphere, with (111) 'Easy' axis parallel to $H_0$,

$$f_0 = 2.8\left(H_0 - \frac{4}{3}\frac{K_1}{M_s}\right) \tag{13}$$

with (100) 'hard' axis parallel to $H_0$,

$$f_0 = 2.8\left(H_0 + 2\frac{k_1}{M_s}\right) \tag{14}$$

where $K_1/M_s$ is first order anisotropy in oersted.

If the sphere is rotated about a (110) crystal axis that is perpendicular to $H_0$, the value of $H_0$, for a given resonance at $f_0$ (MHz), is given by,

$$H_0 = \frac{f_0}{2.8} - \left[2 - \frac{5}{2}\sin\theta - \frac{15}{8}\sin^2\theta\right]\frac{K_1}{M_s} \tag{15}$$

where $\theta$ is the angle between $H_0$ *and that of (100)* axis which becomes parallel to $H_0$, as the sphere is rotated about the (110) axis. Thus, when the sphere is oriented at the ± 27° points,

$$H_0 = \frac{f_0}{2.8}\ \text{MHz} \tag{16}$$

#### *7.3.2 Temperature Effects*

Temperature variations offset the resonance frequency due to variations of $K_1/M_s$ with temperature, if the YIG resonator is in any arbitrary orientation other than the above discussed. $4\pi M_s$ and $\Delta H$ of the material also very over temperature and influence the performance of a YIG device.

## 8. MICROWAVE DEVICES USING SINGLE CRYSTAL YIG

The development of YIG tuned devices, employing single crystal YIG materials came into vogue about two decades ago and has since gained considerable importance due to the phenomenal improvements in the performance capabilities of EW systems and modern electronic instruments. The various functions, such as electronic tuning, frequency selective limiting, frequency discrination, signal distribution, signal isolation, swept signal generation etc., are now being carried out, by these devices, to a high degree of precision and linearity. YIG tuned solid state oscillators, are used as exciters for transmitters, in EW systems, synthesizers and programmable frequency sources in digitally tuned microwave receivers, and transmitters, and sweep generators in microwave instrumentation. Automatic frequency counters, employing YIG tuned harmonic generator-filter, offer stability and accuracy in frequency measurements with digital read out. Panoramic receivers and wide dispersion spectrum analyzers, using these YIG devices, show better display capabilities with elimination of false signal identification and image rejection. Four of the typical applications of YIG devices in systems are described below:

### 8.1 RF Tuners

A typical RF tuner comprises of a YIG tuned filter-oscillator combination driven by suitable electronic drivers with associated isolators and a mixer preamplifier IF filter combination. The operation is based on the 'superhet' principle wherein the input RF signal selected by the YIG filters is down converted to an IF signal for further processing. A signal command enables both the filter and oscillator to be driven to appropriate frequencies, with a fixed IF frequency off-set, throughout the tuning range of the tuner. An instantaneous bandwidth of about 20 MHz over a large dynamic range and linear tunability over octaves or multioctave frequencies has made these YIG tuners vital and critical components for ECM receivers.

### 8.2 YIG Filter Banks

Multi-channel YIG filter bands, either fixed tunable or tunable are in use in multichannel preselectors for applications in microwave communication receivers or channelised receivers. Here, the YIG multicouplers select the frequency pass band which is channelised to the receiver through PIN diode switches. Each frequency channel may be switched on or off in nanoseconds in any sequence or combination for either frequency or time frequency hopping modulated signals. The YIG filter bank offers, as compared to others types, lesser insertion loss at minimum volume and weight.

### 8.3 YIG Tuned Oscillator, Multiplier/Filter

A high spectral purity YIG tuned multiplier/filter for applications as simulators and general local oscillators offers a multi-octave frequency output, by multiplying signals from a 2–7 GHz YIG tuned oscillator and passing these signals through a YIG filter tuned to the desired multiple. The resultant output will be having a high spectral purity of better than – 50 dB in harmonic and sub-harmonic and tunable linearity over a wide frequency band.

### 8.4 Staloc

To overcome mistracking problems due to frequency drift limitations, closed-loop operation of the octave and multioctave filters have been developed using a unique tracking technique called "STALOC" (Self-tracking Automatic Lock-on Circuit). These filters, automatically lock on and track a CW reference frequency at a predetermined offset frequency. These filters find applications in active ECM and jammer set-on receivers.

### 8.5 MSW Devices

An exciting technology has emerged in this decade, which offers wide potential applications, especially signal processing in the UHF and Microwave frequency bands extending to millimetric waves. This promising technology is based on magnetostatic wave (MSW) propagation in magnetically biased epitaxial films of YIG. Some of the important devices are MSW delay lines (Fig. 18) dispersive lines, wideband tunable filters, narrow and filters, tunable oscillators and signal to noise enhancers which are fast finding applications, in military radars and microwave instrumentation.

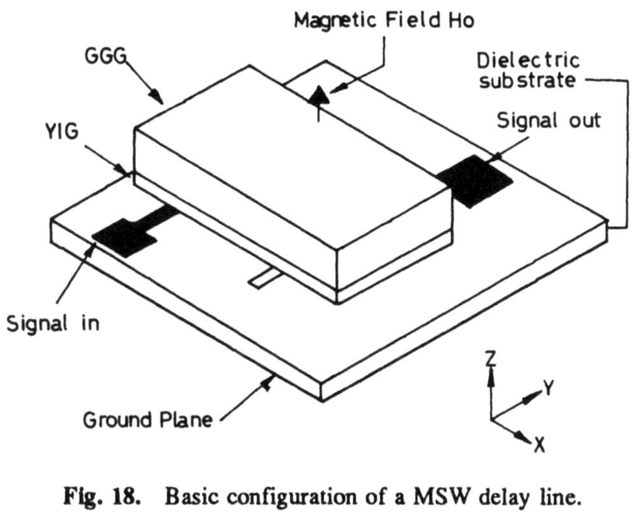


**Fig. 18.** Basic configuration of a MSW delay line.

## 9. TRENDS IN MATERIAL TECHNOLOGY

Today's microwave tunable devices owe much to the phenomenal development of "MICROWAVE MAGNETICS" and associated technologies, over the years. Their applications in microwave systems radar, in communication and Defence applications have enhanced the system capabilities beyond imagination. Direct signal processing of microwaves extending to millimetric waves and multioctave frequency capabilities are the results of innovations in single crystal YIG and allied materials. New materials viz., hexagonal ferrites [39] and single crystal lithium ferrite films [40] are now in the forefront to extend the usefulness of single crystal materials to millimeter wave applications, [41] especially from 18 GHz to 75 GHz.

## REFERENCES

1. F. Bertaut and F. Forrat, Compt. Rend. Acad. Sci., 242, 382 (1956).
2. G. Pauthenet, Doctorate Dissertation, University of Paris (1956).
3. S. Geller and M.A. Gilleo, Acta. Cryst., 10, 239 (1957).
4. W.H. Grodkiewicz, E.F. Dearborn and L.G. Van Uitert, Cryst. Growth, Ed. H.S. Peiser, Pergaman Press, N.Y., 1967, p. 441.
5. J.W. Nielsen and E.F. Dearborn, J. Phys. Chem. Solids, 5, 202 (1958).
6. G.P. Rodrigue "Magnetic materials for millimeter wave applications" IEEE Transactions on magnetics., Sept. 1963, pp. 351–356.
7. G. Winkler and H. Dotsch, 'Hexagonal Ferrites at Millimeter wave length' Proceedings of the 9th European Microwave Conference, 1979, pp. 13–22.
8. P. Roschmann, M-Lemke, W. Tolksdorf and F. Welz 'Aristotropy field and FMC linewidth in single crystal Al, Ga, and Sc substituted Hexagonal ferrites with M structure. Materials Research Bulletin. 19, 1984, pp. 385–392.
9. W.H. Van Aulock, "Handbook of Microwave Ferrite Materials", Academic Press, N.Y. (1965).
10. J. Helszajn, "Principles of Microwave Ferrite Engineering", Wiley-Interscience, N.Y. (1969).
11. E.A. Giess, B.E. Argyle, D.C. Cronemeyer; E. Klokholm, T.R. McGuire, D.F. O'kane, T.S. Plasketl and V. Sadagopan, AIP Conference Proceedings Series N0. 5 (17th conference on Magnetism and Magnetic Materials, Chicago (1971).
12. S. Standley, "Oxide Magnetic Materials", Oxford University Press (1969).
13. F.F.Y. Wang in Treatise on Material Science and Technology, Vol. 2., Academic Press, N.Y. (1973), pp. 273–387.
14. H.J. Van Hook, J. Amer. Ceram. Soc., 44., 208 (1961).
15. W. Tolkdorf, Acta Electronica, 17, 57 (1974).
16. R.A. Laudise, in 'Art and Science of Growing Crystals' Ed. J.J. Gilman, Wiley Interscience, N.Y. (1963).
17. R.C. Linares, J. Applied Physics, 35, 433 (1964).
18. M.J. Kestigian, J. Am. Cram. Soc., 50, 165 (1967).
19. R.A. Laudise and E.D. Kolb, J. Amer, Ceram. Soc., 45, 51 (1962).
20. E.D. Kolb, D.L. Wood, E.G. Spencer and R.A. Laudise, J. Appl. Phys., 38, 1027 (1967).
21. J.E. Pullian, G.R. Archer J.L. and Besser P.J. 'Magnetic Oxide films' IEEE Trans on Magnetics MAG-5, pp. 717–27, 1969.
22. J.D. Adam, Prof. J.H. Collins and J.M. Owens 'Microwave device application of epitaxial magnetic garnets'. The Radio and Electronics Engineer Vol. 45, No. 12, pp. 738–748, Dec. 1975.
23. A.B. Chase and J.A. Osmer, Jour. Cryst. Growth, 5, 239 (1969).
24. V.N. Vertoprakhon, V.D. Zamozhskii and P.V. Klertosv, Sov. Phys. Crystalogr., 13, 113 (1968).
25. S. Geller, G.P. Espinosa and P.B. Grandall, J. Appl. Cryst., 2, 86 (1969).
26. W.J. Croft, Amer. Mineral, 50, 1634 (1965).
27. B. Cockayne, J. Amer, Ceram. Soc., 49, 204 (1966).
28. K.A. Wickersheim, R.A. Lefever and B.M. Hanking, J. Chcmp. Phys., 32, 271 (1960).
29. Measurement of the properties of ferrite materials Miss Sp. Maxwell. The Marconi Review First Quarter 1970, pp. 2–17.
30. C. Kittel; Phys. Review 73, p. 155, 1948.
31. M. Magid. IEEE Trans., IM—13—p. 329, 1964.

32. G.R. Harrison and L.R. Hodges (Jr.), J. Appl. Phys. 33, 1375 (1962).
33. W.L. Bond, Rev. Sci. Instr., 22, 344 (1951).
34. J.W. Jeffery, "Methods in X-ray crystallography", Academic Press (1971).
35. M. Sparks, "Ferromagnetic relaxation theory", McGraw-Hill, N.Y. (1964).
36. A.H. Harrish, "The physical principles in magnetism", Wiley, N.Y. (1965).
37. Lax and Button, Microwave Ferrite and Ferrimagnetics, N.Y., McGraw-Hill 1962.
38. Mathaii, Magnetically tunable band stop filters' IEEE transactions on Microwave theory and Techniques. Vol. MTT-13, pp, 203-212, March 1965.
39. Hexagonal Ferrites for Millimeter Wave Application, D.B. Nicholson, Hewlett-'Packard Journal, 41, Oct 1990, pp. 59–67.
40. Microwave Magnetics and SAW Devices RADC Electromagnetics Sciences Division, Microwave Journal, April 1979, pp. 20–21.
41. D.B. Nicholson, R.J. Matreci and M.J. Levernier, 26.5 to 75 GHz, Preselected mixers based on magnetically tunable barium ferrite filters, Hewlett-Packard Jour., 41, Oct. 1990, 49–58.

# 8

# Radome Materials

G.S. Mani
Institute of Armament Technology
Defence Research and Development Organisation
Ministry of Defence, Pune, India

## 1. INTRODUCTION

### 1.1 Definition

According to IEEE radome is defined as "an enclosure for protecting an antenna from normal effects of its physical environment, generally intended to leave the electrical performance of the antenna unaffected" [1]. In simple terms, a radome is a housing for an antenna. Because of the reason that radome is essentially a protective device, it is required to provide necessary structural strength, but yet, not deteriorate the electromagnetic performance of the antenna under operational conditions.

### 1.2 General Information

Since early 1940s, when airborne radomes started finding applications, tremendous technological advances have been made in the field of radome materials and design aspects. A wide variety of design ideas have been attempted depending on application. A study on the subject of radomes encompasses aspects related to electrical design, mechanical design, fabrication, analysis, testing and other related areas. Various authors have contributed to the literature on the subject [2–9]. Microwave Engineers responsible for developing radome for a specific application should be aware of the above aspects and their implications. In this chapter, materials used for radome are discussed with brief reference to these aspects.

### 1.3 Radome Wall Configurations

An important aspect of radome design is the selection of proper wall configuration. Fig. 1 shows the configurations that can be used for a radome with uniform wall thickness.

#### *1.3.1 Monolithic Wall*

This consists of a single layer of homogeneous radome material. The thin wall radome structure implies the wall thickness to be less than about 1/20th of a

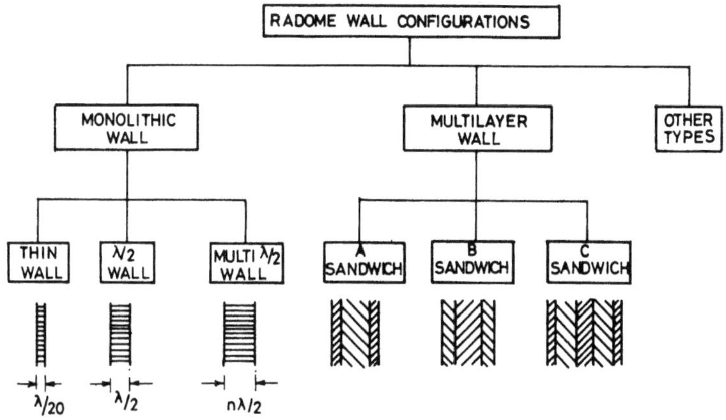


**Fig. 1.** Commonly used radome wall configurations.

wavelength in the material corresponding to highest frequency of operation. Electrically, this can be modelled as a low pass filter and hence radome design based on thin wall configuration can be useful in some of the broadband applications [10]. However, this is subject to structural adequacy of the material in the environment faced by the radome. The configuration can be adapted only if environmental and strength requirements are lenient. It is observed that excellent electrical performance at high angles of incidence can be achieved if the electrical thickness ($d/e$) of the radome is in the region of 0.02. When thin wall configuration is feasible, and considered adequate to meet structural and aerodynamic loads, the radome can have the following features: (a) Light weight; (b) High electrical transmission coefficient; (c) Broadband frequency coverage; (d) Insensitive to changes in polarization; (e) Insensitive to variations in angle of incidence; (f) Small refraction of incident waves; (g) Material parameters not critical; and (h) Manufacturing methods not critical.

If the wall thickness corresponds to one half wavelength in the material, the configuration is called half wavelength configuration. Such configuration exhibits nearly 100% electrical transmission for normal incidence, if the loss tangent of the material is low. Maximum transmission at any desired angle of incidence can be achieved by properly choosing the thickness of the wall. For frequencies between 5 and 25 GHz, the thickness of half wave radome varies in the region of 15 mm to 3 mm for a material with dielectric constant of 4.0. Since for many applications, such thickness may provide adequate strength and rigidity, the half wave radome wall may be considered suitable.

When half wavelength wall configuration is not suitable for any reason, the wall thickness can be made to be a multiple of $\lambda/2$. Such configurations can be a low loss structure at the design frequency. However, the performance of the structure is quite sensitive to loss tangent of the material.

### *1.3.2 Multilayer Wall*

In spite of the variety of available materials, it is often impossible to design a satisfactory radome using single homogeneous sheet of dielectric. The problem may be due to inadequate mechanical strength, stiffness or weather resistance property of the material, though otherwise electrically suitable. It may happen that the thin wall design may have insufficient strength, while half wave design may prove to be excessively heavy or unnecessarily strong or stiff making the design undesirable for an airborne radar use. Sometimes, design of monolithic wall configuration may also lead to impracticable machining tolerances not normally achievable by using available equipment. Under these conditions, multiple layer configuration is adopted for radome wall.

Some of the simple multilayer structures that can be used in radome wall design are A-Sandwich, B-Sandwich and C-Sandwich structures. Out of these, A-Sandwich is most popular, being one of the simplest of the multilayer structures. This has three alternate layers of high and low dielectric constant materials. The core will be of low density, low strength and low dielectric constant material, whereas the skins on either side use high density materials. The main advantage of the configuration is its high strength to weight ratio. Broadband radomes can be designed using A-Sandwich design [11]. However, such designs are more sensitive to variations in incident angle and polarization.

B-Sandwich structure has outer layers of low dielectric constant and a central core of high dielectric constant material. In general, it is found that such B-Sandwich structures have good transmission characteristics over wide range of incident angles. But unfortunately, such structures are not found suitable from structural strength considerations, and hence are not very popular as radome designs.

C-Sandwich is a five layer structure, wherein alternate layers of low and high dielectric constant materials are used. This configuration is observed to have good structural rigidity and also have good electrical characteristics over wide frequency range. However, because of complex design and fabricational aspects involved in such radomes, this has been used only where essential.

### *1.3.3 Other Radome Wall Configurations*

Apart from the above common types, a few other types of radome walls have also been reported in literature. Conducting wires in the form of reactive loads have been added to the radome wall structure [12]. These allow more degrees of freedom in the radome design and can be used to advantage for specific purposes. A grooved wall structure as shown in Fig. 2 has also been reported to give better frequency bandwidth with low depolarization [13]. Gradient dielectric radome wall can also provide broad band properties [14]. Dielectric

**Fig. 2.** Doubly grooved radome wall.

rings or patches have been used to reduce abberations due to antenna scanning within the radome [15].

Apart from the above wall configurations, which were essentially made of dielectric materials, other wall configurations have also been reported. These are the metal space frame radomes capable of quick assembly at site for ground applications [16] and perforated metal radome for missile applications. The latter is reported to have better rain erosion resistance, high strength to weight ratio and low static charge build-up [17].

## 1.4 Radome Design Aspects

Radome design must consider aspects of both mechanical and electrical requirements of the system. Mechanical considerations include structural stability, aerodynamics and the environmental conditions the antenna is likely to be subjected. Electrically, the radome must ensure least distortion in the intrinsic antenna radiation pattern. Both mechanical and electrical aspects of radome design allow much flexibility when considered individually. In a realistic design, however, the high interdependence of the two areas usually cuts the flexibility drastically. Many trade-offs between competing criteria must be made often requiring an interactive, multidisciplinary design approach, as shown in Fig. 3.

### *1.4.1 Aspects of Mechanical Design*

In general, the solution to the mechanical design problems lie in proper selection of radome geometry, materials and fabrication techniques. Radomes installed on aircrafts are often primary structures whose failure can endanger the aircraft. In such cases, the structural design of the radome is of paramount importance and should consider aerodynamic and inertial loads created by aircraft speed and flight profile, radome shape and sudden changes in motion as may be encountered during manoeuvres, landing, breaking or in gusty winds. The radome will be subjected to unsymmetrical pressure patterns due to vehicle manoeuvres, such as yaw, banking and also due to vibrations. These will result in compressive, tensile and shear stresses both in the circumferential and longitudinal directions of an airborne nosecone radome. From typical pressure loading curves, shear and bending moment at different stations on radome can be computed. These, together with other forces, enable estimation of the stresses likely to be developed on the radome surface. The shape, material and fabrication process will be chosen depending on the values of these stresses.

### *1.4.2 Aspects of Electrical Design*

Major aspects requiring attention in the electrical design of radome are transmission efficiency, boresight error, bandwidth requirements and reflections due to radome. In addition, depolarization, emergence of flash lobes, wave trapping, errors due to radome asymmetry are some of the other electrical effects which have to be considered for exhaustive electrical analysis of the radome. In general, proper weightages are given to these parameters and appropriate parameter index as a functions of these variables is computed. Optimisation of the variables is then carried out to yield best parameter index. Optimised values

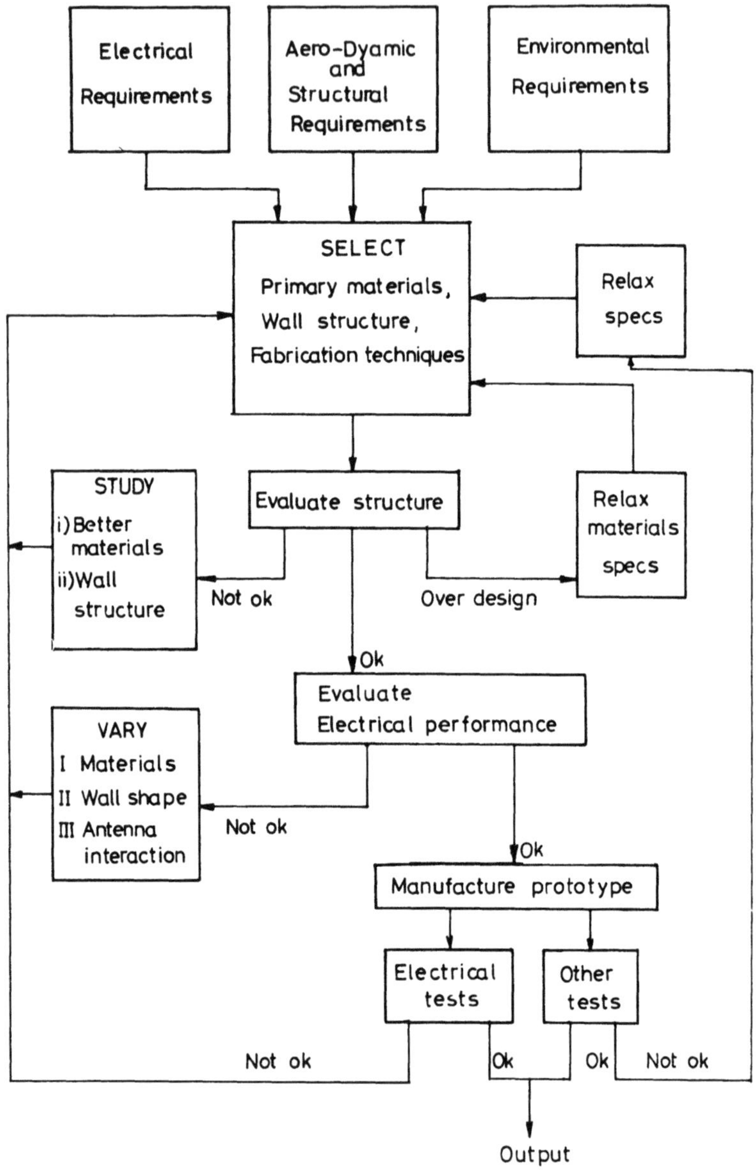


**Fig. 3.** Iterative procedure involved in radome design.

of different variables are then used to fabricate prototype which is then refined for further optimisation.

Electrical performance of the radome largely depends on range of incident angles, which the electromagnetic waves make with the radome wall. In addition, polarization of the antenna enclosed in the radome and the electrical characteristics namely dielectric constant and loss tangent of the material decide the electrical performance of the radome.

Radome shape and location of the antenna relative to the radome dictate the range of incident angles. In the case of ground based or ship-borne radomes, the shape can be chosen on the basis of optimum electrical performance. In case of airborne radomes, it is generally optimised on the basis of aerodynamic considerations. Since shape is an important electrical factor in radome design, it has been found to be convenient to classify radomes on this basis. A normal-incidence radome is identified as the one in which most of the area illuminated by the antenna is approximately normal to the radiated beam of antenna. The transmission properties in such case can be approximated by those exhibited by plane dielectric sheets traversed by plane waves normal to the surface. A large spherical radome, in which the antenna is located at the centre of the sphere will yield performance of an ideal normal-incidence radome. Streamlined radomes, as the name implies, have large fineness ratio (ratio of length to base). Some of the practical shapes used for airborne radome designs are shown in Fig. 4 [3].

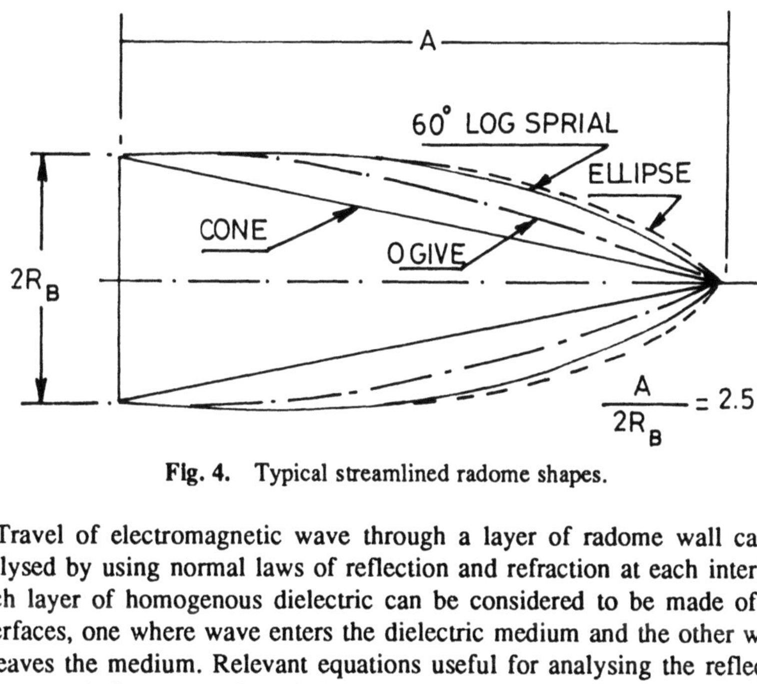


**Fig. 4.** Typical streamlined radome shapes.

Travel of electromagnetic wave through a layer of radome wall can be analysed by using normal laws of reflection and refraction at each interface. Each layer of homogenous dielectric can be considered to be made of two interfaces, one where wave enters the dielectric medium and the other where it leaves the medium. Relevant equations useful for analysing the reflection and transmission properties of the wall configuration are given in Table 1. Geometry involved in the analysis is shown in Fig. 5.

Generally, it is observed that reflection for perpendicular polarization ($R_1$) is larger than that for parallel polarization ($R_{11}$). For normal incidence $R_{11}$ and $R_1$ are equal in magnitude. $R_{11}$ tends to zero at Brewster angle for low loss materials.

Insertion phase shift is defined as the shift in phase introduced by the material for a plane wave incident on its surface. This is dependent on the polarization of the incident wave and the characteristics of the material. Difference in insertion phaseshift for two orthogonal polarizations is an important factor in analysing depolarization characteristics due to radome.

**Table 1. Equations governing radome electrical design**

**Incidence on Flat Dielectric Sheet**

*At Each Interface*

$$\gamma_1 = [\cos\theta - (\varepsilon_r - \sin^2\theta)^{1/2}]/[\cos\theta + (\varepsilon_r - \sin^2\theta)^{1/2}]$$

$$\gamma_{11} = [\varepsilon_r \cos\theta - (\varepsilon_r - \sin^2\theta)^{1/2}]/[\varepsilon_r \cos\theta + (\varepsilon_r - \sin^2\theta)^{1/2}]$$

$$\theta_B = \tan^{-1}[(\varepsilon_r)^{1/2}] \text{ for parallel polarisation}$$

*Homogenous Flat Sheet*

$$R_1 = \frac{\gamma_1\{1 - \exp[-2j\beta_0 d(\varepsilon_r - \sin^2\theta)^{1/2}]\}}{1 - \gamma_1^2 \exp[-2j\beta_0 d(\varepsilon_r - \sin^2\theta)^{1/2}]}$$

$$T_1 = \frac{(1 - \gamma_1^2)\exp[-j\beta_0 d[\cos\theta - (\varepsilon_r - \sin^2\theta)]^{1/2}\}}{1 - \gamma_1^2 \exp[-2j\beta_0 d(\varepsilon_r - \sin^2\theta)^{1/2}]}$$

$$R_{11} = \frac{\gamma_{11}\{1 - \exp[-2j\beta_0 d(\varepsilon_r - \sin^2\theta)^{1/2}]\}}{1 - \gamma_{11}^2 \exp[-2j\beta_0 d(\varepsilon_r - \sin^2\theta)^{1/2}]}$$

$$T_{11} = \frac{(1 - \gamma_{11}^2)\exp[-j\beta_0 d[\cos\theta - (\varepsilon_r - \sin^2\theta)]^{1/2}\}}{1 - \gamma_{11}^2 \exp[-2j\beta_0 d(\varepsilon_r - \sin^2\theta)^{1/2}]}$$

**Monolithic Radome Wall**

*Thin Wall Radome*

Power tr. coeff., $T^2 = 1 - [\pi d(\varepsilon_r - 1)/\lambda_0 \cos\theta]^2$

Diff. in insertion phase-shift $\Delta_{11} - \Delta_1 = \pi d(\varepsilon_r - 1)^2 \sin^2\theta/\lambda_0 \varepsilon_r \cos\theta$

*Multiple Half Wave Radome*

| | |
|---|---|
| Radome Thickness | $d = n\lambda_0/2\,(\varepsilon_r - \sin^2\theta)^{1/2}$ |
| For normal incidence | $d_0 = n\lambda_0/2\,(\varepsilon_r)^{1/2}$ |
| | $T^2 = 1 - [n\pi(\varepsilon_r + 1)\tan\delta/2(\varepsilon_r)^{1/2}]$ |
| For haff wave radome | $n = 1$ |

($\varepsilon_r$: Rel. dielectric constant of dielectric sheet; $\theta$: Angle of incidence; $\theta_B$: Brewster angle; $\gamma$, R: Ref. coeff; T: Tr. coeff; 1: Perpendicular polarisation; 11: Parallel polarisation)

Equations given in Table 1 can be used to plot the electrical performance of radome wall configuration of different materials. Typical curves for some of the wall configurations are shown in Figs. 6 and 7. Such curves help in selection of material for optimum radome performance.

## 2. RADOME MATERIALS

### 2.1 Material Selection

Materials used for fabrication of radomes are primarily selected for adequate strength and good electrical characteristics. The final application or role of the radome in system also greatly influences the selection of material. Whereas the

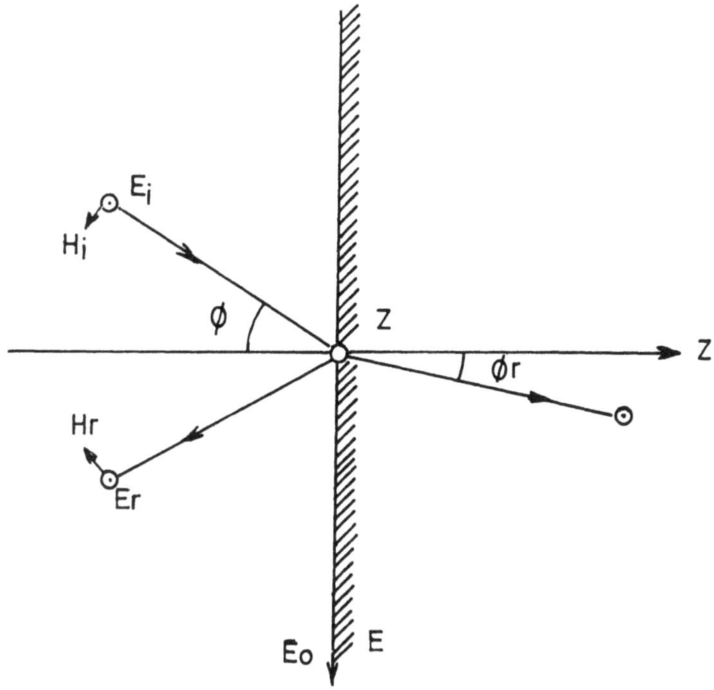


**Fig. 5.** Plane wave reflection and transmission at a dielectric interface.

temperature typically encountered on the surface of ground based radomes can be in the range of – 55 to + 80°C, airborne radomes may be subjected to much higher temperatures, depending on type of aircraft and flight conditions. Fig. 8 shows the estimated equilibrium temperatures on the surface of a 30° conical radome for different aircraft speeds and at different altitudes. For missile radomes, selection of materials should also cater for both thermal shocks and thermal gradients.

Ground radomes will be required to stand pressures resulting from high wind speeds. The pressure of air at velocity $V$ impinging at normal incidence on a flat surface is $P = \rho V^2/2$, where $\rho$ is mass density of air. In general, pressure distribution will depend on the shape of objects in nearby vicinity. In the region 0 to 30 m from ground, the variation in pressure can be about 30%, increasing with height above ground level. In the case of airborne radomes, one has to consider the flight profile and likely manoeuvres of the aircraft. In addition, any pressure differentials must also be considered. Underwater radomes have to be specially designed for these pressure differentials.

Impact strength is another important consideration in the selection of materials. In particular, the airborne radome design should consider impact loads from birds, stones, hails etc. If the radome is mounted on the belly of aircraft, the radome must withstand impact due to sand and dust particles. An important aspect in the design of radomes of this type is to cater for adequate ground clearance during landing and take off. Apart from the above, loads due to ice,

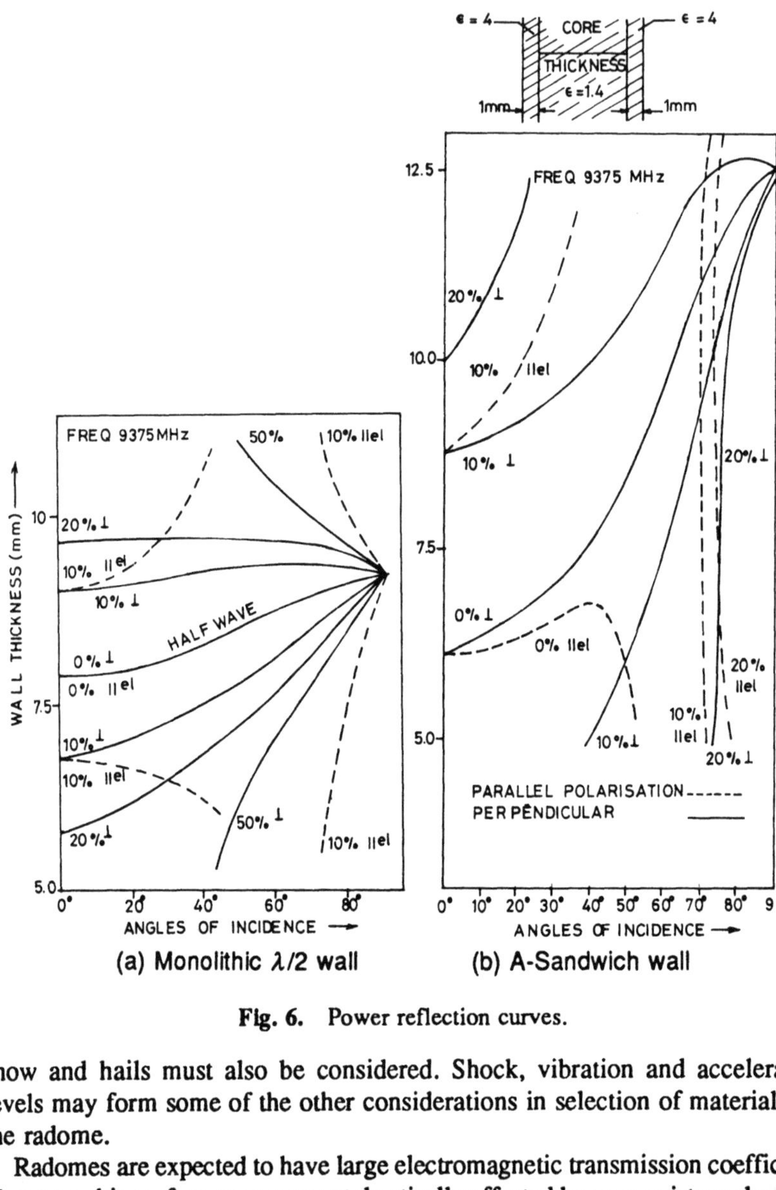


**Fig. 6.** Power reflection curves.

snow and hails must also be considered. Shock, vibration and acceleration levels may form some of the other considerations in selection of materials for the radome.

Radomes are expected to have large electromagnetic transmission coefficient. However, this performance may get drastically affected by any moisture absorbed by the materials forming the radome wall. Dielectric properties of the water vary in the region 80 to 40 over the frequency region of 1 to 20 GHz. The loss tangent in the same frequency region vary from about 0.06 to 0.9. Hence, any possibility of absorption of moisture in the radome wall structure must be totally avoided. Apart from selecting the materials impervious to absorption of moisture, it will be advisable to treat the surface of the radome so that rain water does not stick to surface. If the thickness of water layer over the radome is comparable to wavelength of electromagnetic radiation, the loss will no more be negligible. From practical data collected on ground based radomes,

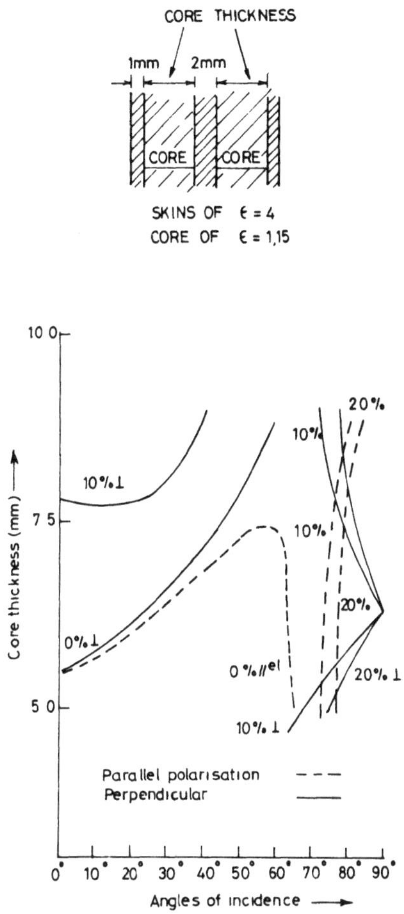


**Fig. 7.** Power reflection of double sandwich.

the thickness of water layer formed on its surface can be expressed by $t = 0.48\ (QR)^{0.32}$, where $Q$ is the rain rate in inches/hour and $R$ the radius of radome in feet [18]. Losses due to such water films can go upto about 2 dB depending on frequency. This emphasises the importance of not only avoiding moisture absorption by the radome materials but also water film formation on surface of radome.

Radomes being an external hardware on any platform, are likely to be subjected to lot of natural or man-created abuse. A major source of such abuse of serious nature arises from corrosion due to chemical fuels or other contaminants with which the airborne or surface radomes are likely to come in contact. In addition radomes for naval use will be required to counter corrosion due to sea environment. Proper material selection to take care of these aspects is essential in radome design. For airborne radomes, static electricity build-up needs to be minimised and this can be done by selecting proper treatment of the outer surface of radome. All coatings used for protection of radome will effect the

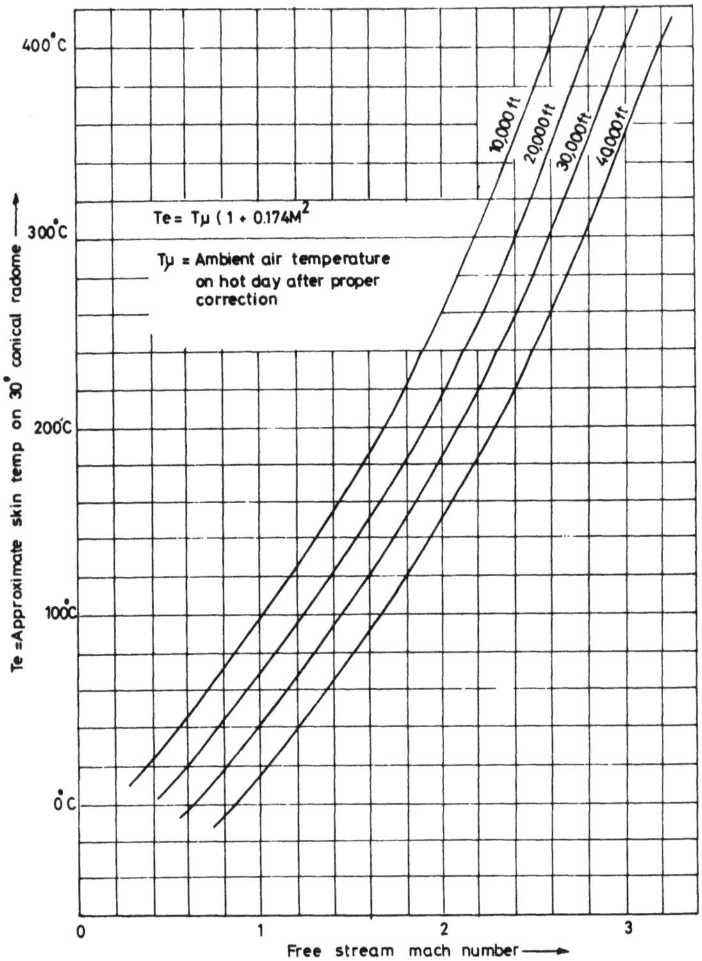


**Fig. 8.** Estimated skin temperature vs machnumber.

electrical performance of the radome, and it is necessary to take their effects while designing the radome.

Electrical parameters of interest in the selection of radome materials are the dielectric constant and loss tangent. Best radome performance is obtained by choosing low loss and low dielectric constant materials. Variation of these values over the range of frequencies of interest dictate the performance of the radome's bandwidth. These electrical characteristics together with the mechanical strength characteristics discussed earlier form the basis for selection of materials for radome design.

## 2.2 Composites

Most of the materials used for radome fall under the category of composite materials. A working definition of composite materials can be "a material system composed of mixture of combination of two or more macro constituents differing in form and/or material composition and that are essentially insoluble in each other" [19].

In principle, composites can be constructed of any combination of two or more materials. But typical composites are those in which a structural constituent called reinforcement is embedded in a matrix. Because the different constituents are intermixed, there is always a contiguous region. It may simply be an interface i.e., a common boundary of the constituents, or a distinct added phase called an interphase. When such interphase is present, there are two interfaces, one between each surface of the interphase and its adjoining constituent as shown in Fig. 9. By suitable combination of the constituents, the composite material can be designed to have the right combination of desired properties. These properties may be related to structural strength, temperature characteristics or the electrical parameters.

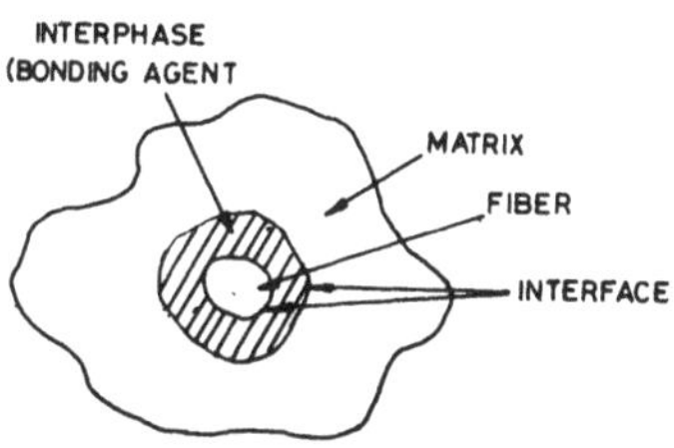


**Fig. 9.** Composites showing constituents and interphases.

The high strength to weight ratio and the capability of tailoring the mechanical and electrical properties of the composites make them very attractive for fabrication of radomes. The constituents of the composites are selected based on the required radome characteristics. Since the radome is required to be electromagnetically transparent, the electrical conductivity of both the reinforcement and the matrix selected for the composite have to be extremely low. In addition they must provide the necessary stiffness to the final composite structure.

### *2.2.1 Fiber Reinforcement*

One of the most common types of reinforcement used in composites is the fiber reinforcement. Glass fibers, aramide fibers and ceramic fibers are the type of fibers generally used in radome fabrication. Out of these glass fibres are most popular.

Whereas mass glass has a strength of 1000 kg/sq cm, the fibrous glass has strength of upto 35,000 kg/sq cm. The stiffness or Young's modulus remain without much change at around $7 \times 10^5$. Basically, the change in strength comes due to less defect in the atomic structure of the fibrous glass.

Different types of glass fibers that can be used as reinforcement are shown in Table 2 [20]. One major problem in characterizing the fibers is that some of the properties noticeably tensile strength and resistivity may depend on many conditions other than the glass itself. For instance, glass fibers drawn singly under clean conditions and tested within minutes of drawing will have

**Table 2. Typical properties of glasses (from [20])**

| Property | E-glass | S-glass | D-glass |
|---|---|---|---|
| Specific gravity | 2.54 | 2.49 | 2.16 |
| Virgin tensile strength (at 72°F) | 3.45 Gpa | 4.6 Gpa | 2.41 Gpa |
| Yield strength (at 1000°F) | 0.83 Gpa | 1.9 Gpa | — |
| Modulus of elasticity (at 72°F) | 724 Gpa | 855 Gpa | — |
| Coeff. of linear expansion/°F | $2.8 \times 10^{-6}$ | $1.6 \times 10^{-6}$ | $1.7 \times 10^{-6}$ |
| Elastic elongation | 4.8% | 5.4% | 4.7% |
| Specific heat | 0.192 | 0.176 | 0.175 |
| Dielectric constant | 6.13 | 5.21 | 4.0 |
| Loss tangent | .004 | .007 | .006 |

considerably higher strength than same type of fiber that has been drawn as a strand, sized, and woven on a reel and stored for some period. Also characteristics such as resistivity may depend very much on extraneous factors such as water content, surface treatment etc. E-glass is a low alkali composition glass primarily for electrical applications. It has 52–56% of $SiO_2$, 12–16% of $Al_2O_3$ and 16–25% of CaO. MgO and $B_2O_3$ vary from 0–6% to 8–13%. S-Glass is a high tensile strength glass which has significant strength to weight ratio, superior strength retention at elevated temperature and high fatigue limit. Typical composition of S-glass is 65% $SiO_2$, 25% $Al_2O_3$ and 10% MgO.

D-glass is an improved dielectric glass developed for high performance radome application. The dielectric constant of D-glass is around 4.0, compared to 6.11 for E-glass and 5.2 for S-glass. The loss tangent is also 0.003, compared to 0.006 for E-glass and 0.007 for S-glass. However, it has lower strength and substantially lower Young's modulus.

Basic raw material for glass fiber is pure molten glass which is extruded through a bushing containing large number of holes. The emergent ends of glass are drawn and sized. Sizing is a process wherein a material is applied on its surface to fill pores and to modify the surface properties to improve adhesion. The range of fiber diameter commonly produced vary from 0.003 to 0.02 mm. Sometimes, conductive additives are added to the fiber to avoid electrostatic charge build-up. Such additives will deteriorate the performance of radomes. As reinforcements, glass fibers can be used in the form of rovings, chopped strands, mats, fabrics and woven rovings.

Aramide fibers under the trade mark of Kevlar is marketed by E.I. Du Pont. It is an organic compound of carbon, hydrogen, oxygen and nitrogen. The relative mechanical properties of Kevlar and other reinforcements generally used in composites is shown in Fig. 10. Kevlar 29 is a low density, high strength aramide fiber useful for ballistic protection. Kevlar 49 is characterised by low density, and high tensile strength and modulus. Kevlar 49 is useful in fabrication of high performance composite applications where light weight, high strength and stiffness are important. It can also resist shattering upon impact and inhibit propagation of cracks. Its stress-strain behaviour is linear upto ultimate failure in tension at 2344 MPa and 1.8% elongation. Its specific

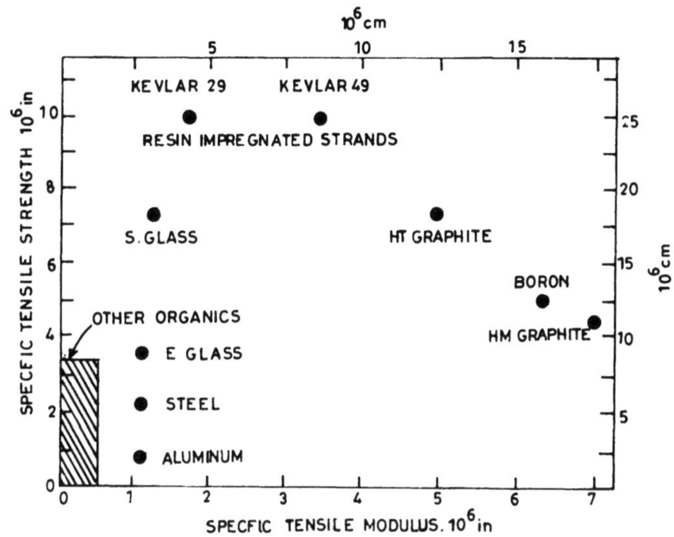


**Fig. 10.** Properties of reinforcing fibers.

modulus is about 4 to 5 times higher than of glass fiber. In spite of these properties, there seems to be some disagreement on its performance under prolonged conditions. Recent reports have indicated that Kevlar does not perform well when used on top of surfaces that are exposed substantially to extremes of weather such as snow and hot sun [21]. Water retention in Kevlar fiber ranges from about 1 to 4% and because of this reason, extreme care should be taken during fabrication of electromagnetic windows using Kevlar. In electrical characteristics, Kevlar compares very well with E-glass fiber. The dielectric constant and loss tangent are about 3.5 and 0.005, respectively.

### 2.2.2 *Resin Matrix*

Selection of resin for use in a particular application depends on a number of factors—compatibility with reinforcement, capability to withstand required temperature and humidity, exposure to ultraviolet rays and usability in the selected fabrication process including necessary tack and drape. Other parameters which determine the selection of resin are its viscosity and wettability to filaments, shrinkage, cure, liberation of volatiles during processing, toxicity and shelf life. The types of resins usable for radome applications are the polyesters, epoxies and polyimides.

Polyester resin is the earliest type of resin which has been popular for use in composites. It is cheap and can be cured under room temperature conditions with proper promoter. It can be used with glass fiber in wet lay-up process or by vacuum impregnation to make laminates. Benzoyl peroxide and amine accelerator can be used to affect room temperature curing. Polyester resin is useful for applications requiring moderate structural strength.

Epoxy resins are characterised by better mechanical properties including good adhesive strength and chemical resistance. They are used for low pressure laminating upto 100 lb/sq in. requiring cure at 350 °F. The maximum usable temperature can be upto 420 °F under intermittent conditions. However, it is felt that there is some degradation of properties after long time exposure to humidity which limits the service temperature to about 275 °F. Epoxy resins are preferred in many applications due to versatility in terms of pot life, curing conditions, viscosity etc. They are used in filament wound radomes and also in sandwich constructions.

Polyimides and polybenzimidazoles are useful for high temperature applications. They have excellent mechanical and electrical properties over wide temperatures and can be used in applications requiring long term exposure over elevated temperatures. For high temperature applications condensation-type polyimides are useful. These resins are however difficult to process and are also costlier than other types by about 5 to 20 times. Polyimide laminates can become water absorbent unless properly sealed.

### 2.2.3 *Other Aspects of Composites*

Performance behaviour of composites not only depend on the materials constituting the composite, but also many other factors—mainly the form and structural arrangement of the constituents and also their interaction. Vast literature is available giving detailed information on these aspects. A few of the major factors which affect performance of radomes are discussed briefly.

#### 2.2.3.1 FIBER-RESIN BOND

The permanence of the mechanical properties of composites, especially their resistance to degradation upon environmental exposure depends on the fiber-resin bond. Sometimes a coupling agent may be used for forming a good bond between the reactive group of the resin and the hydroxyl group which may be formed on the surface of the fiber. A probable explanation is that the coupling agent forms a coating which is more than a molecular monolayer thick and prepares an intermediate zone around the fiber with distinct properties that are beneficial. The ideal coupling agent should provide a low modulus flexible layer at the interface that will improve adhesive strength of the fiber-resin bond and reduce the number of voids in the material.

#### 2.2.3.2 FIBER-MATRIX RATIO

Both quantity and arrangement of fibers affect the composite properties. In general, strength increases directly in relation to amount of reinforcement. The properties of unidirectional composites can be expressed in terms of volume fraction.

The suggested expressions for the electrical conductivity and permittivity in the longitudinal directions are given by mixture rule [22]

$$\sigma_c = V_f \sigma_f + V_m \sigma_m \qquad \text{Conductivity}$$
$$\epsilon_c = V_f \epsilon_f + V_m \epsilon_m \qquad \text{Permittivity}$$

where $V_f$ and $V_m$ are respective fiber and matrix volume fractions ($V_f + V_m = 1$).

Similarly Young's modulus is given by

$$E_c = V_f E_f + V_m E_m \qquad \text{Young's Modulus}$$

However, practical values of permittivity of a composite can better be approximated by using empirical formula [23]:

$$\log \epsilon_c = V_m \log \epsilon_m + V_f \log \epsilon_f$$

Generally, the fiber strength and stiffness greatly exceeds matrix strength and stiffness. Hence, longitudinal mechanical properties are almost entirely governed by the fibers. Compressive strength in the fiber direction is less than the tensile strength, particularly if the fibers are thin or not very straight. For glass reinforcements, the compressive strength may vary from 50 to 75% of the tensile strength. For plastic reinforced with aromatic fibers, however, compressive strength is generally 20 to 25% of tensile strength, due to low compressive strength of the fiber itself.

As regards longitudinal thermal expansion coefficient, the phase with higher modulus tends to impose its expansion coefficient on the final composite property. It can be approximated as [22]

$$\alpha_c = \frac{E_f V_f \alpha_f + E_m V_m \alpha_m}{E_f V_f + E_m V_m} \qquad \text{Linear Thermal Expansion}$$

Strength changes due to causes such as water pick-up will also be governed by similar equation.

The transverse properties are generally governed by the properties of the matrix phase. The transverse strength is mostly lower than that of matrix, being about 50%, but the nature of the fiber matrix interface is particularly important here. Transverse Young's modulus, transverse conductivity and permittivity are also dependent on fiber-volume fraction. The transverse thermal expansion can be approximated by

$$\delta_c = V_f \delta_f + V_m \delta_m \qquad \text{Transverse Thermal Expansion}$$

2.2.3.3 SHORT FIBER REINFORCEMENT

In case the fibers are not continuous, the aspect ratio of the fibers affect the longitudinal properties of the composite. This is because the matrix being less stiff, cannot transfer stress from fiber to fiber. The stiffer the fibers in relation to the matrix, the larger their length to diameter ratio must be if the Young's modulus is to approach that of continuous composite with the same fiber-volume fraction. Same is true for electrical properties such as permittivity and conductivity too; the length of fiber required for satisfactory transfer of flux, current etc. increases with the ratio of the property between the two phases. Aspect ratios of the order of few hundreds for most fibers can give strength close to that obtainable with corresponding continuous fiber composites. Satisfactory Young's modulus can however be obtained by shorter length fibers.

In case bundles of chopped fibers are used in place of individual short fibers, aspect ratio of fiber bundle is more important. Strength and stiffness of such materials are generally not impressive.

### 2.2.3.4 FIBER ORIENTATION

The relationship between strength of the composite and orientation of the reinforcing fiber is shown in Fig. 11. Continuous parallel strands give the highest strength range, bidirectional arrangement gives a middle–strength range and random arrangement gives lowest strength range. Hence by arranging the orientation of fibers it is possible to obtain unidirectional or isotropic properties.

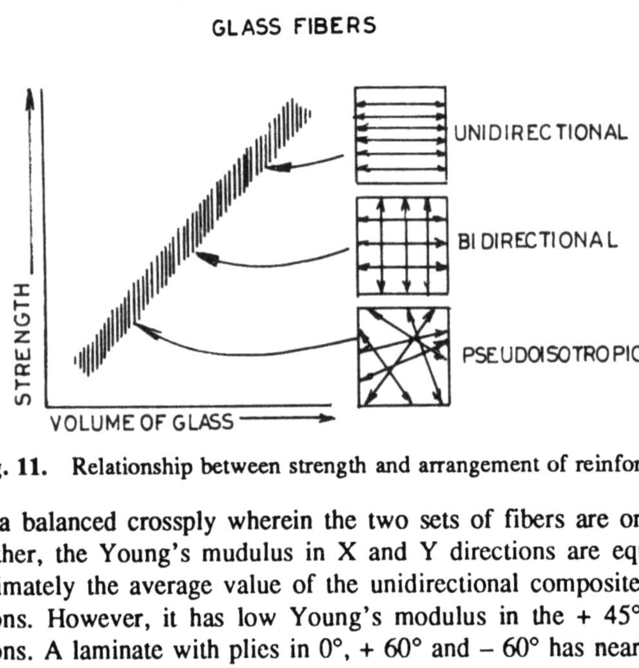


**Fig. 11.** Relationship between strength and arrangement of reinforcement.

For a balanced crossply wherein the two sets of fibers are orthogonal to each other, the Young's mudulus in X and Y directions are equal and has approximately the average value of the unidirectional composite in the two directions. However, it has low Young's modulus in the + 45° and – 45° directions. A laminate with plies in 0°, + 60° and – 60° has nearly the same Young's modulus in every direction. Same is true for electrical characteristics such as permittivity or conductivity in different directions of laminates.

### 2.2.3.5 MULTIDIRECTIONAL WEAVING

Multidirectional woven products approach ideal isotropy in construction, resulting in isotropic physical and mechanical properties. Applications requiring multidirectional woven structures usually are those where extremes in temperatures and highly stressed states are encountered. Very tough fibers such as ceramic or quartz are only used in such applications. Desired composite properties can be properly tailored into the structure during the weaving process. Variation in structural requirements in any of three orthogonal directions can be altered by properly altering volume fraction of yarn in the direction. The variations can be from 10 to 60% in the desired direction. In most of the cases, yarn bundles are used in each direction instead of individual filament or yarn. Blocks are woven with yarn reinforcements as per requirements which can be used for the final application. It is also possible to have multiyarn system where different types of yarns can be used in different directions. Preforms of such hybrid multidirectional types find applications in space. Schematic of 3D, 4D and 5D arrays are shown in Fig. 12. The 7D woven blocks can have yarns in 3 orthogonal directions and 4 diagonal directions.

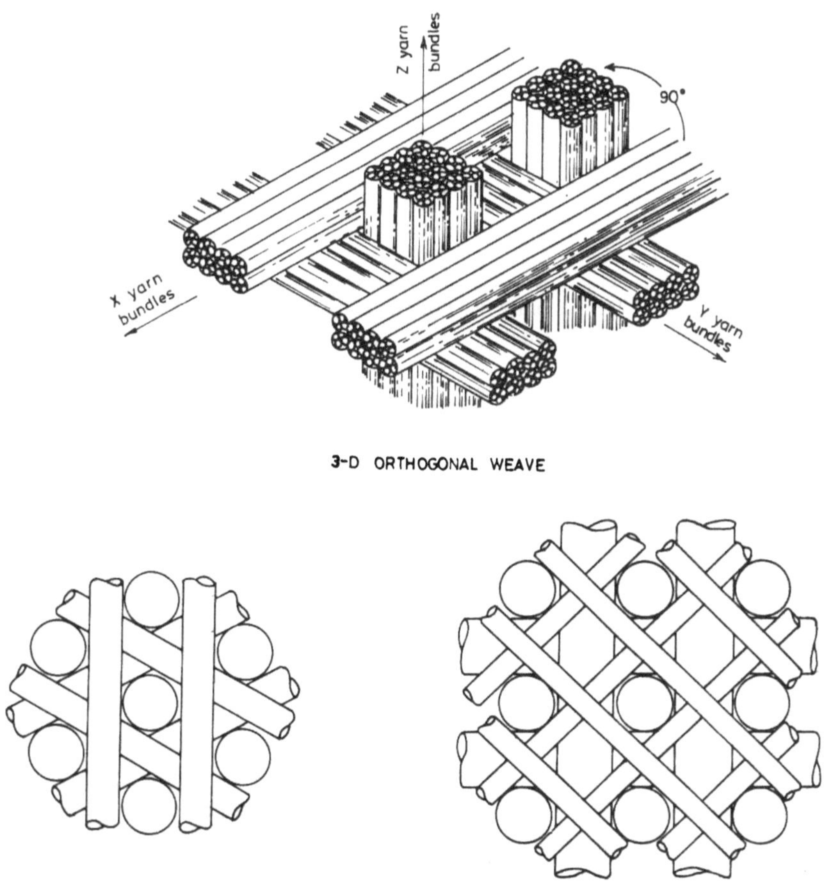


3-D ORTHOGONAL WEAVE

4-D Hexagonal array (Schematic)

5-D Weave Schematic

**Fig. 12.** Multidimensional woven structures.

2.2.3.6 Fabrication methods

Since the performance of a radome is very much dependent on the type of process used in its fabrication, it is necessary for the microwave engineer responsible for radome design to be aware of different fabricational methods and their relative merits. This is shown in Table 3.

Filament winding technique is a popular method since it lends itself to automatic equipment. Selective orientation of reinforcing fibers is possible in this method and hence it offers good design flexibility. Also good control over resin to glass ratio can be obtained in this method. Because of this, uniform dielectric constant of the laminated material and good repeatability can be obtained in this method. By proper adjustment of circumferential and loop windings, mechanical properties in selective directions can be achieved if required.

**Table 3. Manufacturing methods of radomes (from [3])**

| Radome wall configuration | Vacuum bag | Autoclave | Pressure bag | Filament winding | Matched die moulding | Foam-in-place |
|---|---|---|---|---|---|---|
| A. *Solid wall* | | | | | | |
| 1. Halfwave | 1 | 2 | 3 | 1 | 3 | 4 |
| 2. Thin-wall | 1 | 1 | 3 | 1 | 1 | 4 |
| B. *Multilayer* | | | | | | |
| 1. A-Sandwich | 1 | 3 | 3 | 3 | 4 | 2 |
| 2. C-Sandwich | 1 | 3 | 3 | 3 | 4 | 3 |
| 3. Multilayer | 1 | 3 | 3 | 4 | 4 | 4 |

1: Most widely used; 2: Frequently used (dictated by type of materials); 3: Little used (dictated by specialised requirements); 4: Not applicable.

Matched die moulding uses a pair of dies. Fabrication of these dies and their alignment is an important step, since final dimension and finish depend on these dies. Matched die moulding can give accurate and repeatable products of high strength and electrical homogeneity. The technique involves high initial cost outlay in terms of tools and dies. Modification of moulding tools to change radome shape or thickness is difficult and expensive. Matched die moulding technique is suitable for fabrication of radomes based on foam-in-place core or sharply curved thick honeycomb.

In vacuum bag moulding process, dry glass fabric is wetted with resin and laid on male mould. A plastic film bag is placed over the lay-up and after proper sealing, the space between lay-up and plastic bag is evacuated by connecting to a vacuum source. Atmospheric pressure is utilised in moulding the required part. Generally the required thickness is built up by laying a number of layers of reinforcing fabric. It is important to use resin with proper viscosity, so that the resin and air can be easily squeezed from the lay up. This is a low cost technique but getting uniform resin content over complete surface requires skill and practice. This technique is useful only for radomes which are not subjected to high stresses.

Principle used in autoclave technique is similar to vacuum bag moulding but it uses much greater pressures. Also the moulding can be carried out under elevated temperature condition depending on resin requirement. Special autoclave equipment needs heavy capital investment. Preimpregnated cloth called prepregs can be used in this technique and can give accurate, repeatable products with consistent performance.

## 2.3 Honeycomb Core Materials [24]

Typical core materials used in sandwich radomes have low dielectric constant of less than 1.5 and loss tangent less than about 0.02. Most common of them is the non-metallic honeycomb structural material. The basic A-sandwich construction using honeycomb core is shown in Fig. 13. The honeycomb core material is made of hexagonal shaped cells. The cell sizes usually used range

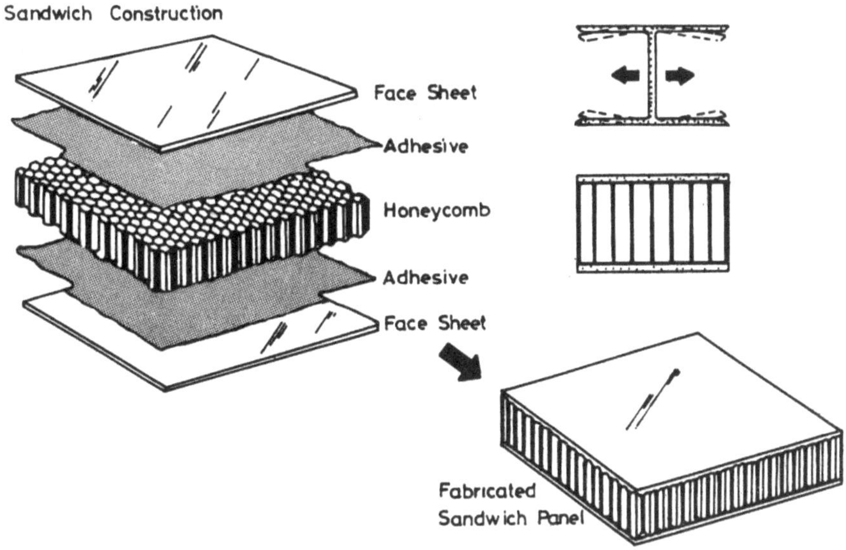


**Fig. 13.** A-sandwich panel construction with honeycomb core.

from 3/16" to 1/2" with densities ranging from 2 to 10 lb/cft. The core to skin adhesive rigidly joins the sandwich components and allows the complete unit to act as single entity with high torsional rigidity. The rigidity obtained by the sandwich construction can be understood by comparing it with an I-beam. The facings of sandwich panel acts as flanges and the core as the web of the I-beam. While loading on one of the facings is in the compression, the other is in tension. The core resists the shear load and increases the stiffness of the structure. In comparison to the web of I-beam, the honeycomb core gives continuous support of the facings. Fig. 14 shows the comparison of sandwich structures with normal monolithic structures.

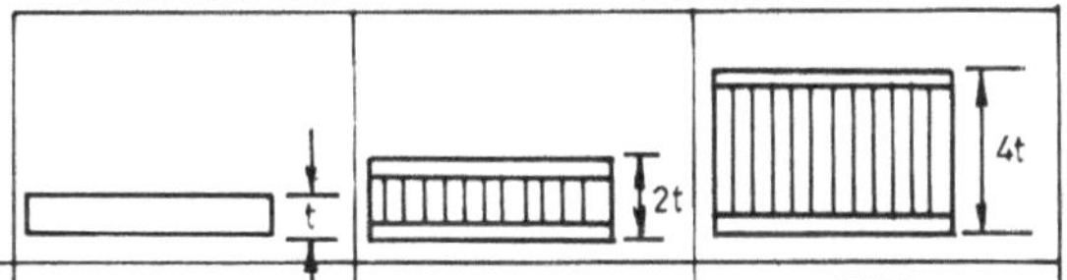


| | Monolithic Wall | Sandwich Wall | |
|---|---|---|---|
| Relative Stiffness(D) | 100 | 700 | 3700 |
| Relative Strength | 100 | 350 | 925 |
| Relative Weight | 100 | 103 | 106 |

**Fig. 14.** Sandwich vs monolithic wall-strength comparison.

Glass fabrics, Nomex paper and Kevlar are some of the materials used for honeycomb core in radome applications. Dielectric constant of different types of honeycomb as a function of density is shown in Fig. 15. Apart from normal standard hexagonal cells, over-expanded and flexible cell honeycombs shown in Fig. 16 are useful in fabricating intricately shaped radome structures. The

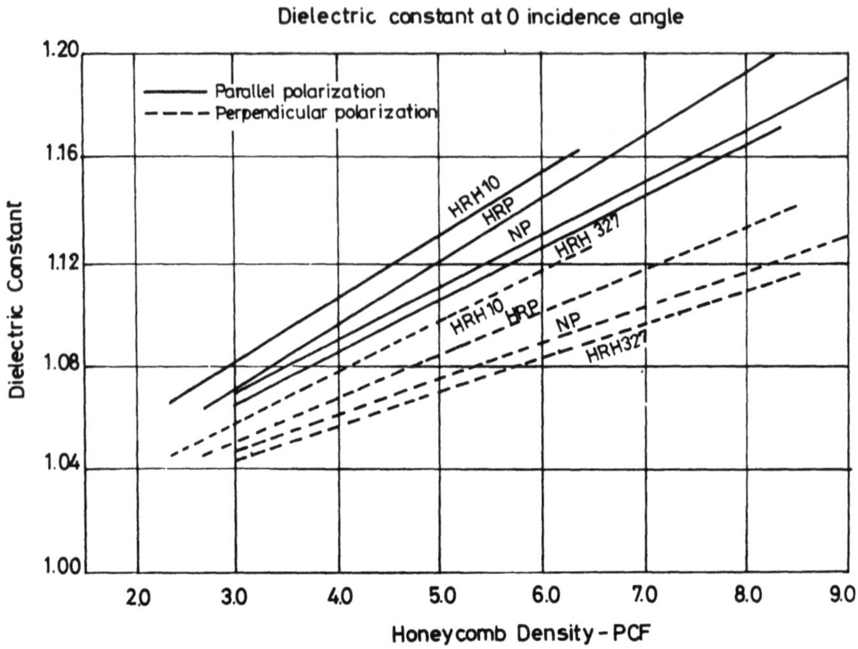


**Fig. 15.** Dielectric constant vs honeycomb density.

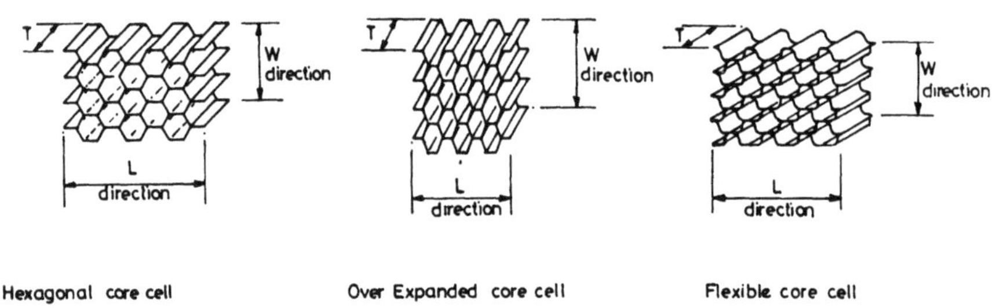


**Fig. 16.** Types of honeycomb core cells.

over-expanded configuration is a hexagonal honeycomb which has been over expanded in *W* direction providing a rectangular cell configuration. This facilitates curving or forming in the L-direction. This process tends to increase *W* shear properties when compared to normal hexagonal honeycomb. The flex core cell configuration provides good formability into compound curvature with reduced antielastic curvature, and without buckling the cell walls. Higher shear strengths are possible with these flexible core honeycomb compared to hexagonal honeycomb structures where curvatures of very tight radii are to be formed.

Typically the Nomex aramide fiber, the main constituent of the non-metallic honeycomb softens at about 450 °F, and hence the properties drop off rapidly at this temperature. However, when returned to ambient conditions, the honeycomb is expected to regain its original strength. Typical properties of honeycomb strength at higher temperatures is shown in Fig. 17.

An important aspect to be considered during fabrication of sandwich radomes

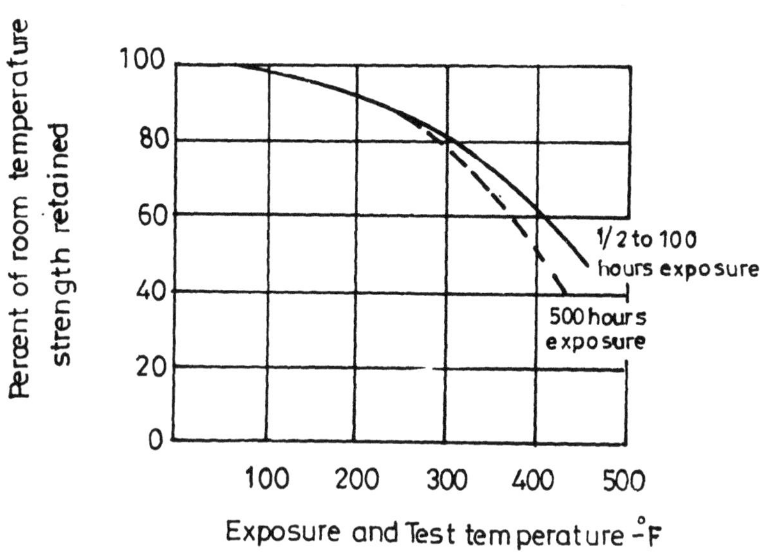


**Fig. 17.** Temperature effects on honeycomb.

using honeycomb cores is the need to properly bond the core to the skins. In order to achieve a good attachment, the adhesive must have the unique qualities of surface wetting and controlled flow during early stages of cure. This controlled flow prevents the adhesive from flowing down the cell wall and leaving a low strength top skin attachment. Adequate attention must also be paid for closing the sandwich structure at the edges, to avoid moisture pickup through these edges.

Transfer of heat through a sandwich panel is dependent upon the basic principles of convection, conduction and radiation. Non-metallic honeycomb cores used for radomes minimise heat flow from the outer skin. When the structure is to act also as a heat shield, flow of heat can further be reduced by filling the cells, so that convection within the cells can be eliminated. Special foams are sometimes used, where possible water accumulation in the cells due to skin damage must be prevented.

## 2.4 Foam Materials

### *2.4.1 Rigid Foam*

Rigid foams with closed cells are useful as core material in sandwich radomes. Depending on structural strength required for the radome, the density of the foam can be chosen. Low molecular weight poliyols are used with isocyanates, resulting in high degree of cross linkage for obtaining the high strength foam. Control of density is done by controlling water added during crosslinking reaction. High temperature polyurethane foam prepared from TDI and alkyol triallycyaniviate copolymer has higher strength [25]. Typical densities used in radomes are of the order 50 to 200 kg/cu m. Compressive strengths of these foams are nearly twice that of normal foams. Typical strength and dielectric constant properties of the foam are given in Fig. 18.

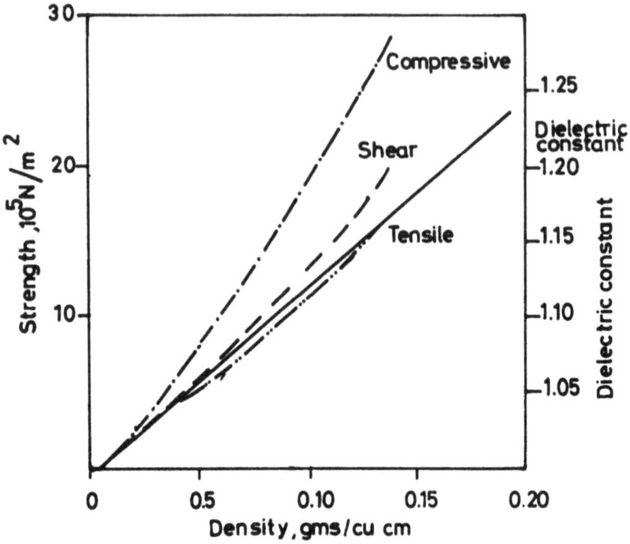


**Fig. 18.** Properties of rigid polyurethane foam.

Intricately shaped structural sandwich radomes can be moulded with foam-in-place core. In such cases, the foam is built up between the skins in the required form and density. However, extreme care has to be taken to ensure that foaming action is proper and fills the complete volume in which the sandwich structure is to be realised. Costly tooling and heavy metallic moulds are required to withstand high pressures which are likely to be built up during the foaming process.

Alternatively, the required shape can be moulded from plain sheets by heating it upto the foam softening temperature. This temperature is a function of the foam materials and density. Uniform heating in thermostatically controlled air circulated oven to attain equilibrium temperature is preferred. The sheet is then removed from the oven and placed on preheated male metal mould. A female mould conforming to the outside profile of the core shape is then pressed over the foam sheet. This technique is called thermo forming and intricate doubly curved rigid foam cores for sandwich radome can be fabricated by this technique. Skins are then bonded to the shaped core to make the sandwich structure. Radome of tangent ogive and other shapes fabricated by this method have shown good electrical and mechanical performance [26].

### *2.4.2 Syntactic Foam*

Syntactic foams are produced from low density glass microbaloons. The hollow glass microspheres range in density from 5 to 50 lb/cft and in size from 0.02 to 0.2 mm diameter. These microbaloons are mixed with proper resin to yield low density foams used as core. The syntactic foam serves to make the sandwich structure lightweight, increase its stiffness and improve crack resistance. Typical dielectric constant of this type of foam range from 1.5 to 2.5, densities from 0.5 to 1 g/cu cm and loss tangent from 0.005 to 0.01. Syntactic foam can have more ordered cells so that sandwiches with thinner skins can be built improving the electrical efficiency [27]. Both additive and condensation polymides can be used in conjunction with syntactic foam to build high performance radomes. Unpublished reports have shown good phase characteristics for radomes built using syntactic foam.

## 2.5 Materials for Ground Based Radomes

Ground based radomes enclosing large antennas would normally be spherical or part-spherical structures. The structures can either be air supported or can be rigid. The air support inflatable radomes can be lightweight and designed for mobile stations. They will require continuous air pressure to maintain support of the radomes. The risk factor for these radomes is rather high since the structure can collapse due to accidental fabric failure or due to abnormal wind or snow conditions. The rigid radomes are generally made of panels joined together by means of metallic or dielectric members. For easy fabrication, all the panels are designed to be identical and can be either flat or curved. The panels can be of thin wall, halfwave or sandwich configuration.

For air supported radomes, thin rubberised fabrics or thin plastics such as dacron can be used. Thin low loss glass fabrics coated on both sides with teflon can also be utilised. These materials have dielectric constants of the order 3 to 3.5 and loss tangents of about 0.01. Typical thickness range from 0.5 to 1.5 mm. Outer coating of hypalon or tedlar is used to give better weather resistance characteristics. RAYDEL is a composite fabric of high modulus reinforcing yarns, coupled with a matrix based on perfluoropolymer resin [28]. It is a registered trademark product of Chemical Fabrics Corporation and has been used for air supported and pretensional radome designs. With a dielectric constant of 2.7 and loss tangent of 0.005 at 24 GHz, it can be used to design low loss broadband ground based radomes. The special resin used in its formulation gives it the capacity to perform well even in adverse rain conditions.

Stretched polypropylene film material has dielectric constant of 2.3 and loss tangent of 0.003 and can be used for making homogeneous tuned wall radomes [29]. It has excellent fragment-defeat properties, is self bonding under heat and pressure, and can provide high level of ballistic protection against fragmentation munitions. In order to maintain high ballistic properties, the film panels must be constructed free from entrapped air and moisture with uniform interply and proper polymer orientation. Polyurethane foam glass fiber laminate panels with suitable flange attachment can also be used as panels. Where sandwich structures are essential from strength consideration, A-Sandwich with proper cores are used. Wherever metallic supports for membranes are used, the design has to take care of the effects due to aperture blockage.

## 2.6 Materials for Missile Radomes

### *2.6.1 General Considerations*

Seeker missiles are the precision guided weapons which have enormous potential in modern warfare. The fully autonomous active seeker has many operational advantages but presents great design challenges, compared to semi-active or passive types. Fig. 19 shows a simplified block diagram of seeker electronics. Highly integrated transmitter-receiver modules having small size and weight are sometimes gimbally mounted along with the antenna inside the seeker

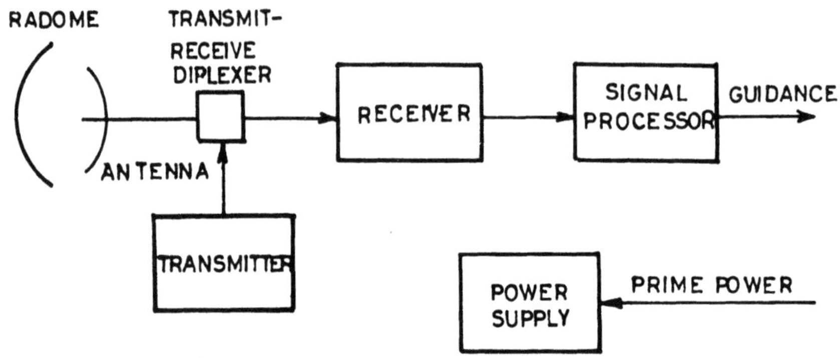


**Fig. 19.** Typical missile seeker head electronics.

head. The output is the guidance data which is fed to the autopilot for guiding the missile path.

Sensor antenna and radome form an important subsystem of the missile seeker. Compared to the microwave antenna, millimetric wave sensors offer better resolution and adverse weather capacity. Millimetric wave sensors are also being considered for re-entry interceptors for exo-atmospheric operations. Seekers operating on radiometric principle require wider rf bandwidth sensors. For better target discrimination, multi-mode seekers are used where information from more than one sensor is used for guidance. In all cases, radome design must be compatible with sensors.

### *2.6.2 Radome Requirements*

Radome for missiles must satisfy electrical, mechanical and aerodynamic requirements. Halfwave radome with hemispherical dome offers good electrical performance, but the aerodynamic drag may not be acceptable. Halfwave radome may cater for 5 to 20% frequency bandwidth, depending on material dielectric constant and fineness ratio. Broadband antiradiation homer antenna will require multilayer radome wall.

Amplitude distortion caused by radome material results in transmission loss and subsequent degradation of system sensitivity. Phase distortion introduced by radome causes error in boresight and boresight slope. Proportional navigational guidance used in some missiles is sensitive to target position error slope and any error introduced in boresight slope by the radome will affect the overall performance. General guideline for acceptable values for missile seeker radomes are transmission loss of 1 dB, boresight error of 1.0 milliradian and boresight error slope of 0.05 deg/deg. Frequency modulation is sometimes preferred to conventional pulse modulation in missile seekers since it offers high average power and good short range performance. In such cases, transmitter-receiver isolation and front-end VSWR become important design demands. Reflections from radome must be minimised in such cases for all antenna scan angles.

Seekers employing common aperture for dual mode sensors present many challenges in radome design. In such cases, the radome design will be required to be optimised at several wavelengths which may be widely separated.

### *2.6.3 Materials*

Conventional resin glass composites can be used only for low speed missiles, which do not demand operation in high temperatures. For speeds of the order of Mach 3 or above, it is necessary to use inorganic materials based on silicon, glass, or ceramics. Apart from capability to withstand high temperature without deterioration in performance, these materials can be designed to provide adequate

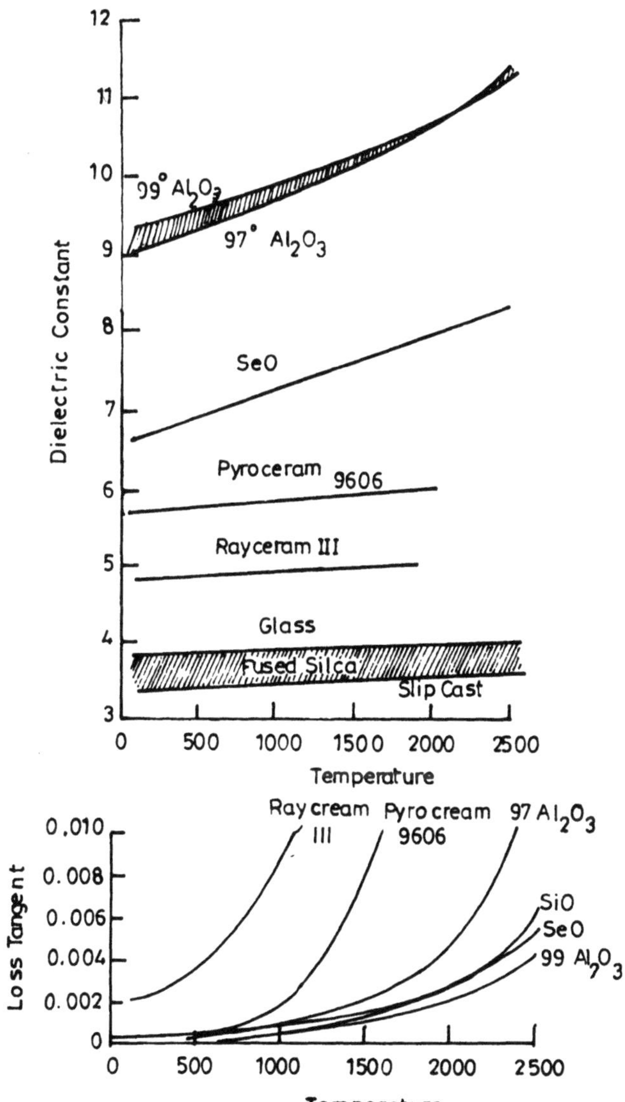


**Fig. 20.** Electrical characteristics of some inorganic materials.

protection against rain erosion. Electrical properties of some of the inorganic materials useful in radome fabrication are shown in Fig. 20.

### 2.6.3.1 SOLID WALL RADOME MATERIALS

Two important types of materials which can be used for solid wall high temperature radomes are 99+% pure alumina and glass ceramics. Pyroceram 9606 is a glass ceramic available from Corning Glass Co. useful as a high temperature missile radome material. Some of the main features of alumina and glass ceramic materials are shown in Table 4. Whereas dielectric constant

**Table 4. Typical properties of alumina and glass ceramic materials**

| | Alumina | Glass ceramic (Pyroceram 9606) |
|---|---|---|
| Specific gravity | 3.7 to 3.9 g/cu cm | 2.62 g/cu cm |
| *Variation over 25 to 500°C* | | |
| Dielectric constant | 9.6 to 10.3 | remains fairly constant at 5.7 |
| Loss tangent | 0.0001 to 0.0005 | 0.0002 to 0.001 |
| Flexural strength | 270–250 × $10^6$ N/m$^2$ | 235–200 × $10^6$ N/m$^2$ |
| Young's modulus | 380–350 × $10^9$ N/m$^2$ | 120 N/m$^2$ |

of alumina is about 9.6, that of glass ceramic is 5.65. Hence for halfwave design, physical thickness of glass ceramic radomes is larger than of alumina. Due to increased physical thickness, its mechanical properties are adequate in most cases, even though some of the mechanical strength characteristics of glass ceramic materials are poorer. Also, though density of glass ceramic is less, it has only marginal weight advantage over similar alumina design. Both materials being hard, can withstand rain erosion and thermal shocks well. For alumina, thermal differentials in excess of 400°C can result in failure. Above 500°C, mechanical strength of pyroceram deteriorates more drastically than of alumina. In its pure form, permittivity of alumina increases from 9.6 at room temperature to 11.4 at 1000°C. By adding titanate, it can remain nearly constant at about 11.6 over complete temperature range.

### 2.6.3.2 SILICA BASED RADOME MATERIALS

Fused silica formed by sintering slip casting process is the less dense form of silica. At a specific density of 2.2 g/cu cm, its dielectric constant value of 3.5 remains fairly constant over large temperature range. However, being porus in nature, methods of preventing water absorption have to be incorporated in making radomes out of fused silica. The material has also poor rain erosion resistance.

Silicon nitride has excellent strength and erosion resistance characteristics. The dielectric constant is dependent on the density of the material varying from about 8 for a density of 3.2 g/cu cm to 5.7 for a density of 2.4 g/cu cm. Reaction sintered silicon nitride can be prepared to give much lower densities. Gas evolution taking place during this process allows formation of porus structures. The amount of porosity, and thus dielectric constant is dependant on composition and process parameters. However, a hard impervious coating is required for these low density material to provide resistance to water absorption and to sand/rain erosion. The loss tangent can generally be kept less than 0.005 in all cases. It is observed that variations in dielectric constant with temperature in silicon nitride are intrinsic in nature and little improvement can be achieved by process variation. On the other hand, the loss is predominantly controlled by intergranular phases and impurities. Monolithic radomes using reaction bonded silicon nitride can be used for narrow band applications.

By having a two layer structure of low or medium density silicon nitride core as a thickbase, with a thin skin of higher density silicon nitride, it is possible to achieve acceptable radome performance over a wide range of frequencies. The high density outer skin can provide erosion resistance and can also form a suitable base for moisture sealent.

Chemical vapour deposited silicon nitride (CVDSN) has the potential as a multiple frequency window material and can be used in conjunction with sensors operating at different wavelengths.

### *2.6.4 Frangible Glass as Radome Material*

Frangible materials, as referred here, are those which shatter into a multitude of small fragments in precisely predictable fashion. Chemically strengthened glasses and glass ceramics can be designed to break with very small but concentrated amount of energy. One of the methods by which frangibility of materials can be exploited in the field of missiles is by employing a frangible dome over a dielectric hemispherical radome. The outer frangible dome can be aerodynamically shaped so that it can protect the inner parts of the missile from the rigours of atmospheric flight in lower atmosphere. Then, as the missile reaches less dense regions, the frangible dome would be triggered to be blown off, exposing the simple low dielectric hemispherical dome. Being electrically superior, the hemisphere can perform much better and guide the missile to the target. Thus the frangible dome could function as a protective cover for the electrically superior radome of low mechanical strength. By properly choosing the glass ceramic material for the outer dome it can be designed to not only give protection to the inner radome, but also enable a strong signal to guide the missile on its initial flight path.

## 2.7 Materials for Simple Electromagnetic Windows

Sometimes, simple rf transparent window will be required to protect the antenna, with no major structural requirements or aerodynamic loads. Thermal plastics can be used as EM windows for such applications. Typical thermal plastic materials that are available commercially are given in Table 5, along with their main electrical characteristics. These materials are generally used in thin wall configuration. Thermal plastics can be moulded into the required shape by vacuum forming or injection moulding and this offers economically viable solution when large quantities are involved.

Noryl is a polyphenylene oxide and is useful for small ECM antennas such as cavity backed Archemedian spirals. Since thin wall configuration is used, the axial ratio of the original antenna can be preserved. Open ended waveguides in large phased arrays can be protected by having Noryl protection caps. Teflon

**Table 5. Some thermal plastics useful as EM windows**

| | Manufacturer | Dielectric constant | Loss tangent |
|---|---|---|---|
| NORYL (Polyphenylene oxide) | General electric | 2.58 | .005 |
| KYDOX (PVC Acrylic) | Rohn and Hass | 3.44 | .008 |
| TEFLON (Tetrafluoro ethylene) | E.I. Dupont | 2.1 | .0005 |

and Lexan are other thermoplastics which can be used in the form of thin flat sheets or in the contour conforming to the platform. The dielectric constant of Teflon is so low, that larger thickness of protective cover can be used compared to glass laminates. It can be used even at subsonic speeds, if certain amount of erosion can be tolerated.

## 2.8 PROTECTIVE FILMS AND COATINGS FOR RADOMES

In order to protect radomes in certain operational environments, it is advisable to use protective films or coatings. Some of the important coatings of relevance for radomes are for protection against erosion due to rain, sand and dust, moisture pick-up, static electricity build-up, ultraviolet radiations etc.

### *2.8.1 Protection Against Rain and Sand Erosion*

To combat erosion due to rain, and sand, neoprene coating of about 0.015" (0.4 mm) thickness on radomes have been used in the past. The dielectric constant of Neoprene is about 2.8 and its effect must be accounted for in predicting the radome performance. Under prolonged heavy rain conditions at high speeds, radomes with these coatings start eroding. Hence such radomes require periodic recoats.

Polyurethane is another useful coating material which can be used for reducing erosion due to rain. These are created by reacting isocyanatic groups containing an active hydrogen, such as primary or secondary amines or other methanes. Such a coat used as top coat on a three layer coating system could have 5 to 8 years of service. It is generally used over an epoxy, polyamide or polysulfide primer. Special coating and curing procedures are generally recommended for these coatings. Successive coats should be applied when the previous coat is set-to-touch but not tack-free. Polyurethane coatings are about 4 times tougher than neoprene coating systems [30].

Long exposure to ultraviolet radiations can cause deterioration in case of ground radomes, especially if they are made of foam. A white protective coating of epoxy or acrylic loaded with titanium dioxide can give protection against these radiations.

## 3. EFFECTS OF MATERIAL VARIABLES AND TOLERANCES

For proper engineering design, it is essential to have a knowledge of the effects of variables and tolerances on the final performance of the radome. A few of them having direct bearing on the electrical aspects of radome are discussed briefly here.

### 3.1 Errors in Manufacturing Thickness

Theoretical power reflection percentages for different errors in thickness for a halfwave radome sheet is shown in Fig. 21. These curves are for a flat monolithic dielectric sheet of dielectric constant 4.0 designed to operate 9375 MHz. An error in thickness of 0.0005" results in 1% power reflection at 60° angle of

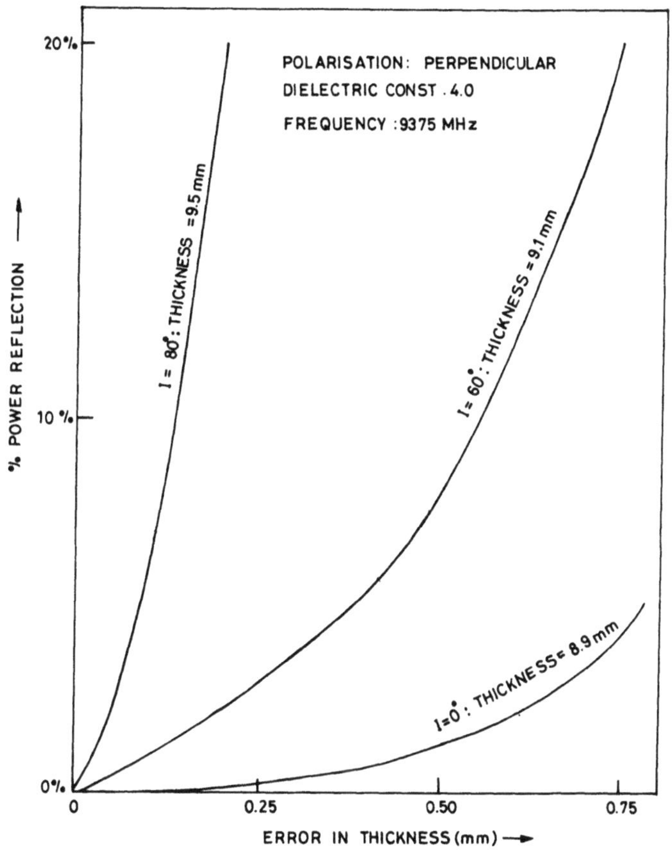


**Fig. 21.** Effect of solid laminate thickness errors on power reflection.

incidence; this value rises to 8% at 80°. For a corresponding sheet at *Q* band (37.5 GHz) these values are 5% and 28% at 60° and 80°. Need for control of wall thickness for streamlined radomes having large angles of incidence is obvious. Even for sandwich radomes, small errors in thickness of skins or core are not often serious at low angles of incidence. But if the shape of the radome is such that high angles of incidence are likely to be encountered, strict tolerances on sheet thickness is essential.

### 3.2 Thickness Change Due to Erosion

Sometimes, eventhough tight quality control measures have been imposed to ensure proper wall thickness, deterioration in performance over a period of time may be noticed due to erosion of the radome wall. Airborne radomes flying at supersonic speeds in heavy rain are subjected to erosion, and subsequent poor electrical performance. A change of the order of 0.005" is not unreasonable for a neoprene coated radome. In such cases, power reflection coefficient of 5% at X-band and upto 20% at *Q* band for 80° radome can be expected. Periodic maintenance with recoating is the possible solution for overcoming this.

### 3.3 Thermal Expansion of Radome

Radome thickness can vary due to the thermal expansion. It is estimated that a change of about 100°C normally encountered during a supersonic flight can result in about 4% rise in reflection of power at X-band. More serious problem, than the increase in wall thickness due to thermal expansion, is the differential expansion between materials used in different parts of a radome. Stiffener rings, ablation boots and similar parts of a radome are generally made of metal alloys whose thermal expansion coefficient are different from materials used in radome fabricaton. Careful engineering design is required for countering such problems.

### 3.4 Effects Due to Variation in Dielectric Constant

Dielectric constant is the single most important parameter to affect the radome electrical performance. Hence this is required to be well controlled during the manufacture and the life-time of radome. Fig. 22 gives some idea of the effect of variation in dielectric constant on a resonant halfwave radome. The change in the value of dielectric constant of radome wall can be either due to processes

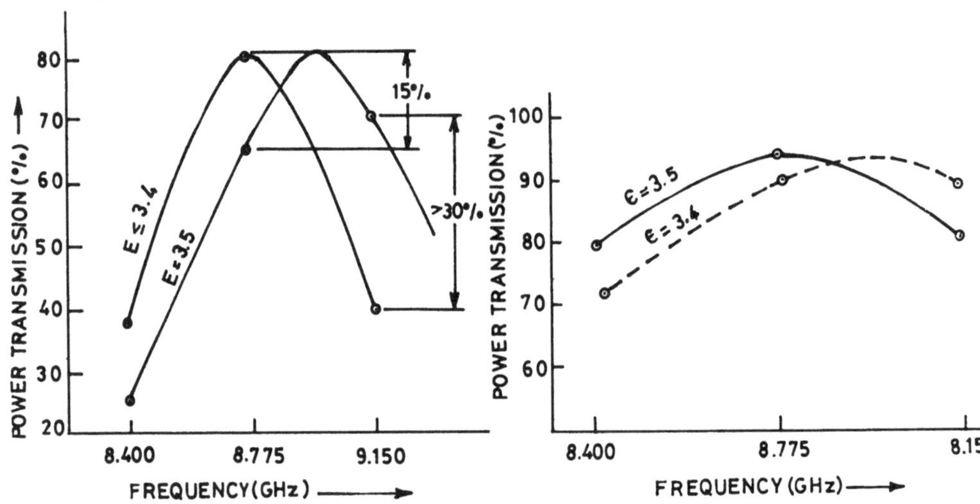


**Fig. 22.** Effect of variation in dielectric constant on radome performance.

or materials involved during fabrication, or can be due to external environmental effects. Strict control over the characteristics of resins, reinforcement and volume ratios during the fabrication of composites is essential to maintain the final electrical performance. Parameters such as fiber orientation which have been discussed earlier are also important.

The curing procedure involved in the fabrication of radome requires application of simultaneous heat and pressure in proper sequence. Whereas heat is used to facilitate and control chemical reaction, pressure is applied to squeeze any excess resin and to minimize void content. Selection of proper curing process depends on materials involved. Various related aspects such as resin flow, pressure distribution during curing, compaction and formation of voids have

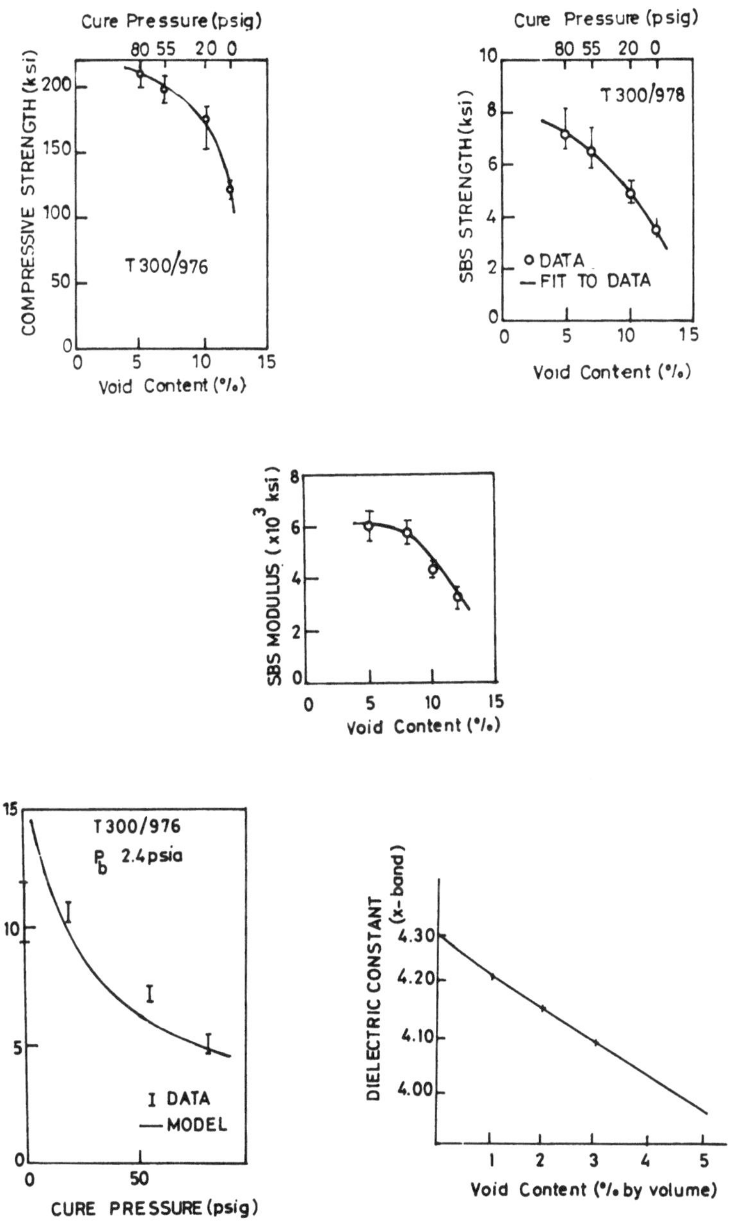


**Fig. 23.** Relationship between mechanical properties and curing parameters.

been investigated in detail through suitable model studies [31, 32]. Experimental measurements to relate the void content with cure pressure are done through photomicrographic analysis [33]. The relationship between cure pressure, void content and compression strength for a thermoset resin composite system is given in Fig. 23. Factors during operation, such as temperature variations and any moisture absorption would result in change of dielectric constant. If moisture content is high, pressures produced by heated vapours of air trapped in the voids can cause variations in its electrical characteristics, as well as result in physical change. This aspect must be viewed seriously if sandwich radomes using honeycomb core is subjected to large temperature variations.

### 3.5 Minimising Radome Errors

Though some of the errors pointed out earlier cannot be totally eliminated, it is essential that adequate precautions are taken at all stages including selection of materials, design, fabrication and operational maintenance.

At the material selection stage, it is advisable to select proper compatible resins and fibers taking into account the strength requirements. It is essential to have a perspective idea of the environmental and other stresses to which the radome is likely to be subjected during its life-time. Errors due to irregular properties of basic radome materials can be avoided by checking these values just before use. Apart from mere selection of materials, attention must be paid to proper control of fabrication and curing methods to ensure homogeneity, proper bonding and to minimise voids. Matched die moulding with heavy pressures can keep dimensional thickness errors during manufacture within tolerable limits. Use of woven socks or preforms of required shape can eliminate increase in thickness at joints or overlaps.

To avoid ingress of moisture, proper sealing at the edges is essential. The humidity problem might be minimised by keeping the radome covered while not in use. For airborne radomes, this will leave the radome surface dry and free of surface moisture before take off. It would be advisable to inspect visually for any damage or erosion periodically, so that corrective measures like repair or total replacement can be taken before catastrophic failure takes place.

## 4. RADOME MATERIAL MEASUREMENTS

Standard measurement techniques which are applicable for characterising mechanical and electrical properties of materials are also directly applicable for radome materials. Excellent review of electrical characterisation of materials with lots of reference is given in literature [34]. There are however certain specific types of measurements which are relevant to radomes during fabrication and development and these are dealt briefly in the following paragraphs.

### 4.1 Flat Panel Tests

It is advisable that properties of all constituent materials are measured before fabrication of radome. A proper record of these results will help in analysing any malfunction of radome subsequently. For aiding this, it is common practice to make flat panels using the same materials under the same working conditions as are used for actual radome fabrication. All other parameters are also kept identical so that the flat panels truly characterise the radome wall.

Electrically, the flat panel is tested for transmission loss and insertion phase by following free space method. A typical set-up wherein measurements can be carried out using a network analyzer is shown in Fig. 24. Corrections to accommodate effects due to multiple reflections can be incorporated to improve accuracy [35]. By rotating the sample panel, transmission coefficient can be

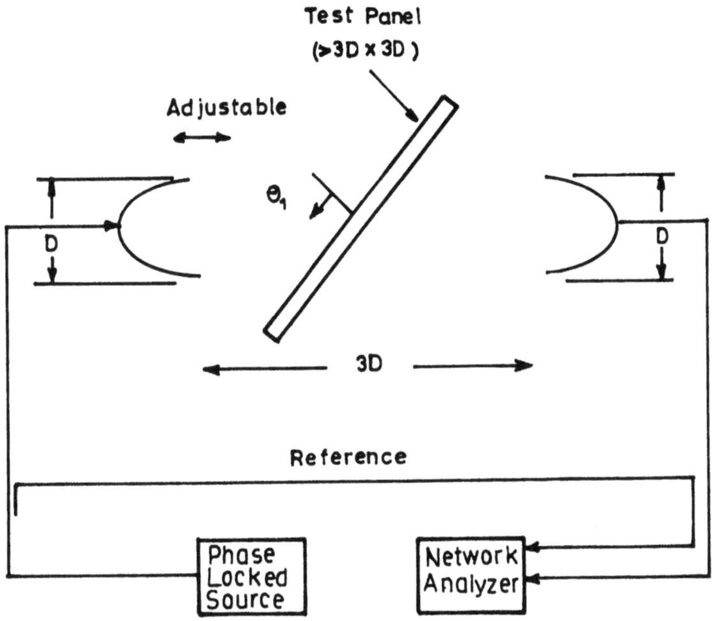


**Fig. 24.** Free space panel measurement set-up.

obtained as a function of incident angle. Curve fitting can be used to estimate value of dielectric constant within an error of 0.004 to 0.009 and loss tangent within 0.00003 [36].

### 4.2 Pattern Comparison Method

The ultimate purpose of radome is to ensure minimum distortion to antenna radiation pattern. Hence one of the practical methods of checking the performance of radome is to plot radiation characteristics of the antenna with and without the radome. By noting the differences in major features of the two radiation patterns, effects such as loss in gain, beam broadening, boresight shift, depolarisation due to radome can be studied and the radome accepted for qualification. Though lot of automation can be built in, the basic set-up for the measurement is shown in Fig. 25. Reflections caused by the radome, can also be checked by following the principle of substitution. These set-ups can also be used to optimise the location of the antenna within the radome for certain critical applications.

### 4.3 Performance Tests as Useful Production Tools

During the production of highly sophisticated radomes of complex design, it is

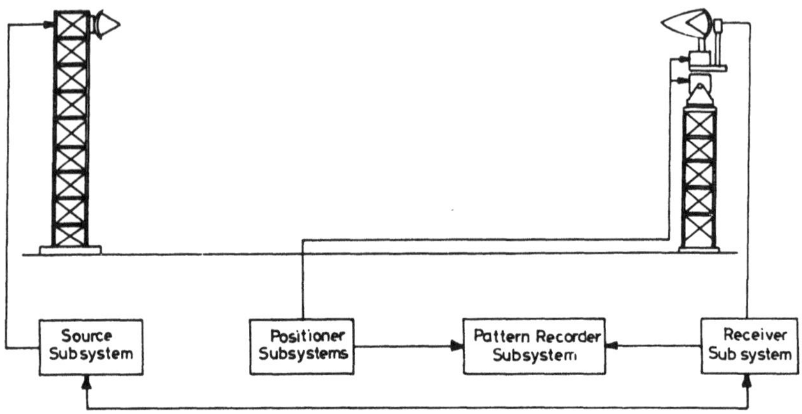


**Fig. 25.** Pattern comparison method for evaluating radome.

necessary to verify the correctness of the steps followed in the fabrication. Any manufacturing error due to oversight or fall in quality of workmanship may prove to be costly to be rectified when detected at the final stage. Even after production, it is worth checking the performance of the radome using some gauges prior to shipment and installation with antenna assembly. These gadgets are normally evolved from simple original concepts to suit the specific requirements.

### *4.3.1 Concept Based on Loading of Resonant Circuit*

Two instruments called Osculating Cavity Dielectrometer (OCD) and Osculating Electrode Dielectrometer (OED) were developed to permit monitoring of fabrication for E–3A AWACS radomes [37]. Since the radome design involved C-sandwich with step-tapered glassply lay-ups, and machine tapered honeycomb cores, precision monitoring of wall thickness was required. Both OCD and OED use rf oscillators with rf resonant circuits. The basic principles of OCD and OED are shown in Fig. 26. OED uses interdigital electrodes, and the OCD a 1" diameter antenna aperture through which the rf interacts with the radome

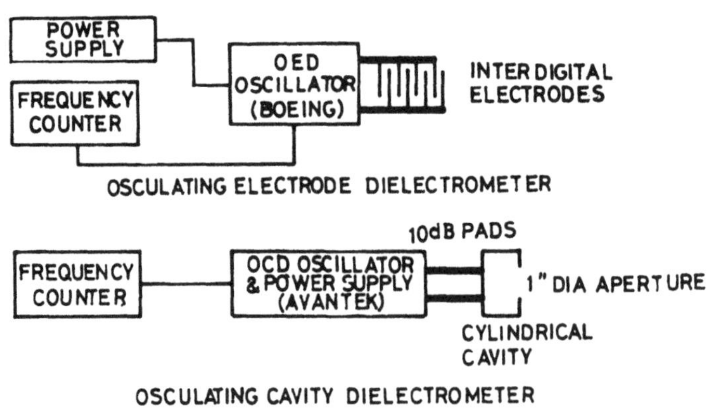


**Fig. 26.** OCD/OED equipment set-up.

skin. The oscillator frequency shift is directly related to electrical thickness of the material. The OED is placed closed to radome surface and conforms with radome curvature, whereas the OCD is spaced approximately 1/8" above the radome surface. OED is more sensitive to radome surface roughness, but less sensitive to radome curvature. Tests are carried out at different locations on the radome surface to check uniformity.

### *4.3.2 Concepts Based on Reflections from Radome*

Microwave stethoscope based on reflection principle is used to monitor reflections of power as shown in Fig. 27. The reflected power over the test frequency band can be plotted as dB below 100% reflection and can serve to monitor the electrical thickness of radome. Comparison made with theoretical reflections or

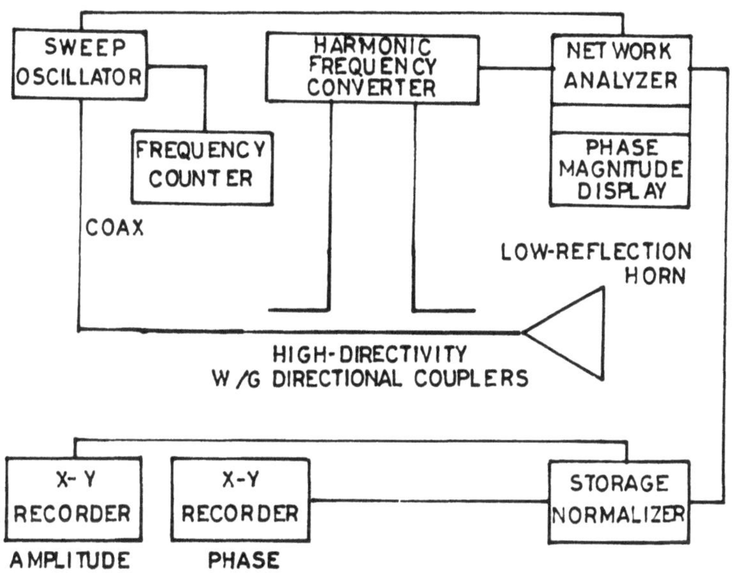


**Fig. 27.** Microwave horn stethoscope.

calibrated panels can qualify the radome. The same principle can also be used for assessing the quality of anti-static coatings and also any defects in radome wall such as moisture pick-up or wall thickness variations due to erosion [38].

A simple gauge based on same principle uses short circuit stubs and shaped grooves on its flange to make measurements more reliable and accurate [39]. A flexible conducting foil is held in intimate contact with the internal surface of radome and acts as a back plate of the cavity. The gauge has been found to be useful for pre-shipment checks.

## 5. USEFUL SPECIFICATIONS, STANDARDS, TEST METHODS AND DATABASES

Some of the specifications, standards and test methods connected with radome materials and related aspects are as follows:

## 5.1 Specifications

### *5.1.1 Materials*

| | | |
|---|---|---|
| L-P-383 | : | Plastic Material, Polyester Resin, Glass Fiber Base, Low Pressure Laminated |
| MIL–C–7439 | : | Coating System, Elastomeric, Rain Erosion Resistant, and Rain Erosion Resistant with Anti-Static Treatment, for Exterior Aircraft and Missile Plastic Parts |
| MIL–R–7575 | : | Resin, Polyester, Low Pressure Laminating |
| MIL–C–8073 | : | Core Material, Plastic Honeycomb, Laminated Glass Fiber Base, for Aircraft Structural Applications |
| MIL–C–8087 | : | Core Material, Foamed in-Place Polyester Disocyanate Type, Interchangeability and Replaceability of Component Parts for Aircraft and Missile |
| MIL–C–9084 | : | Cloth, Glass, Finished, for Polyester Resin Laminates |
| MIL–R–9299 | : | Resin, Phenolic, Laminating |
| MIL–R–9300 | : | Resin, Epoxy, Low Pressure Laminating |
| MIL–R–25042 | : | Resin, Polyester, High Temperature Resistant, Low Pressure Laminating |
| MIL–P–25395 | : | Plastic Materials, Heat Resistant, Low Pressure, Laminated Glass Fiber Base, Polyester Resin |
| MIL–P–25421 | : | Plastic Materials, Glass Fiber Base—Epoxy Resin, Low Pressure Laminated |
| MIL–R–25506 | : | Resin, Silicone, Low Pressure Laminating |
| MIL–P–25515 | : | Plastic Materials, Phenolic Resin, Glass Fiber Base, Laminated |
| MIL–P–25518 | : | Plastic Materials, Silicone Resin, Glass Fiber Base, Low Pressure Laminated |
| MIL–C–27315 | : | Coating Systems, Elastomeric, Thermally Reflective and Rain Erosion Resistant |
| MIL–C–81773 | : | Coating, Polyurethane, Aliphatic Weather Resistant |
| MIL–C–83231 | : | Coatings, Polyurethene, Rain Erosion, Resistant for Exterior Aircraft and Missile Plastic Parts |
| MIL–C–83286 | : | Coating, Urethane, Aliphatic Isocyanate, for Aerospace Applications |
| MIL–Y–83370 | : | Yarn, Roving and Cloth, High Modulus, Organic Fiber |

### *5.1.2 Construction and Other Aspects*

| | | |
|---|---|---|
| MIL–A–8860 | : | Airplane Strength and Rigidity, General Specification for |
| MIL–A–8869 | : | Airplane Strength and Rigidity, Special Weapons Effects |
| MIL–B–5087 | : | Bonding, Electrical and Lightning Protection, for Aerospace Systems |
| MIL–M–8856 | : | Missiles, Guided: Strength and Rigidity: General Specification for |
| MIL–P–9400 | : | Plastic Laminate Materials and Sandwich Construction, Glass Fiber Base, Low Pressure Aircraft Structural, Process Specification Requirements |
| MIL–R–7705B | : | Radomes, General, Specifications |
| MIL–S–9041 | : | Sandwich Construction, Plastic Resin, Glass Fabric Base, Laminated Facings and Honeycomb Core for Aircraft Structural Applications |

| | | |
|---|---|---|
| MIL–S–25392 | : | Sandwich Construction, Plastic Resin, Glass Fabric Base, Laminated Facings and Polyurethane Foamed-in-Place Core for Aircraft Structural Applications |

## 5.2 Standards and Test Methods

| | | |
|---|---|---|
| FED–STD–406 | : | Plastics, methods of testing |
| MIL–STD–401 | : | Sandwich constructions and core materials; general test methods |
| MIL–STD–810 | : | Environmental test methods |
| ARTC–4 | : | Electrical test procedures for Radomes and Radome materials |
| ASTMD 3039–76 | : | Test method for tensile properties of fiber-resin composites |
| ASTM D 790–84 | : | Test method for flexural properties of unreinforced and reinforced plastics and electrical insulating material. |
| ASTM D 3355–74 | : | Test method for fiber content of unidirectional fiber-resin composites by electrical resistivity |
| ANSI/ASTM–D–2344–84 | : | Standard test method for apparent interlaminar shear strength of parallel fiber composites by short beam method. |

## 5.3 Other Information [40]

| | | |
|---|---|---|
| PLASPE | : | An engineering and marketing database from 'plastics technology' magazine: can be accessed by subscribers using a terminal and modem via telephone |
| HIMP | : | High temperature material properties data bank operated by Centre for Information and numerical data analysis and synthesis (CINDAS) of Purdue University, in conjunction with US Dept. of Defence. |
| EPOS | : | Engineering plastics on screen–A data base on mechanical, thermal, electrical and other properties of injection mouldable materials from ICI, UK. |
| POLYFACT | : | An on-line menu driven interactively formatted database provided by E.I. Dupont. |

## REFERENCES


1. IEEE standard 145–1983, Institute of Electrical and Electronics Engineers Inc, 345, East 47th Street, New York.
2. T.E. Tice (ed) "Techniques of Airborne Radome Design", AFAL—TR-66-391, Vols. 1 and 2, Dec. 1966.
3. J.D. Walton, Jr (ed) "Radome Engineering Handbook Design and Principles", New York: Marcel Dekker, 1976.
4. W. Cady, M. Karelitz, and L. Turner, "Radar Scanners and Radomes.", New York: McGraw-Hill, 1948.
5. A.W. Rudge, K. Milne, A.D. Oliver and P. Knight, "Handbook of antenna design", (Ch. 14, Vol 2). London: Peter Peregrinus, 1983.
6. R.C. Johnson and H. Jasik, in "Antenna Engineering Handbook", (Ch. 44). New York: McGraw-Hill, 1984.

7. M.I. Skolnik, "Radar Handbook", (Ch 14)., New York: McGraw-Hill, 1970.
8. Proc. Symposia on Electromagnetic Windows, Georgia Inst. of Technology, Atlanta-Held on number of years starting from 1955.
9. International Conferences on Electromagnetic Windows, Paris-Held on number of Years starting from 1967.
10. G.S. Mani et al, "Electrical Design Report of 1-4 GHz Radome", DLRL report no. DLRL. : R : 86 : 020 (Restricted), March 1986.
11. G.S. Mani and Ramsingh, "Design and fabrication of radomes for a broadband spiral antenna", DLRL technical notes (Restricted), Vol 2, March 1981.
12. R.H. Cary, "Some novel techniques for avoiding obscurations", IEE Conf. Publication NO. 155, 1971.
13. E.L. Rope and G.P. Tricoles, "Anisotropic Dielectrics, Tilted grooves on flat sheets and an axially symmetric radomes", IEEE International Symp. Antenna Propag. Dig., pp 610–611, June 1979.
14. D.L. Loyet, "Broadband radome design techniques", Proc. 13th EM Window Symposium, Georgia Institute of Technology, pp 169–173, 1976.
15. D.C. Seller, "Abberation of dielectric patches and rings", Proc. Int. Conf. on EM Windows, Paris, 1967.
16. A. Cohen, P. Davis, S.C. Nilo and J.F. Orabona, "A 150 feet metal space frame radome", Proc, OSU-WALDC Radome Symp. WADC Tech. Report 57–314.
17. Pelton and Munk, "A Streamlined Metallic Radome", IEEE Trans. Ant. and Prop., Vol. A, pp. 22, no. 5, pp. 799–803, 1974.
18. A. Cohen and A.P. Sondski, "Effect of rain on satellite communication earth terminals and rigid radomes", Microwave Journal, Vol 9, no. 9, pp 111–121, Sep. 1966.
19. M.M. Schwartz, "Composite Materials Handbook.", New York: McGraw-Hill, 1983.
20. G. Lubin, "Handbook of Advanced Composites", New York: Van Nostrand Reinhold, 1982.
21. "Kevlar Use", Aviation Week and Space Technology, pp. 37, April 7, 1986.
22. J. Cook, "Reinforcements for plastics intended for electrical and electronic applications", Journal of Naval Sciences, Vol. 1, No. 3, pp. 221–233, July 1975.
23. R.H. Cary, "Avionic Radome Materials", AGARD report no. AGARD AR-75.
24. "The basics on bonded sandwich construction", Hexcel Corporation Book Part no. TSB 120 and TSB 124.
25. G.M. Brydon and R.H. Cary, "Some Radome materials with superior dielectric and temperature characteristics", Proc. Int. Conf. on EM Windows, Paris, 1967.
26. G.S. Mani, et al, "Development of Broadband Radomes for Electronic Equipment Pod", DLRL report No. DLRL R : 86 : 024 (Restricted), April 1986.
27. M.S. Cray and M.G. Taylor, "Syntactic foam—A new method for radome construction", Presentation given at Fourth Int. Conf. on EM Windows, Barcol, France.
28. M.B. Punnett, "Raydel—Microwave transmissive fabric radome composites", Proc Int. Conf. on EM Windows, Georgia Inst. of tech, pp. 117–123, 1978.
29. Joseph J. Prifti, "Hardened Tuned-wall plastic radome for military radar",
30. "Astrocoat-Sterling erosion resistant coating system", Sterling Lacquer Mfg. Co., 3150, Brannon Avenue, St. Louis, Mo63139, USA.
31. G.S. Springer, "A model for curing process of epoxy matrix composites", Progress in Science and Engineering of Composites, Japan Society of Composite Materials pp 25–35, 1982.

32. T.G. Gutowski, "A resin flow/fiber deformation model for composites", Advances in Technology in Materials and Processes, SAMPE, pp. 925–934, 1985.
33. J.M. Tanged et al, "Effects of cure pressure on resin flow, voids and mechanical properties", Jour. of Composite Materials, Vol. 21, pp. 421–440, May 1987.
34. M.N. Afsar, J.R. Birch and R.N. Clarke, "The measurements of properties of materials", Proc. IEEE, Vol 74, No. 1, pp. 183–199, January 1986.
35. W.W. Ho, "High temperature millimetric dielectric characterization of radome materials", Proc. SPIE—Int. Soc. opt. Engrs, Vol. 362, pp. 190–195, 1982.
36. F.I. Shimabukuro et al, "A quasi-optical method for measuring complex permittivity of materials", IEEEE Trans. on Microwave Theory and Tech, Vol MTT—32, no. 7, pp. 659–665, 1984.
37. T. Larry Norin and Luis L. Oh, "Monitoring AWACS radar radome fabrication", Proc. 14th EM Window Symposium, Georgia Inst. of Tech., pp 137–145, 1978.
38. N.R. Ray, "A portable radome tester", Proc. 12th Symp. on EM Windows, pp 117–124, June 1974.
39. Leon J. Ricardei and Seymong Sutkin, "A microwave radome performance verification gauge", IEEE Trans. on Ant. and Prop., Vol. 39, no. 2, pp. 131–142, Feb 1991.
40. Gladius Lewis, "Selection of Engineering Materials", New Jersey, Prentice-Hall, 1990.

# 9

# High Frequency Applications of High-$T_c$ Superconductors

C.M. Srivastava
Advanced Centre for Research in Electronics and Department of Physics
Indian Institute of Technology, Bombay, India

## 1. INTRODUCTION

The current investigations on the technological impact of the high temperature superconductors (HTSC) indicate that the most promising applications of these materials are in the high-frequency devices. The replacement of waveguides and coaxials circuits by microwave integrated circuits more than two decades ago heralded a revolution in the field of microwave circuits. The main reasons for the widespread use of microwave integrated circuits are increased reliability, improved reproducibility and significant reduction in size, weight and cost. These gains are of vital importance in space and satellite electronic circuits. Further in military applications, where equipments are subjected to severe shock and vibrations, MIC devices provide better mechanical integrity than waveguides bolted together.

Even before the advent of HTSC superconductors were used in some special high frequency circuits and cavities [1, 2, 3]. The reason is that lower conductor losses result in lower noise, higher speed and wider bandwidth. Amongst the major systems applications is the use of superconducting microwave cavities for accelerating charged particles. The possibility of using HTSC in super-collider experiment is currently being examined.

Several successful attempts [4, 5, 6] have been made to develop passive microwave devices like resonators, filters and delay lines using HTSC. These devices are in the stripline configurations. A stripline is a thin narrow strip of metal which is placed a short distance above a metal ground plane. The strip is placed parallel to the ground plane and is separated from it by a thin dielectric slab. The transmission occurs via the transverse electromagnetic (TEM) mode much in the same manner as in a two wire line. A stripline is lossier than a waveguide but it is smaller and cheaper and hence is extensively used.

The losses in stripline primarily originate from the dielectric and the conductor. With improvement in the microwave dielectric materials the conductor losses in stripline devices today dominate even when the best quality copper or gold is used. With the advent of HTSC and use of liquid nitrogen as coolant for the device it is possible to reduce the conductor losses and enhance the quality factor of the device in a cost effective manner.

The possibility of complete elimination of refrigeration below ambient temperature exists in a space environment where temperature close to 100 K can be obtained from radiative cooling. If $T_c$ can be further pushed by about 30 K from the existing maximum of 125 K for thallium based compounds it should be possible to use many of these devices in aerospace industry without additional refrigeration.

Although today several high $T_c$ superconducting compounds are known, most of the high frequency applications have been attempted on $YBa_2Cu_3O_7$(YBCO) films. Only recently some results on Tl-Ca-Ba-Cu-O thin films have also been reported.

There are however some problems which need to be solved before the rapid deployment of the HTSC microwave device is possible. The fabrication of these devices require development of good quality epitaxial thin films with high $J_c$, low $R_s$, long term stability, methods of patterning without affecting the superconducting properties, ohmic contacts and large size substrates of good quality. The device performance is critically dependent on these parameters of the film. Further, the type of correlation which exists between the structural and microwave properties of HTSC films is not clear at present. It is therefore not possible to predict the device performance from the measurements on the unpatterened films.

In this article, we discuss the electromagnetic properties of good conductors and superconductors. Specially, we examine the dependence of the propagation characteristics on surface resistivity for a microstripline. We then describe some of the microwave components which have been successfully fabricated using HTSC materials. Finally, we give the current trends and future potential applications.

## 2. ELECTROMAGNETIC PROPERTIES

### 2.1 Normal Metals—Anomolous skin effect

The dc resistivity of a good conductor like copper changes by 3 orders of magnitude on cooling from 300 to 0 K. However, in the same temperature range the high frequency surface resistance decreases only by a factor of 7. This is due to the anomolous skin effect. At very low temperatures the mean-free-path $l$, of pure annealed copper is about 1 mm. The magnetic field $H$ inside the conductor is parallel to the surface and though it is much larger than the electric field its intensity gets rapidly attenuated inside the conductor ($I \propto e^{-z/\delta}$), where $\delta$ is the skin depth and is given by

$$\delta = [2/\mu_c \, \omega \sigma]^{1/2} \tag{1}$$

Here $\mu_c$ is the permeability of the conductor, $\omega$ is the frequency and $\sigma$ is the conductivity. The magnetic field at low temperatures thus penetrates a distance much less than $l \sim 1$ mm ($\delta << l$). The surface electrons which are the active carriers of charge transport cannot then interact with the external rf field and exchange energy. This leads to anomolous skin effect and results in high value for the high-frequency surface resistance compared to the d.c. resistance.

In the region where anomolous skin effect is absent ($\delta > l$), the normal state surface resistance is given by

$$R_s = \frac{1}{\sigma\delta} = (\pi f \mu_c \rho)^{1/2} \tag{2}$$

For copper at 300 K, $\rho = 1.7\ \mu\Omega$-cm and $\mu_c \sim \mu_o = 4\pi \times 10^{-7}$ H/m. For 10 GHz we obtain $R_s = 40$ m$\Omega$. At 77 K, $\sigma_{cu}$ (77)/$\sigma_{cu}$ (300) ~ 16. We then have at 77 K, $R_s \sim 10$ m$\Omega$.

The dependence of surface resistance on temperature and frequency for copper has been measured by Hammond et al [7]. It is found that $R_s$ varies as $\omega^{1/2}$ for the metals in the normal state. Further, since $\sigma$ increases as temperature is lowered, $R_s$ decreases with temperature at constant frequency.

### 2.2 Superconductors—Surface impedance

In superconductors the electrons at the surface cannot travel deep into the materials and are confined within a sheath of thickness $\lambda$, where $\lambda$ is the Landon penetration depth. The microwave properties of superconductors can only be understood through quantum mechanics. The surface impedance $Z_s = R_s + j X_s$ depends on the mechanism of charge transport in the superconductor. In the extreme local limit the surface impedance is given by Ginsburg [8]

$$Z_s(\omega) = (1 + j) \left[\frac{\omega\mu_0}{2\sigma_{sc}}\right]^{1/2} \tag{3}$$

where

$$\sigma_{sc} = \sigma_1 - j\sigma_2 \tag{3a}$$

is the conductivity of the superconductor. Using the BCS weak coupling theory [9] Mattis and Bardeen [10] have obtained the expressions for $\sigma_1$ and $\sigma_2$. Since these expressions involve complicated integrations over the quasi-particle energy states we adopt a simplified semi-classical model which to some extent reproduces the results of quantum mechanical calculations.

For superconductors, zero resistivity leads to the acceleration equation

$$m\dot{\mathbf{v}} = e\mathbf{E}$$

Since $\mathbf{J} = ne\mathbf{v}$, where $n$ is the number of charge carriers per unit volume, we can write this equation as

$$\dot{\mathbf{J}} = \frac{ne^2}{m}\mathbf{E} \tag{4}$$

We consider the two fluid model [11] in which the current is assumed to be a superposition of a normal (resistive) $\mathbf{J}_n$ and a superconductive component $\mathbf{J}_s$

$$\mathbf{J} = \mathbf{J}_n + \mathbf{J}_s \tag{5}$$

with

$$\mathbf{J}_n = \sigma \mathbf{E} \tag{5a}$$

where $\sigma$ is the normal state conductivity of the metal. Then, neglecting displacement current

$$\nabla \times \mathbf{H} = \sigma \mathbf{E} + \mathbf{J}_s \tag{6}$$

and so

$$\nabla \times \nabla \times \mathbf{H} = \sigma \nabla \times \mathbf{E} + \nabla \times \mathbf{J}_s$$

or

$$-\nabla^2 \mathbf{H} = \sigma \nabla \times \mathbf{E} + \nabla \times \mathbf{J}_s \tag{7}$$

From Eq. (4)

$$\nabla \times \dot{\mathbf{J}}_s = \frac{ne^2}{m} \nabla \times \mathbf{E} = -\frac{ne^2 \mu}{m} \dot{\mathbf{H}} \tag{8}$$

Using Eq. (7) and (8), we obtain

$$\nabla^2 \mathbf{H} = \sigma \mu \dot{\mathbf{H}} + \frac{1}{\lambda^2} \mathbf{H} \tag{9}$$

where $\lambda = (m/ne^2\mu)^{1/2}$ is the Landon penetration depth. If we assume

$$H \sim H_0 \exp(j\omega t) \exp(-\gamma z) \tag{10}$$

$$\gamma^2 = \frac{1}{\lambda^2} + \frac{2j}{\delta^2} \tag{11}$$

where $\delta$ is the skin depth (Eq. 1) and $\gamma$ is the propagation constant in the superconductor.

The surface impedance is defined by the ratio of the electric field $E_{\parallel}$ to the magnetic field $H_{\parallel}$ tangential to the conductor surface

$$Z_s = \frac{E_{\parallel}}{H_{\parallel}} = R_s + jX_s = \frac{j\omega\mu}{\gamma} \tag{12}$$

In the normal state ($T > T_c$) $\lambda = \infty$ and Eq. (11) gives

$$\gamma = \frac{1}{\delta} + \frac{j}{\delta} = (j\omega\mu\sigma)^{1/2} \tag{13}$$

Using Eqs. (12) and (13) we obtain the expression for $R_s$ which agrees with Eq. (2).

For the superconductor ( $T < T_c$), $\lambda$ is finite and can be very small at low temperatures compared to $\delta$, so Eq. (11) determines the characteristics of microwave propagation. Substituting from Eq. (11) into Eq. (12), we obtain

$$R_s = \frac{\omega\mu\lambda}{\sqrt{2}} \left[ \frac{\left(1 + \frac{4\lambda^4}{\delta^4}\right) - 1}{1 + \frac{4\lambda^4}{\delta^4}} \right]^{1/2} \tag{14}$$

$$\sim \frac{\omega^2 \sigma \lambda^3 \mu}{2} \qquad (\lambda << \delta) \tag{15}$$

$$X_s = \frac{\omega \mu \lambda}{\sqrt{2}} \left[ \frac{\left(1 + \frac{4\lambda^4}{\delta^4}\right)^{1/2} + 1}{1 + \frac{4\lambda^4}{\delta^4}} \right]^{1/2} \tag{16}$$

$$\sim \omega \mu \lambda \qquad (\lambda << \delta) \tag{17}$$

From Eqs. (15) and (16), we obtain

$$\delta = \frac{R_s^2 + X_s^2}{\omega \mu (R_s X_s)^{1/2}} \tag{18}$$

and

$$\lambda = \frac{R_s^2 + X_s^2}{\omega \mu (X_s^2 - R_s^2)^{1/2}} \tag{19}$$

The conclusion that $R_s$ is proportional to $\omega^2$ and $X_s$ is proportional to ω has been derived by Hartwig [12].

Many studies of the temperature dependence of surface resistivity of superconducting films have been reported [7, 13, 14]. This data can be compared with the expression for $R_s$ in Eq. (15). According to two fluid model $\sigma \propto t^4$ and $\lambda^2 \propto (1 - t^4)^{-1}$, where $t = T/T_c$. So $R_s$ can be expressed as [1]

$$R_s \sim \frac{\mu^2 \sigma_n (T_c) \lambda_0^3 \omega^2}{2} \frac{t^4}{(1 - t^4)} \tag{20}$$

The theory by Mattis and Bardeen [10] gives for $T < T_c$,

$$R_s(\omega, T) \, \alpha \, \frac{\omega^2 \exp(-\Delta / k_B T)}{T} \tag{21}$$

Measurements of $R_s$ on YBCO films by Fathy et al [15] indicate that $R_s$ varies as $\omega^2$ as expected. However, Kobayashi et al [13] found that both $R_s$ and $X_s$ in YBCO films vary as $\omega^{0.5}$. Recent measurements of microwave losses in YBCO films by Pakulis et al [14] show that the temperature dependence of $R_s$ given by Eq. (20) is not applicable and the loss is proportional to $t/\sqrt{1 - t}$ throughout the temperature range, $0 < t < 1$. A complete understanding of the temperature and frequency dependence of microwave losses in high $T_c$ ceramic superconductors is not yet available probably due to the presence of intergranular losses which dominate at low temperature.

Figure 1 shows the surface resistance of a thin film of Tl-Ca-Ba-Cu-O measured by Hammond et al [7] as a function of temperature at 9.55 GHz. This is compared with the $R_s$ for a similar copper film. At 77 K the Tl-film has at least 20 times lower loss than copper for fields upto 10 Gauss.

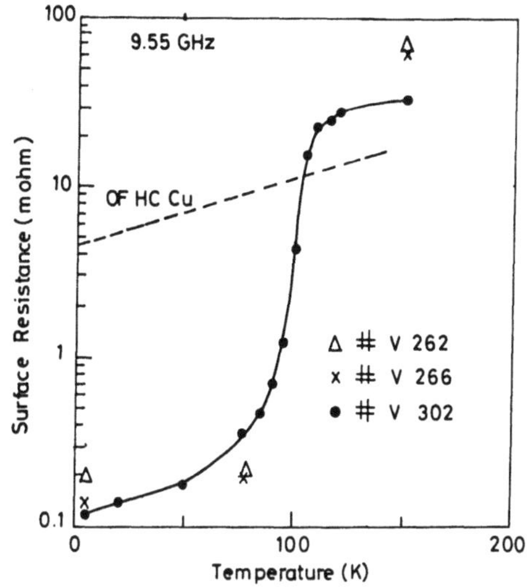


**Fig. 1.** The measured surface resistance vs temperature at low power for three Tl-Ca-Ba-Cu-O thin films. The complete temperature scan for film # V302 is shown. The other two films were measured at 4.2, 77 and 150 K. At 77 K the $R_s$ for all three films is at least 20 times smaller than copper at the same temperature [7].

## 2.3 Residual Losses

The surface resistance of most superconductors can be expressed as

$$R_{exp} = R(\omega, T) + R_{res} \tag{22}$$

where $R(\omega, T)$ is given by Eq. (20) or (21). $R_{res}$ is a sample dependent constant whose origin is yet not known. Amongst several possible causes of residual resistance are the presence of normal conducting materials, tunneling across cracks, generation by phonons by the rf field and grain boundary scattering of charge carriers. It is observed that for $T_c/T > 5$, $R_{res}$ invariably dominates.

# 3. ATTENUATION AND DISPERSION IN MICROSTRIP LINES

The analysis of the propagation of the electromagnetic fields in a stripline using superconductors has been carried out by Kautz [16]. Improved analysis including the effect of dispersion on the transmission line performance has been given by Whitaker et al [17], Kwon et al [18] and Ekholm and McKnight [19].

In this section, we discuss the theory of microstrip transmission line and calculate the propagation constant $\gamma$ which has its real and imaginary parts

$$\gamma = \alpha + j\beta \tag{23}$$

where $\alpha$ is the attenuation constant and $\beta$ is the phase factor. The geometry of the microstrip line is shown in Fig. 2. $\alpha$ is affected by both, conductor and the dielectric losses, but $\beta$ is primarily determined by the line geometry and depends only to a small extent on the conductor. It is convenient to take

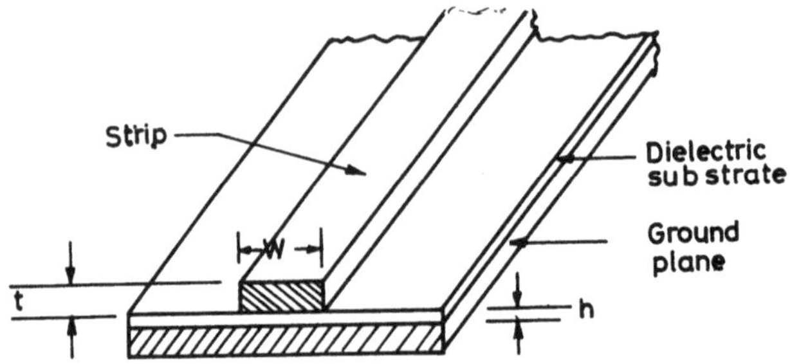


Fig. 2. Geometry of a microstrip line. The ground plane and the strip may or may not be of the same metal. The dielectric substrate is a low microwave loss material. '$w$' is the width and '$t$' is the thickness of the strip and '$h$' is the thickness of the dielectric substrate.

$$\alpha = \alpha_c + \alpha_d$$

$$\beta = \beta_c + \beta_m \tag{24}$$

where the subscripts c, d and m stand for conductor, dielectric and modal contributions to $\alpha$ and $\beta$.

Using the transmission line equation

$$\frac{d^2V}{dx^2} = \gamma^2 V \tag{25}$$

for the propagation of a sinusoidal voltage $V(x, \omega)e^{j\omega t}$ on the microstrip line, it can be shown that $\gamma$ is approximately given by

$$\gamma = [ZY]^{1/2} = \frac{Z}{Z_0} \tag{26}$$

where $Z$ is the series impedance, $Y$ is the shunt admittance and $Z_0$ is the characteristic impedance. For the microstrip line [19]

$$Z = j\omega\mu_0\, g_1 + 2Z_s g_2 \tag{27}$$

$$Y = \omega\varepsilon_0\,(j\varepsilon_r^{\text{eff}} + \varepsilon_r \tan\delta)/g_1 \tag{28}$$

where $\varepsilon_0$ and $\mu_0$ are the permittivity and permeability of free space, $Z_s$ is the surface impedance, $\varepsilon_r^{\text{eff}}$ is the effective dielectric constant, $\varepsilon_r$ and $\tan\delta$ are the relative permittivity and loss tangent of the substrate and $g_1$ and $g_2$ are functions depending on the geometry of the line.

The attenuation constant $\alpha_c$ depends on $Z_0$ and the surface resistivity, $Z_s$ of the ground plane and the strip. If both are made of the same material

$$\alpha_c = R_{\text{eal}} \left\{\frac{Z_s}{Z_o}\right\} g_2 \quad \text{m}^{-1} \tag{29}$$

Detailed analysis of the losses in microstrips using normal metals has been given by Pucel et al [20]. Caulton [21] has given a simplified formula for such striplines

$$\alpha_c = \frac{R_s}{Z_0}\left(\frac{1}{w} + \frac{1}{2h + w}\right) \quad \mathrm{m}^{-1} \tag{30}$$

Here $R_s$ is the surface resistivity of the conductor, $w$ is the width of the strip and $h$ is the thickness of the dielectric substrate.

The characteristic impedance $Z_0$ is given in terms of $\varepsilon_r$, $w$ and $h$ by an empirical relation obtained by Wheeler [22]

$$Z_0 = \frac{42.4}{(1 + \varepsilon_r)^{1/2}} \ln\left\{1 + \left(\frac{4h}{w}\right)\left[\left(\frac{14 + 8/\varepsilon_r}{11}\right)\left(\frac{4h}{w}\right) + \sqrt{\left(\frac{14 + 8/\varepsilon_r}{11}\right)^2\left(\frac{4h}{w}\right)^2 + \frac{1 + 1/\varepsilon_r}{2}\pi^2}\,\right]\right\} \tag{31}$$

This gives similar results as the simplified expressions obtained for $Z_0$ by Schneider [23] for the two limits of wide ($w/h > 1$) and narrow ($w/h < 1$) microstrip lines,

$$\frac{1}{\sqrt{\varepsilon_{\mathrm{eft}}}} \frac{120\pi}{\frac{w}{h} + 2.42 - 0.44\frac{h}{w} + \left(1 - \frac{h}{w}\right)^6} \qquad \left(\frac{w}{h} > 1\right) \tag{32}$$

$$Z_0 = \frac{1}{\sqrt{\varepsilon_{\mathrm{eff}}}} 60 \ln\left(\frac{8h}{w} + \frac{w}{4h}\right) \qquad \left(\frac{w}{h} \leq 1\right) \tag{33}$$

where

$$\varepsilon_{\mathrm{eff}} = \frac{\varepsilon_r + 1}{2} + \frac{\varepsilon_r - 1}{2}\left(1 + \frac{10h}{w}\right)^{-1/2} \tag{34}$$

The effective dielectric constant is given by Eq. (31) only upto the cut-off frequency

$$f_c = \frac{c}{4h\,(\varepsilon_r - 1)^{1/2}} \tag{35}$$

Eq. (35) gives the region ($0 < f < f_c$) in which the lowest order longitudinal $TE_1$ mode for the microstrip dominates. For $f > f_c$, $\varepsilon_{\mathrm{eff}}$ becomes frequency dependent and is not given by Eq. (34).

Dispersion at high frequencies has been analyzed by Mittra and Itoh [24] and Atwater [25]. At very high frequencies the transmission line behaves as a waveguide with the wave completely confined to the dielectric region. Consequently the effective dielectric constant $\varepsilon_{\mathrm{eff}}$ approaches $\varepsilon_r$ as the frequency is increased to a very high value.

It is convenient to relate $\varepsilon_{\mathrm{eff}}$ to a dielectric filling factor $q$ by

$$q = \frac{\varepsilon_{\mathrm{eff}} - 1}{\varepsilon_r - 1} \tag{36}$$

For $f > f_c$ as $\varepsilon_{\mathrm{eff}}$ increases with frequency and approaches $\varepsilon_r$, $q \to 1$.

For a 25 mil alumina substrate with $\varepsilon_r = 10$, $f_c$ from Eq. (35) is 40 GHz. With $h/w = 1$ Eq. (34) gives $\varepsilon_{eff} = 6.86$. This gives $q = 0.65$. Yamashita et al [26] have discussed numerical methods for the estimate of $\varepsilon_{eff}$ beyond $f_c$. This is needed to estimate the modal phase factor $\beta_m$ in Eq. (24)

$$\beta_m = \frac{2\pi f}{c}\sqrt{\varepsilon_{eff}(f)} \tag{37}$$

The expression for phase factor due to the conductor is

$$\beta_s = \text{Im}\left\{\frac{Z_s}{Z_0}\right\} g_2 \quad \text{m}^{-1} \tag{38}$$

For a superconductor from Eqs. (12) and (17),

$$\text{Im}\left\{\frac{Z_s}{Z_0}\right\} = \frac{\omega\mu\lambda}{Z_0}$$

For $f = 10$ GHz, $\lambda = 5000$ Å and $Z_0 = 50\ \Omega$ since $g_2 \sim 1$ we obtain $\beta_c \sim 1$ rad/m. For $\varepsilon_{eff}(f) \sim 16$, from Eq. (37), $\beta_m \sim 10^2$ rad/m. Thus $\beta_m >> \beta_c$.

It can be shown that in general $\beta_c << \beta_m$. It is therefore not necessary to include $\beta_c$ in the calculation of the phase velocity. We then have

$$v_{ph} = \frac{2\pi f}{\beta_m} = \frac{c}{\sqrt{\varepsilon_{eff}(f)}} \tag{39}$$

Ekholm and McKnight [19] have calculated the phase velocity as a function of frequency for a microstrip line with $w/h = 1$, $h = 0.5$ mm and $\varepsilon_r = 27.0$. They have used numerical approximation of Yamashita et al [26] to estimate $\varepsilon_{eff}(f)$. From Eq. (35) $f_c$ is nearly 20 GHz. Their result for $v_{ph}$ vs $f$ is shown in Fig. 3. The decrease in $v_{ph}$ occurs due to increase in $\varepsilon_{eff}(f)$ in the frequency band between 1 and 100 GHz.

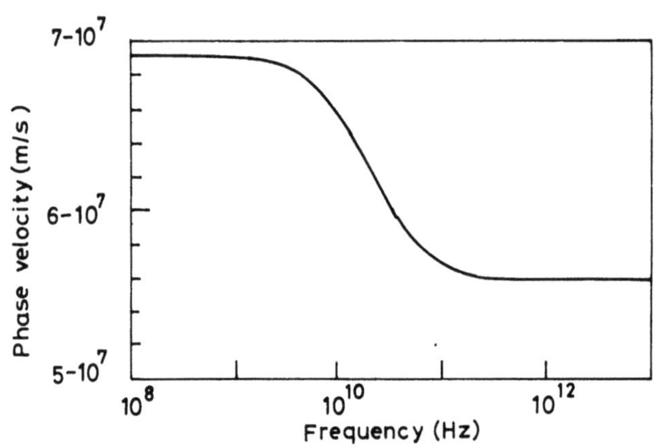


**Fig. 3** Phase velocity as a function of frequency for the microstrip with $w = h >> t$ and $h = 0.5$ mm, $\varepsilon_r = 27$ [19].

From our approximate analysis the low frequency phase velocity limit should be $c/\sqrt{\varepsilon_{eff}}$, where $\varepsilon_{eff}$ is given by Eq. (34) while the high frequency phase velocity limit should be $c/\sqrt{\varepsilon_r}$. These two values are $7.55 \times 10^7$ m/s and $5.77 \times 10^7$ m/s respectively which are close to the values given in Fig. 3 obtained from a more accurate analysis.

It may be noted that the dispersion in phase velocity arising from $\varepsilon_{eff}(f)$ spreads over a large frequency band, 1 to 100 GHz. If a pulse propagates on the microstrip line which has dominant frequency components in this region severe distortion in the shape of the pulse will occur due to this dispersion.

The transition from the higher to lower phase velocity occurs near 10 GHz. At this frequency the wavelength within the substrate is $\lambda/\sqrt{\varepsilon_r} \sim 0.57$ cm and is comparable to the transverse dimension of the microtrip. The dispersion therefore is primarily arising from geometry related modal effects which determines the redistribution of energy from the $TE_1$ mode to higher order modes for $f > f_c$. On the other hand the attenuation in the system is determined both by the contributions by the conductor and the dielectric.

The loss in the dielectric substrate which fills the region between the conductors in the transmission line is small. An approximate expression for $\alpha_d$ in Eq. (24) is given by [19]

$$\alpha_d \approx \pi \frac{\varepsilon_r \tan\delta}{\sqrt{\varepsilon_{eff}}} \frac{1}{\lambda_0} \quad \text{m}^{-1} \tag{40}$$

where $\lambda_0$ is the free space wavelength and $\tan\delta$ is the loss tangent of the dielectric.

Lanthanum aluminate ($LaAlO_3$) and lanthanum gallate ($LaGaO_3$) have been extensively used as substrates in HTSC microstrip line. At 77 K these materials have [27] $\varepsilon_r \sim 23$ and $\tan\delta \sim 2 \times 10^{-4}$. Using Eqs. (34) and (40) for stripline with $w/h = 1$, we obtain $\alpha_d = 1.08$ dB/m.

For a copper microstrip line on $LaAlO_3$ substrate at 80 K and 9.55 GHz, $R_s$ has a value [7] nearly of 10 mΩ. With $w/h = 0.25$ and $w = 0.5$ mm, we get a value of about 56 Ω for $Z_0$ from Eq. (33) and obtain $\alpha_c \sim 3.5$ dB/m from Eq. (30).

Compared to copper at 80 K Tl-Ca-Ba-Cu-O film deposited on $LaAlO_3$ has $R_s \sim 0.5$ mΩ [7]. In this case then $\alpha_c$ would be 20 times smaller than the copper microstripline. It would then also be smaller than the attenuation $\alpha_d$ produced by the dielectric.

### 3.1 Superconducting Microstrip Line and Waveguide Structure—Potential and problems

There are several transmission line structures which have been used in the microwave circuit integration. Some of these [28] are microstrip, suspended stripline, inverted stripline, slot line, coplanar waveguide, trapped inverted stripline, Fin line and image line. Amongst these microstrip is the most commonly used structure. Specially the superconducting devices are generally made in this structure due to its simplicity.

An analysis of the attenuation and dispersion of microstrip line of YBCO on yttria-stabilised zirconia substrates on the basis of the theory developed in

Sec. 3 shows that at 60 K the HSTC line should be significantly less attenuating than the coper line at the same temperature. The application of HTSC microstrip lines at frequencies above 10 GHz appears to be most useful on account of very high attenuation of cooled copper at these frequencies.

Winter and Rose [29] have studied the transmission properties of high $T_c$ superconductor waveguide. Their results show that these have the potential of 100 GHz of band-width for transmission over long distances with low attenuation. They also show that the millimeter waveguide would have the advantage over optical systems since they can have much lower carrier frequency which would result in wider dynamic range.

Despite attractive advantages widespread use of high $T_c$ superconductors in high-frequency devices have not been possible due primarily to the difficulties of fabrication of reproducible high quality films for microstrip lines. Likewise HTSC microwave cavities for accelerating charged particles have been investigated but the progress is slow due to difficulties arising from the techniques used for their fabrication.

One of the major problems arising from imperfection in structure due to the methods used for synthesis is the discrepancy between theoretical and experimental results which in the case of attenuation constant may be of a few orders of magnitude. The disagreement may also be due to the inadequacy of the theoretical models like those used in Sec. 3 [30] for the analysis of the properties of HTSC materials.

### 3.2 Substrate for Microstrip Lines

The substrates which are often used in HTSC microstrip lines are MgO, $SrTiO_3$, $Al_2O_3$, $LaGaO_3$ and $LaAlO_3$. Their properties are given in Table 1.

**Table 1. Microstrip line substrate materials**

| Material | $\varepsilon_r$ | tan $\delta$ | $f_c$ (GHz) | Ref. |
|---|---|---|---|---|
| MgO | 9.65 | $4 \times 10^{-4}$ | 51 | [31] |
| $Al_2O_3$ | 8.30 | $1 \times 10^{-4}$ | 55 | [32] |
| $LaGaO_3$ | 23.1 | $2 \times 10^{-4}$ | 32 | [33] |
| $LaAlO_3$ | 23.2 | $5 \times 10^{-4}$ | 32 | [34] |
| $ZrO_2$ : Y* | 27.0 | $6 \times 10^{-4}$ | 29.0 | [35] |
| $SrTiO_3$ | 1875.0 | $10^2$ | 3.6 | [33] |

*Yttria-stabilized zirconia.

The most commonly used HTSC material is TBCO. The films are generally prepared by laser ablation deposition followed by post-deposition thermal processing.

In Table 1, we have also given reference to the work in which these substrates have been used to make a microstrip line. We have also indicated in the table the cut-off frequency $f_c$, obtained from Eq. (35) with $h = 0.5$ mm. At frequencies greater than $f_c$ the microstrip line supports a variety of waveguide modes in addition to the quasi-TEM mode of low frequencies. This has important effects on the distribution of the pulse shape during transmission if it comprises of frequency components above $f_c$.

### 3.3 Thin Film Microstrip Resonators

A 35 GHz ring resonator has been fabricated by Bhasin et al [36] from laser ablated YBCO films on $LaAlO_3$ substrates. The structure of the resonator is given in Fig. 4. Below 60 K the HTSC strip performed better than the gold. At 25 K the measured $Q$ of HTSC resonator was 600 which was 1.5 times the $Q$ of identical resonator made of gold.

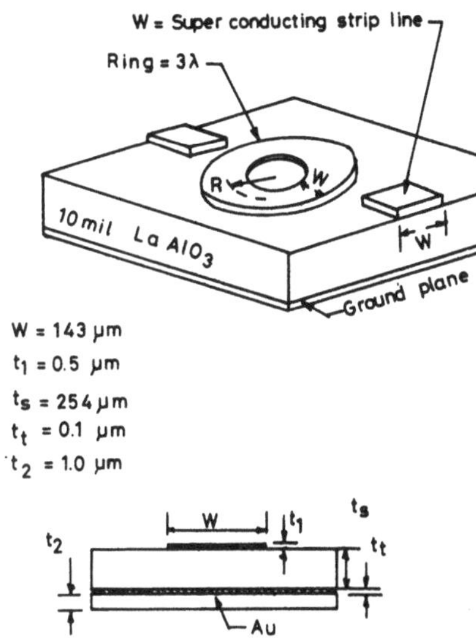


**Fig. 4.** 35 GHz ring resonator structure fabricated from laser ablated YBCO thin film on lanthanum aluminate substrate. The film is c-axis aligned and its $T_c$ is 89.8 K. The metal ground plane is deposited by first evaporating 100 Å of Ti for adhesion followed by 1 micron of gold [36].

A high $Q$ Tl-Ca-Ba-Cu-O thin film microstrip resonator has been fabricated by Hammond et al [7]. They measure a loaded $Q$ of 7300 at 2.6 GHz and 77 K which is 20 times higher than of an identical silver resonator. On increasing the frequency to 7.3 GHz the $Q$ at 77 K of HTSC resonator was 6000 which was better by a factor of 10 than of the silver resonator. The HTSC resonator $Q$s decreased monotonically as the input power was increased. At effective power levels in the resonator upto 100 Watts in frequency range 2 to 7 GHz the $Q$ was at least three times higher than the silver resonator at all frequencies.

The analysis of $Q$ is performed [37] in the coplanar geometry in terms of the quality factor $Q_c$ due to conductor losses, $Q_r$ to radiation and $Q_d$ to losses due to dielectric substrate

$$\frac{1}{Q} = \frac{1}{Q_c} + \frac{1}{Q_r} + \frac{1}{Q_d} \tag{41}$$

Radiation losses are reduced by choosing smaller lateral waveguide dimensions.

But this lowers $Q_c$ and hence an optimization procedure is required. $Q_r$ generally is in the range of 10,000. $Q_d$ is approximately given by $(\tan\ \delta)^{-1}$. From Table 1, $Q_d$ is thus in the range of 3000 to 10,000. The conductor quality factor can be described as

$$Q_c = g/R_s \tag{42}$$

where $g$ is a geometry dependent factor which can be calculated and $R_s$ is the surface resistance. $Q_c$ depends on the structural imperfections of the HTSC films such as grain boundaries, twinning, point defects, surface roughness and some intrinsic parameters which are so far unknown and contribute to $R_{res}$ of Eq. (22).

### 3.4 Superconducting Filters

Currently, narrow band MIC filters cannot be used in communication systems on account of their low unloaded $Q$ (50 to 500) compared to the coaxial and waveguide structures for which $Q$ is two orders of magnitude larger (5000 to 10,000). A low $Q$ filter introduces high midband loss and degradation in selectivity and gain at the band edges. It is expected that the replacement of copper by HTSC will permit MIC structures whose performance will be comparable to those of waveguides at least at low power levels. We have seen in the previous section that microstrip resonators with HTSC film can achieve $Q > 7000$.

Bonetti and Williams [27] have developed techniques which help in the design and development of microstrip filters on high $\varepsilon_r$ substrates. They have used this technique to design and fabricate a 4-pole, C-band, 150 MHz band-pass filter with a 0.05 ripple Chebychev response built on a lanthanum galate substrate. Experimental results at liquid nitrogen show excellent agreement with the predicted performance.

## 4. HIGH-$T_c$ THIN FILM MICROWAVE DETECTORS AND MIXERS

It is noted in Sec. 3 that HTSC microstrip transmission line can provide distortion free propagation of electrical transients with a bandwidth of more than 100 GHz. The HTSC films used in these applications have to be of very high quality deposited on a substrate generally with c-axis orientation. In contrast for active superconductive devices which can operate at liquid nitrogen temperatures use has invariably been made of granular films. Such films have been successfully tested as sensitive microwave and far-infrared detectors and mixers [38, 39]

Konopka et al [40] have fabricated granular YBCO and Bl-Ca-Sr-Cu-O (BCSCO) detectors which operate over the frequency range of 24 to 110 GHz at 77 K. The sensitivity achieved at 110 GHz is comparable to that of the crystalline detectors.

Mixing experiments of two microwave signals in YBCO-on-MgO and BCSCO-on-MgO detectors in the 25 GHz range were carried out successfully [40]. YBCO mixer output was linear at the signal levels between –10 and –50 dBm. The mixing action disappeared at 80 K when the onset of superconductivity in the film was 85 K. By tuning both signal and local oscillator microwave sources,

IF frequencies could be varied from 50 MHz to 5 GHz. This could be achieved without any decrease in the mixer output upto 3 GHz. The detector response time was estimated to be less than 40 ps.

High $T_c$ films with substantial granularity are produced either by dc magnetron sputtering or by chemical deposition from nitrate precursors.

The physical processes responsible for detection and mixing effects are only partially known. The granular films comprise of a collection of superconducting grains interconnected by grain boundary Josephson junctions and regions of the normal non-superconducting phase. The general behaviour of the detector has to be analyzed on the basis of this complex structure. Absence of a full understanding of the physical mechanism makes the device optimization extremely difficult.

The detection sensitivity increases with the decrease in temperature but there is a limit to the detectors lowest-usable temperature (typically 50 K) on account of the complicated transport mechanism at low temperature in which quantum effects dominate.

## 5. MICROWAVE SUPERCONDUCTING CAVITIES FOR PARTICLE ACCELERATORS

We have seen that in the microwave frequency range superconductors remain substantially free of resistance at frequencies upto 100 GHz. This is particularly true of type I superconductors. Even for type II superconductors, the Meissner effect may persist metastably above $H_{c1}$ to a superheating critical field $H_{sh}$ because it takes a finite time to nucleate a flux line. Experimentally, for type I superconductors $H_{sh}$ is found close to $H/\sqrt{K}$, where $K$ is the Abrikosov constant and is given by $K = \lambda/\xi$. Here, $\lambda$ is the London penetration depth and $\xi$ is the coherence length. For the extreme type II superconductor, theory predicts $H_{sh} \sim 0.75\ H_c >> H_{c1}$. The ability of superconductors to support lossless high frequency surface currents below $H_{sh}$ is exploited in many devices, the most important amongst these is the electron linear accelarator called Linac.

The principle of Linac is simple. A cluster of electrons are accelerated by passing them through a series of resonant cavities. The time taken for the electron to travel from one cavity to the next is the half period of the rf source. The electric fields in the successive cavities act in the same sense upon the particle because the velocity is such that it takes just the right amount of time to reach the cavity when the electric field is oriented for forward acceleration [1].

The electrons are introduced in linac at high speeds close to velocity of light. Let the initial energy of the electrons be 1 MeV. It is proposed to accelerate them to 1 GeV. If the resonant cavity has a volume $V$ of 1/30 $m^3$ and the field gradient $E$ in the cavity is 8 MV/m, nearly 400 resonant cavities are needed to reach this energy. It is, therefore, necessary to have cavities with high $Q$ to reduce the operational cost.

The peak power $E_p$, stored in the cavity is

$$E_p = 1/2\ \varepsilon_0 E^2 V \quad (43$$

where the permittivity of free space, $\varepsilon_0 = \left(\frac{1}{36\pi}\right) \times 10^{-9}$ Fm$^{-1}$. With $E = 8$ MV/m and $V = 1/30$ m$^3$, $E_p \sim 9.5$ J per cavity. The energy loss per cavity is given by

$$W = \frac{E_p f}{Q} \tag{44}$$

For copper cavity, $Q \sim 6 \times 10^4$. If $f \sim 1$ GHz, $W \sim 170$ kW. The total power consumption for 400 cavities is then 68 MW. At present such high powers can be produced at high frequencies only in the pulse mode. If the surface resistance can be reduced by three to four orders of magnitude the $Q$ can be substantially improved and the power dissipation reduced to a level that continuous operation of a Linac would be possible. This possibly can be achieved with the help of a HTSC superconductors at 77 K.

We note that the performance of the microwave cavity is limited by two parameters—the critical field ($H_c$) and the surface resistance ($R_s$). It is in fact the superheating critical field $H_{sh}$, which determines the peak power handling capability of the cavity without loosing the superconducting properties through the relation

$$E_{max} \sim H_{sh} \left(\frac{\mu_0}{\varepsilon_0}\right)^{1/2} \tag{45}$$

For Nb, $H_{sh} \sim 1600$ Oe, so $E_{max} \sim 60$ MV/m. In HTSC material, $H_{sh}$ is much higher. For example, in YBCO the thermodynamic critical field is 27,000 Oe and the superheating critical field is 20,000 Oe. One can, therefore, take for HTSC that $H_{sh}$ is 10 times higher than Nb so the accelerating voltage capability of a HTSC cavity increases to 600 mV/m. This leads to substantial reduction in the number of cavities required to ahieve the same energy.

The improvement in $Q$ due to reduction in surface resistance when HTSC cavity is used leads to lower power dissipation and thereby to lower cost of operation. The $Q$ of the cavity is given by

$$Q = \frac{\mu_0 V f}{R_s A} \tag{46}$$

where $f$ is the frequency and $V/A$ is the volume to surface ratio of the cavity. Assuming $V/A \sim 0.1$ m, $f = 1$ GHz, $R_s \sim 10^{-7}\,\Omega$ approximate for a superconductor for $T_c/T > 5$, we obtain $Q \sim 10^8$. This is four orders of magnitude larger than for a copper cavity.

Attempts to use HTSC cavities in Linac are in progress [41] and it is likely that soon attractive economic benefits will be realised in this application.

## 6. CONCLUSION

The high frequency properties of high temperature superconductors have been discussed and in particular their surface impedance has been examined as a function of frequency and temperature. The possible advantages of microstrip transmission lines made of high temperature superconductors over copper or gold are described through a detailed study of the attenuation and dispersion of microwave transients at 77 K. It is shown that the performance of HTSC passive microwave components show great promise for improved performance over their metallic counterparts. However, there may be some problems associated with the fabrication of good quality, reproducible thin films of HTSC on dielectric substrates which may delay early utilization of these devices in microwave systems. A brief outline of application of granular superconductivity observed in HTSC films to active devices like detectors and mixers is given. Finally the possible utilization of HTSC cavities in linear particle accelerator has been discussed.

## REFERENCES


1. J.E.C. Williams, 'Superconductivity and its applications', Pion Limited, London, NW2, 1970.
2. V.L. Newhouse, 'Superconducting devices' in 'Superconductivity' Vol. II, Ed. R.D. Parks. Marcel Dekker Inc, N.Y. 1969.
3. T. van Duzer and C.W. Turner, 'Principles of superconducting Devices and Circuits', Elsevier, N.Y. Chapter 2. 1981.
4. P.H. Carr, 'Potential microwave applications of high temperature superconductors', Microwave Journal, p. 91-93, Dec. 1987.
5. D.E. Oates, A.C. Anderson and B.S. Shih, 'Superconducting stripline resonators and high $T_c$ materials, IEEE MTT-S Digest p. 627-630, 1989.
6. J.M. Pond and C.M. Krowne, 'Slow wave properties of superconducting microstrip transmission lines', IEEE MTT-S Digest p. 449–452, 1988.
7. R.B. Hammond, G.V. Negrete, M.S. Schmidt, M.J. Moskowitz, M.M. Eddy, D.D. Strother and D.L. Skoglund, 'Superconducting Tl-Ca-Ba-Cu-O Thin film microstrip resonator and its power handling performance at 77 K', IEEE MTT-S Digest, p. 867–870, 1990.
8. D.M. Ginsburg, 'Calculation of the electromagnetic absorption edge in superconducting alloys', Phys. Rev. **151**, pp. 241–245, 1966.
9. J. Bardeen, L.N. Cooper and J.R. Schrieffer, 'Theory of Superconductivity', Phys. Rev. **108** 1175 (1957).
10. C.C. Mattis and J. Bardeen, 'Theory of anomolous skin effect in normal and superconducting metals', Phys. Rev. **111**, 412–417, 1958.
11. F. London, 'Superfluids, Macroscopic Theory of superconductivity', Vol. II, New York, Dover publications, 1960.
12. W.H. Hartwig, 'Superconducting resonators and devices', Proc. IEEE, **61**, 1 pp. 58–70, Jan. 1973.
13. Y. Kobayashi, T. Imai and H. Kayano, 'Microwave measurement of surface impedance of High-$T_c$ superconductors', IEEE MTT-S Digest, p. 281–284 1990.
14. E.J. Pakulis, R.L. Sandstrom, P. Chaudhari and R.B. Laibouitz, 'Temperature dependance of microwave losses in Y–Ba-Cu-O films', Appl. Phys. Lett, **57** (9) 940–942 1990.
15. A. Fathy, D. Kalokitis, V. Pendrick, E. Belohoubek, T. Venkatesan, L. Nazar, J.B. Barner, A. Findikoglu, A. Inam, X.D. Wu, X.X.Xi and W.L. McLean,

'Microwave Properties and Modelling of High-$T_c$ Superconducting Thin Film Meander Line', IEEE MTT-S Digest p. 859–862, 1990.
16. R.L. Kautz, 'Picosecond pulses on superconducting striplines', J. Appl. Phys. **49** pp. 308–314, 1977.
17. J.F. Whitaker, T.B. Norris, G.A. Mourou and T.Y. Hsiang, 'Pulse dispersion and shaping in microstrip lines', IEEE Trans. Microwave Theory Tech. Vol. MTT–35, pp. 41–47 1987.
18. O.K. Kwon, B.W. Langley, R.F.W. Pease and M.R. Beasley 'Superconductors as very high-speed system level interconnects', IEEE Electron Device Lett. Vol. EDL-8, pp. 582–585, 1987.
19. E.B. Ekholm and S.W. McKnight, 'Attenuation and Dispersion for High-$T_c$ Superconducting Microstrip Lines', IEEE Trans. Microwave Theory Tech. MTT-38, 387–395, 1990.
20. R.A. Pucel, D.J. Masse and C.P. Hartnig, 'Losses in Microstrip', IEEE, Trans. Microwave Theory Tech. MTT-16, 342–346, 1968.
21. M. Coulton, 'Film Technology in Microwave integrated circuits', Proc. IEEE. **59**, pp. 1481–1489, 1971.
22. H.A. Wheeler, 'Transmission line propagation of a strip on a dielectric sheet on a plane'—Trans. IEEE, Microwave Theory Tech., MTT-25, 631–647, 1977.
23. M.V. Sneider, 'Microstrips for microwave integrated circuits', B.S.T. Jour. **48**, 1421–1244, 1969.
24. R. Mittra and T. Itoh, 'A new technique for the analysis of the dispersion characteristic of microstrip lines', IEEE Trans. Microwave Theory Tech. MTT-19, pp. 47–56, 1971.
25. H.A. Atwater, 'Tests of microstrip dispersion formulas', IEEE Trans. Microwave Theory Tech. Vol. 36, pp. 619–621, 1988.
26. E. Yamashita and K. Atsuki, 'Analysis of microstrip-like transmission line by nonuniform discretization of integral equations', IEEE Trans. Microwave Theory and Tech. MTT-24, pp. 195–200, 1976.
27. R.R. Bonetti and A.E. Williams, 'Preliminary design steps for thin-film superconducing fields', IEEE MTT-S Digest, 273–276, 1990.
28. D. Venkatsubbiah, 'Practical aspects of passive components while operating from dc to microwave frequencies' in Microwave Technology and Applications' Eds. RVS Sitaram, G.P. Srivastava and P.G. Nair, Tata McGraw-Hill Publ. Co., New Delhi 1988, pp. 186–194.
29. J.H. Winters and C. Rose, 'High-$T_c$ superconductor waveguides: Theory and applications' IEEE, Trans. Microwave Theory Tech. MTT-39, pp. 617-623, 1991.
30. T.E. van Deventer, P.B. Katechi, J.Y. Josefowicz and D.B. Rensch, 'High frequency characterisation of high-temperature superconducting thin film lines', IEEE MTT-S Digest, 285-288, 1990.
31. M. Wintergill, J. Fontanella, C. Andeen and D. Schucle, 'The temperature variation of dielectric constant of pure CaF, SrF, BaF and MgO', J. Appl. Phys. **129** 1561, 1963.
32. K. Char, N. Newman, S.M. Garrison, R.W. Barton, R.C. Taber, S.S. Landerman and R.D. Jacowitz, 'Microwave surface resistance of epitaxial YBa Cu O thin film on supphire', Appl. Phys. Lett. **57**, 409, 1990.
33. D.W. Cooke, E.R. Gray, R.J. Houlton, H.H.S. Javadi, Maez, B.L. Bennett, B. Rusnak, E.A. Meyer, P.N. Arendt, J.G. Berry, D.R. Brown, F.H. Garzon, I.D. Raistrick, A.D. Rollett, B. Bolmaro, N.F. Elliott, N. Klein, G. Muller, S. Orbach, H. Piel, J.Y. Josefowicz, D.B. Rensch, L. Drabeck and G. Gruner, 'Surface

resistance of YBa Cu O films deposited on LaGaO substrate', Physica C 162–164, p. 1537, 1989.
34. K.H. Young, G.V. Negrete, R.B. Hammond, A. Inam, R. Ramesh, D.L. Hart and Y. Yonezawa, 'Clear correlations observed between $YBa_2Cu_3O$-$\delta$ thin film properties and GHz microwave resonator performance', Appl. Phys. Lett. **58** p. 1789, 1991.
35. A.C. Anderson, B.Y. Tsam, J.W. Steimbeck and M.S. Dilorio, 'RF surface resistance of YBCO thin films', MIT Lincoln Labotatory Quaterly Tech. Rep. Mar. 11, 1988.
36. K.B. Bhasin, C.M. Chorey, J.D. Warner, R.R. Romanofsky and V.O.Heinen, 'Performance and Modeling of superconducting ring resonators at millimeter-wave frequencies', IEEE MTT-S Digest, pp. 269–272, 1990.
37. R. Dill, J. Otto, G. Riha, P. Russer, L. Schultz, G. Solkner, A.A. Valenzuela and E. Wolfgang, 'Test chip for high temperature superconductor passive devices', IEEE MTT-S Digest, pp. 863–866, 1990.
38. J. Konopka, R. Sobolewski, A. Konopka and S.J. Lewandowski, 'Microwave detection and mixing in Y-Ba-Cu-O thin films at liquid nitrogen temperatures', Appl. Phys. Lett., **53**, pp. 796–798, 1988.
39. M. Leung, P.R. Broussard, J.H. Claassen, M. Osofsky, S.A. Wolf and U.Strom, 'Optical detection in thin granular films of Y-Ba-Cu-O at temperature between 4.2 and 100 K' Appl. Phys. Lett. **51**, pp. 2046–2048, 1987.
40. J. Konopka, R. Sobolewski, G. Jung, W. Kula, P. Gierlowski, A. Konopka and S.J. Lewandowski, 'Microwave detectors based on granular high-$T_c$ thin films', IEEE Trans. Microwave Theory Tech. **38**, pp. 160–165, 1990.
41. H. Padamsee et al, 'Microwave superconductivity for particle accelerators—How the high $T_c$ superconductivity measures up' in 'Superconductivity and its applications', Eds. H.S. Kwok and D.T. Shaw Elsevier, New York (1988) p. 249.

MIX
Papier aus verantwortungsvollen Quellen
Paper from responsible sources
FSC® C105338

If you have any concerns about our products,
you can contact us on
**ProductSafety@springernature.com**

In case Publisher is established outside the EU,
the EU authorized representative is:
**Springer Nature Customer Service Center GmbH**
**Europaplatz 3, 69115 Heidelberg, Germany**

Printed by Libri Plureos GmbH
in Hamburg, Germany